Square Foot & UNIFORMAT Assemblies Estimating *Third Edition*

RSMeans

RSMeans
Square Foot & UNIFORMAT Assemblies Estimating

Third Edition

RSMeans

Copyright © 2001
Construction Publishers & Consultants
63 Smiths Lane
Kingston, MA 02364-0800
781-422-5000
www.rsmeans.com

Managing Editor: Mary Greene. Editors: Andrea Sillah and Barbara Balboni.
Production Manager: Michael Kokernak. Production Coordinator: Marion Schofield.
Composition: Sheryl Rose. Proofreader: Robin Richardson.
Book and cover design: Norman R. Forgit.

10 9 8 7 6 5 4 3 2

Library of Congress Cataloging in Publication Data

ISBN: 978-0-87629-018-7

Table of Contents

Contributors to This Edition

Barbara Balboni, primary editor of this book, is a Construction Cost Engineer/Editor at R.S. Means Co., Inc. She is the Senior Editor of Means *Square Foot Costs, Assemblies Cost Data,* and *Interior Cost Data,* and maintains several components of the Means cost database. Ms. Balboni also conducts seminars on square foot and unit price estimating, and performs market research and consulting projects in construction cost estimating. Prior to her tenure at R.S. Means, she was a Designer/Project Manager for Architects Lowrey, Blanchard Architectural Associates, Inc., and the McKenna Group, Ltd., where she was responsible for all phases of architectural projects from programming and schematic design through construction. She is a member of the American Institute of Architects and the Boston Society of Architects.

Brian Bowen, who reviewed and provided valuable input on this book, is the former President of Hanscomb, Inc., an international construction consulting firm headquartered in Atlanta that specializes in providing project and construction management services, including budget and cost control, value engineering, scheduling, and quality control. Prior to his retirement in 2000, Mr. Bowen worked for Hanscomb for 40 years and served as President of Hanscomb Worldwide since 1995. Mr. Bowen also assisted the General Services Administration and the American Institute of Architects in the development of the first UniFormat classification system in the 1970s. Throughout his career he has been active on task forces and committees of organizations including the American Institute of Architects and the American Society for Testing and Materials, where his work included conducting and developing training courses and authoring cost estimating publications. He is a member of the Society of American Value Engineers, the American Association of Cost Engineers, and the Royal Institution of Chartered Surveyors.

Robert P. Charette reviewed this book and wrote the UNIFORMAT II section of the Appendix. Mr. Charette, a professional engineer, is the former Director of Value Engineering and Life Cycle Costing Services for Hanscomb Consultants in Canada and the United States. He also co-chaired, with Brian Bowen, the ASTM Building Economics

Subcommittee that developed the UNIFORMAT II classification system. Mr. Charette has lectured and presented seminars on value engineering, life-cycle costing, elemental estimating, and UNIFORMAT II at colleges including the Harvard Graduate School of Design, McGill University, and the University of Montreal, and is currently an associate adjunct professor at Concordia University. He is a member of the Architectural Engineering Institute of the American Society of Civil Engineers, the Society of American Value Engineers, the Design-Build Institute of America, and the Association of Project Managers.

Patricia L. Jackson, PE, reviewed this book and provided helpful commentary. Ms. Jackson is President and owner of Jackson A&E Associates, Inc., in the Dallas, Texas area, a firm that offers owners representation and program management for the construction of new facilities, building expansion, and renovation projects nationwide. She is an architect and licensed professional engineer with 25 years' experience in project management and engineering. Prior to her work with Jackson A&E, Ms. Jackson was Vice President of project management for Aguirre Corporation of Dallas, Texas; Director of Engineering Operations and Development for R.S. Means Co., Inc.; and Manager of Technical Programs for the Construction Specifications Institute, among other positions. She is a member of the American Concrete Institute, the American Institute of Steel Construction, the American Society of Civil Engineers, the National Society of Professional Engineers, and the Construction Specifications Institute.

Authors and Editors of Previous Editions

Billy J. Cox, an original author of this book, is a former Engineer/Editor for R.S. Means Co., Inc., and a commissioned officer in the United States Air Force. Mr. Cox served as Base Architect for Webb Air Force Base and Keesler Air Force Base, where he was responsible for design and construction management projects. He also was an assistant professor of architecture at the Air Force Institute of Technology.

F. William Horsely, co-author of earlier editions of this book, was a Consulting Construction Engineer for R.S. Means Co., Inc. and author of *Means Scheduling Manual.* Mr. Horsely conducted scheduling, project management, and estimating seminars throughout the country and taught courses for the Pre-Cast Concrete Institute, the Portland Cement Association, the Bridgeport Engineering Institute, and Yale and Wesleyan Universities.

John H. Ferguson, PE, editor of the second edition of this book, is the Manager of Engineering Operations at R.S. Means Co., Inc. Mr. Ferguson oversees the development, production, technical editing, and quality of Means annual cost guides, consulting and estimating projects, and seminars. He is a member of the American Society of Civil Engineers and the National Society of Professional Engineers.

Introduction

This book explains square foot and assemblies estimating techniques for use in developing conceptual or appraisal estimates. Accurate early project cost information is needed by many parties involved in the design and construction industry, including real estate developers, builders, owners, designers, appraisers, and attorneys.

Often in the earliest stages of a project little information is available other than the proposed type of building, a probable location for the project, and total project size. A reliable cost estimate can be difficult to attain with so few details. Yet, that is when an estimate has the most influence—during the initial project planning. This book will show you how to assemble reliable square foot costs to develop a conceptual, or budget, estimate. Since conceptual estimate figures are used to establish a project's feasibility and budget, it is critical that the numbers achieve a certain level of credibility and accuracy.

Estimators must begin their work in much the same way as building designers. From the owner's or developer's description of the project's basic parameters, an estimator must create a conceptual estimate, and the architect and/or engineer must create a conceptual design. Building codes, zoning regulations, and published references are good sources of information to help define the project's size and other requirements and parameters. Costs can be derived from available square foot cost data. However, care must be taken when using cost data from more than one source, as the projects the data is based on should be the same, in most if not all respects, to the project being estimated.

The Assemblies Approach

Assemblies (or systems) estimates break the building down into its major components—structural system, exterior wall, roofing, and so forth. By pricing the building component by component, it is no longer necessary to find an exact building match to derive a credible project cost. The advantage of using the assemblies estimating technique is improved accuracy without a significant increase in estimating time.

The first step in any assemblies estimate is understanding the factors that affect the project's cost. For example, how do increased design loads and varying soil conditions affect the total cost of a building? It's one thing to take the total project costs for a number of buildings, divide them by their square foot dimensions, and develop ballpark figures for cost projections. It's an entirely different matter to know just why each of those buildings has its own unique price tag. Cost is not related to size alone. The assemblies square foot estimating process allows estimators to quickly and efficiently analyze all the major elements of the project—and compare alternatives—while developing a total project cost.

An assemblies estimate can be used during the schematic design and design development phases of a project to further refine the estimate and compare and evaluate the cost impact of various building systems. The estimator can assign and price specific systems for each area of construction.

The term "assemblies square foot estimate" refers to an estimate that divides the costs of each element in an assemblies estimate by the building square foot area. The result is square foot cost for each of the various elements of the building.

For an estimate to be well organized, it must have a clear direction from the beginning. One of the biggest advantages of assemblies estimating is that it simplifies the complex task of organizing and including the various parts of a construction project. The assemblies approach follows the sequence of construction activities, beginning with excavation and foundation, through the major structural elements, exterior closure, interior finishes, mechanical and electrical systems, and, finally, sitework and landscaping.

Reliable assemblies estimates must be based on more than a simplistic analysis of completed projects. They should be defined in relation to specific building components likely to be used in the proposed project. The costs of systems used in the estimate must be analyzed with realistic data. For instance, foundation and structural elements should be selected after analysis of the dead and live loads on the building.

Using the assemblies approach is fast and logical, and makes more sense in the final analysis than guesswork or poorly-matched standardized cost information. The relatively small amount of time required to complete an assemblies estimate is well invested in a credible and realistic estimate of probable project costs.

How to Use This Book

Means Square Foot & UNIFORMAT Assemblies Estimating Methods is a practical guide to a type of estimating used for budgeting projects at a time when only preliminary project information is available. At this stage, the plans have not been developed, and only basic project parameters are known. This book explains how to use square foot and assemblies estimating techniques correctly and advantageously under these circumstances.

Chapters 1 and 2 show how to create conceptual estimates for space planning, construction with cost limitations, and projects with size and site limitations. The chapters that follow are organized by UNIFORMAT II divisions and explain, step-by-step, the procedures for creating an assemblies square foot estimate.

The assemblies estimating method reorganizes the familiar 50 divisions of the CSI MasterFormat system into the seven alpha-numeric UNIFORMAT II divisions that closely reflect the order in which a building is constructed.

- A: Substructure
- B: Shell
  - B10: Superstructure
  - B20: Exterior Closure
  - B30: Roofing
- C: Interiors
- D: Services
  - D10: Conveying
  - D20: Plumbing
  - D30: HVAC
  - D40: Fire Protection
  - D50: Electrical
- E: Equipment & Furnishings
- F: Special Construction & Demolition
- G: Building Sitework

Using the assemblies method, it is not necessary to know every detail to establish a price for a building element, such as the foundation. You

The Estimate Example

need to know only building dimensions, proposed location including important site conditions, and major project requirements.

Running throughout this book is a sample estimate for a three-story office building, which will provide a detailed example of how all the pieces of cost information for the estimate go together. This example closely follows the organizational pattern of the assemblies format. The three-story office building estimate appears at the end of Chapters 3-16, set off from the body of the text with a black outline and a gray background.

The three-story office building estimate follows the process of preparing a cost estimate step-by-step using assemblies square foot costs, and shows how all the costs are derived. Using this format, an estimate can be developed based on assemblies costs from any reliable source, including published data; the owner's records of recent, similar projects; the contractor's and subcontractors' data; or the architect's or engineer's records. *Means Assemblies Cost Data* has been used exclusively for the office building estimate example, and relevant pages have been reproduced to show the exact source of the cost information.

The office building estimate summary pages list the costs for each UNIFORMAT II division. These forms are cumulative, with new cost items added to the summary shown for the previous chapter.

The second part of this "How to Use" section illustrates the components of *Means Assemblies Cost Data* pages and the estimate summary forms.

After all assemblies are selected and tabulated, the estimate must be totaled and all mark-ups added. This is a critical part of the estimate, which does not always get the attention it deserves. Chapter 16 describes the process of adding general conditions, sales taxes, contingencies, overhead and profit, and design fees to the estimate, and adjusting costs to a specific location.

Note: For consistency, the 2001 and 2002 editions of Means Assemblies Cost Data *are referenced throughout the examples in this book. The line numbers, tables, and reference numbers mentioned can be found in subsequent editions of* Means Assemblies Cost Data. *The numbering system used by Means was changed from UniFormat to UNIFORMAT II beginning in 2002, and the assembly headings shown in this book are from the 2002 edition. Costs are based on 2001* Means Assemblies cost database.

The prices in Means Assemblies Cost Data *are averaged for 30 major cities in the United States. This data has been accumulated from actual job costs, and material dealers' quotations have been combined with negotiated labor rates. Hundreds of contractors, subcontractors, and manufacturers have furnished cost information on their products.*

Variations in wage rates, labor efficiency, union restrictions, and material prices will result in local fluctuations. Local, regional, or national shortages of construction materials can severely influence material costs as well as cause considerable job delays with a corresponding increase in indirect job costs. Sales tax is not included in material prices. Prices include the installing contractor's overhead and profit (O&P). The prices are those that would be quoted to either the general or prime contractor. An allowance for the general or prime contractor's mark-up, overhead, supervision, and management should be added to the prices in this book. The usual range for this item is 5% to 15%. A figure of 10% is the most typical allowance.

How to Use the Assemblies Cost Tables

The following is a detailed explanation of a sample Assemblies Cost Table. Most Assembly Tables are separated into three parts: 1) an illustration of the system to be estimated; 2) the components and related costs of a typical system; and 3) the costs for similar systems with dimensional and/or size variations. Next to each bold number below is the item being described with the appropriate component of the sample entry following in parenthesis. In most cases, if the work is to be subcontracted, the general contractor will need to add an additional markup (R.S. Means suggests using 10%) to the "Total" figures.

1 System/Line Numbers (B1010-223-2000)

Each Assemblies Cost Line has been assigned a unique identification number based on the UNIFORMAT II Elemental Classification system.

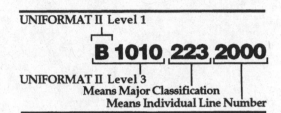

UNIFORMAT II Level 1

B 1010 , 223 , 2000

UNIFORMAT II Level 3
Means Major Classification
Means Individual Line Number

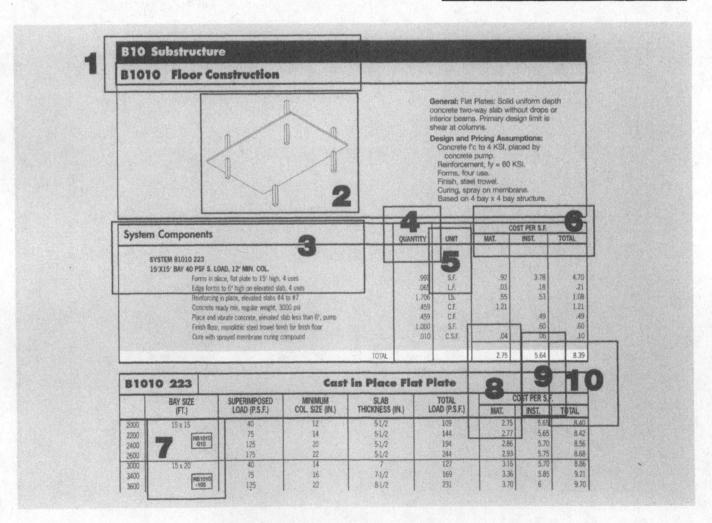

1 B10 Substructure

B1010 Floor Construction

2

General: Flat Plates: Solid uniform depth concrete two-way slab without drops or interior beams. Primary design limit is shear at columns.

Design and Pricing Assumptions:
Concrete f'c to 4 KSI, placed by concrete pump.
Reinforcement, fy = 60 KSI.
Forms, four use.
Finish, steel trowel.
Curing, spray on membrane.
Based on 4 bay x 4 bay structure.

System Components **3**

SYSTEM B1010 223
15'X15' BAY 40 PSF S. LOAD, 12" MIN. COL.

	QUANTITY **4**	UNIT **5**	MAT.	INST.	TOTAL **6**
			COST PER S.F.		
Forms in place, flat plate to 15' high, 4 uses	.990	S.F.	.92	3.78	4.70
Edge forms to 6" high on elevated slab, 4 uses	.065	L.F.	.03	.18	.21
Reinforcing in place, elevated slabs #4 to #7	1.706	Lb.	.55	.53	1.08
Concrete ready mix, regular weight, 3000 psi	.459	C.F.	1.21		1.21
Place and vibrate concrete, elevated slab less than 6", pump	.459	C.F.		.49	.49
Finish floor, monolithic steel trowel finish for finish floor	1.000	S.F.		.60	.60
Cure with sprayed membrane curing compound	.010	C.S.F.	.04	.06	.10
TOTAL			2.75	5.64	8.39

B1010 223			Cast in Place Flat Plate			**8** MAT.	**9** INST.	**10** TOTAL
	BAY SIZE (FT.) **7**	SUPERIMPOSED LOAD (P.S.F.)	MINIMUM COL. SIZE (IN.)	SLAB THICKNESS (IN.)	TOTAL LOAD (P.S.F.)	COST PER S.F.		
2000	15 x 15	40	12	5-1/2	109	2.75	5.65	8.40
2200	RB1010 -010	75	14	5-1/2	144	2.77	5.65	8.42
2400		125	20	5-1/2	194	2.86	5.70	8.56
2600		175	22	5-1/2	244	2.93	5.75	8.68
3000	15 x 20	40	14	7	127	3.16	5.70	8.86
3400	RB1010 -105	75	16	7-1/2	169	3.36	5.85	9.21
3600		125	22	8-1/2	231	3.70	6	9.70

2 Illustration

At the top of most assembly tables is an illustration, a brief description, and the design criteria used to develop the cost.

3 System Components

The components of a typical system are listed separately to show what has been included in the development of the total system price. The table below contains prices for other similar systems with dimensional and/or size variations.

4 Quantity

This is the number of line item units required for one system unit. For example, we assume that it will take 1.706 pounds of reinforcing on a square foot basis.

5 Unit of Measure for Each Item

The abbreviated designation indicates the unit of measure, as defined by industry standards, upon which the price of the component is based. For example, reinforcing is priced by lb. (pound) while concrete is priced by C.F. (cubic foot).

6 Unit of Measure for Each System (Cost per S.F.)

Costs shown in the three right hand columns have been adjusted by the component quantity and unit of measure for the entire system. In this example, "Cost per S.F." is the unit of measure for this system or "assembly."

7 Reference Number Information

| RB1010 -010 | You'll see reference numbers shown in bold rectangles at the beginning of some major classifications. These refer to related items in |

the Reference Section, visually identified by a vertical gray bar on the edge of pages.

The relation may be: (1) an estimating procedure that should be read before estimating, (2) an alternate pricing method, or (3) technical information.

The "R" designates the Reference Section. The numbers refer to the UNIFORMAT II classification system.

Example: The rectangle number above is directing you to refer to the reference number RB1010-010. This particular reference number shows comparative costs of floor systems.

8 Materials (2.75)

This column contains the Materials Cost of each component. These cost figures are bare costs plus 10% for profit.

9 Installation (5.64)

Installation includes labor and equipment plus the installing contractor's overhead and profit. Equipment costs are the bare rental costs plus 10% for profit. The labor overhead and profit is defined on the inside back cover of this book.

10 Total (8.39)

The figure in this column is the sum of the material and installation costs.

Material Cost	+	Installation Cost	=	Total
$2.75	+	$5.64	=	$8.39

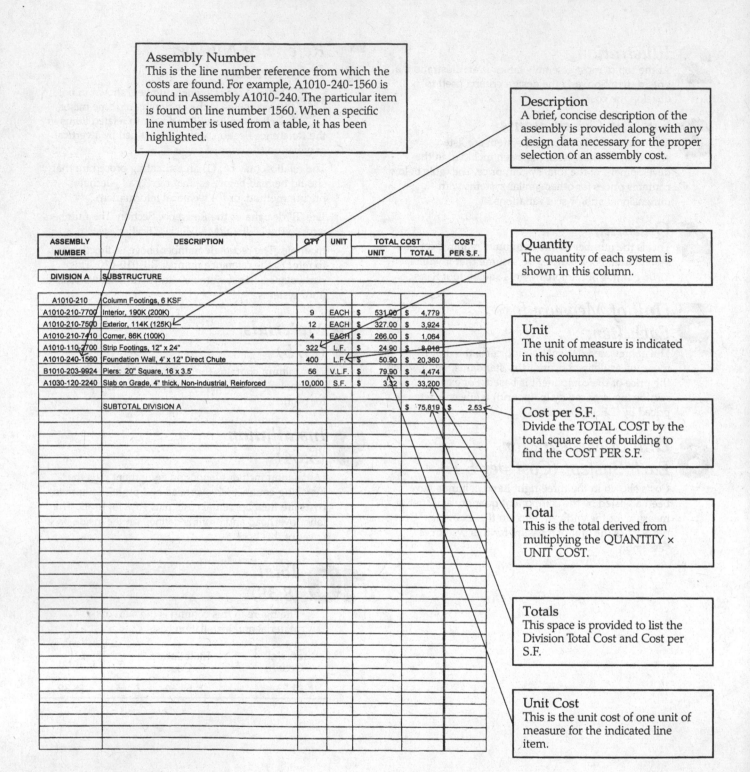

Assembly Number
This is the line number reference from which the costs are found. For example, A1010-240-1560 is found in Assembly A1010-240. The particular item is found on line number 1560. When a specific line number is used from a table, it has been highlighted.

Description
A brief, concise description of the assembly is provided along with any design data necessary for the proper selection of an assembly cost.

Quantity
The quantity of each system is shown in this column.

Unit
The unit of measure is indicated in this column.

Cost per S.F.
Divide the TOTAL COST by the total square feet of building to find the COST PER S.F.

Total
This is the total derived from multiplying the QUANTITY × UNIT COST.

Totals
This space is provided to list the Division Total Cost and Cost per S.F.

Unit Cost
This is the unit cost of one unit of measure for the indicated line item.

ASSEMBLY NUMBER	DESCRIPTION	QTY	UNIT	TOTAL COST UNIT	TOTAL COST TOTAL	COST PER S.F.
DIVISION A	SUBSTRUCTURE					
A1010-210	Column Footings, 6 KSF					
A1010-210-7700	Interior, 190K (200K)	9	EACH	$ 531.00	$ 4,779	
A1010-210-7500	Exterior, 114K (125K)	12	EACH	$ 327.00	$ 3,924	
A1010-210-7410	Corner, 86K (100K)	4	EACH	$ 266.00	$ 1,064	
A1010-110-2700	Strip Footings, 12" x 24"	322	L.F.	$ 24.90	$ 8,018	
A1010-240-1560	Foundation Wall, 4' x 12" Direct Chute	400	L.F.	$ 50.90	$ 20,360	
B1010-203-9924	Piers: 20" Square, 16 x 3.5'	56	V.L.F.	$ 79.90	$ 4,474	
A1030-120-2240	Slab on Grade, 4" thick, Non-industrial, Reinforced	10,000	S.F.	$ 3.32	$ 33,200	
	SUBTOTAL DIVISION A				$ 75,819	$ 2.53

xviii

Chapter 1

Types of Estimates

What does it cost to construct an office building today? About $90 per square foot, give or take a few dollars depending on the location, right?

In a nutshell, that type of guess is what square foot cost estimating is all about. It is common in the construction industry to think of costs in those terms, and general figures for basic building types tend to be an accepted standard at any given time. Most contractors have these kinds of numbers in mind based on their experience over the years in a particular type of construction. Other industry professionals, such as architects and appraisers, also have an idea of what a certain type of house or office building will cost, depending on its size. Square foot cost estimating develops specific techniques for determining the estimated construction cost of a building based on its size and other selected criteria.

Experience teaches that general square foot cost figures come from a number of sources:

- Recent experience or cost figures taken from jobs of similar construction.
- Historical costs with appropriate mark-ups.
- Published costs for similar building types.
- Unit price estimates.
- Assemblies (or systems) estimates.
- Order of magnitude estimates.

To put square foot estimating in context, the next section defines the four major types of estimates.

Four Types of Estimates

Most buildings use similar materials to create the whole structure. Installation is often accomplished using typical methods, and contractors generally use a standard type of labor force, either union or open-shop. That is the basic idea behind estimating the cost of a construction project. Almost any two structures can be compared in relation to the basic elements they have in common. This principle is one of the cornerstones of estimating.

Several different methods can be used to arrive at the estimated cost of a particular project. To better understand these methods and their

1

appropriate use, a description of the four primary types of estimates follows, starting with the most accurate and detailed and ending with the quickest and least accurate. They are:

- Unit Price
- Assemblies (or Systems)
- Square Foot and Cubic Foot
- Order of Magnitude

Unit Price Estimates

Unit price estimates involve a careful breakdown of all the elements that go into the building process. Each unit item is matched to a specific quantity. To complete a unit price estimate, it is necessary to know the specified or required prices for all material, equipment, and labor. In general, this type of estimate is based on a complete set of working drawings and specifications, and can be started before the drawings are complete. The accuracy of unit price estimates is generally + or −5% with a complete set of construction documents.

Because of the extreme level of detail involved, unit price estimating is a lengthy and costly process. The basic purpose of this type of estimate is to develop prices for a designed structure and to arrive at an estimated cost for construction. It is usually not economically feasible or even possible to use unit price estimating in the conceptual stages of a building construction project. Typically, unit price estimates are performed for a project that is being negotiated or competitively bid.

Unit price estimates conform to the 16 MasterFormat divisions adopted by the Construction Specifications Institute (CSI), which are widely used by the building construction industry. These divisions have been developed to help maintain consistency in the classification of construction information. They are particularly useful when comparisons must be made between construction projects that have little else in common. Cost manuals used for unit price estimates such as *Means Building Construction Cost Data* divide all data into 16 CSI divisions, and are assembled one division at a time. Figure 1.1 shows the 16 CSI MasterFormat divisions.

The unit price estimating process begins with the *quantity takeoff*, a detailed analysis of the material and equipment required to construct a project. The estimator examines the plans for the project and notes the quantities, then matches them to the appropriate unit prices. For instance, if a wall in a building requires 50 cubic yards of concrete, that quantity is multiplied by the appropriate unit price for concrete to produce the cost for the concrete in the wall. If the unit price for 3,000 psi concrete is $70.00 per cubic yard, the estimated cost of the concrete itself (the cost of the material only) would be $3,500. This price is called an "extended cost." Such costs can be calculated for any part of a construction project: material, equipment, labor, and, eventually, total project cost including general conditions, overhead, and profit. Totals of all these costs, organized according to CSI format, are added to calculate the total unit price estimate.

Assemblies Estimate

When a project is in the conceptual stage, building professionals can use assemblies estimating (sometimes known as "systems" estimating) to consider and price different systems or elements of the structure. Without having an idea of the cost of building systems, a design professional could spend many weeks designing a building only to find that the design and the budget are incompatible when the estimate is done. Assemblies estimates provide a quick but fairly comprehensive way to study the cost impact of building systems as they relate to the project budget.

Both designers and building owners find it increasingly important to consider the cost relationship between major components in the structure. The assemblies approach to preliminary cost estimating involves grouping several trades and/or work items and tasks. Assemblies estimates do not require a detailed design. Instead, they are based on the size of the structure and other parameters as they pertain to a particular site and requirements of the owner. The degree of accuracy of an assemblies estimate is generally + or – 10%.

For example, a foundation usually requires excavation, formwork, reinforcing, concrete placement and finish, and backfill. To create a unit price estimate, it would be necessary to evaluate several separate line items for each of these activities to arrive at a cost. In an assemblies estimate, the work is combined into a single package. The cost of individual work tasks required to construct one assembly (e.g., a complete footing) is represented as one line item cost, including materials, labor, and equipment.

A close examination of a building system reveals that each item normally included in a unit price estimate finds a new slot in the assemblies estimate. While the unit price approach assigns costs for each piece of material, each equipment item, and all installation costs to a specific CSI division, an assemblies estimate essentially reshuffles the deck, reorganizing certain items that were formerly grouped into a single trade breakdown or division.

Breaking a Project Down into Components

One logical approach to developing an assemblies estimate is to break down the structure into several convenient components. A common

Division Number	MasterFormat
1	General Requirements
2	Site Construction
3	Concrete
4	Masonry
5	Metals
6	Wood & Plastics
7	Thermal & Moisture Protection
8	Doors & Windows
9	Finishes
10	Specialities
11	Equipment
12	Furnishings
13	Special Construction
14	Conveying Systems
15	Mechanical
16	Electrical

Figure 1.1

method is to consider the project components in the order they will be constructed. This method follows the ASTM E 1557 "Standard Classification for Building Elements and Related Sitework—UNIFORMAT II," a standardized way of classifying building elements that was issued in 1993. For the purpose of this book, the square foot estimating process follows UNIFORMAT II's system of organization.

UNIFORMAT II Divisions

A. Substructure:

A10 Foundations: Standard Foundations, Special Foundations, Slab on Grade

A20 Basement Construction: Basement Excavation, Basement Walls

B. Shell:

B10 Superstructure: Floor Construction, Roof Construction

B20 Exterior Closure: Exterior Walls, Exterior Windows, Exterior Doors

B30 Roofing: Roof Coverings, Roof Openings

C. Interiors:

C10 Interior Construction: Partitions, Interior Doors, Specialties

C20 Stairs: Stair Construction, Stair Finishes

C30 Interior Finishes: Wall Finishes, Floor Finishes, Ceiling Finishes

D. Services:

D10 Conveying: Elevators & Lifts, Escalators & Moving Walks, Other Conveying Systems

D20 Plumbing: Plumbing Fixtures, Domestic Water Distribution, Sanitary Waste, Rain Water Drainage, Other Plumbing Systems

D30 HVAC: Energy Supply, Heat Generating Systems, Cooling Generating Systems, Distribution Systems, Terminal & Package Units, Controls & Instrumentation, Systems Testing & Balancing, Other HVAC Systems & Equipment

D40 Fire Protection: Sprinklers, Standpipes, Fire Protection Specialties, Other Fire Protection Systems

D50 Electrical: Electrical Service & Distribution, Lighting & Branch Wiring, Communication & Security, Other Electrical Systems

E. Equipment & Furnishings:

E10 Equipment: Commercial Equipment, Institutional Equipment, Vehicular Equipment, Other Equipment

E20 Furnishings: Fixed Furnishings, Movable Furnishings

F. Special Construction & Demolition:

F10 Special Construction: Special Structures, Integrated Construction, Special Construction Systems, Special Facilities, Special Controls and Instrumentation

F20 Selective Building Demolition: Building Elements Demolition, Hazardous Components Abatement

G. Building Sitework:

G10 Site Preparation: Site Clearing, Site Demolition & Relocations, Site Earthwork, Hazardous Waste Remediation

G20 Site Improvements: Roadways, Parking Lots, Pedestrian Paving, Site Development, Landscaping

G30 Site Mechanical Utilities: Water Supply, Sanitary Sewer, Storm

Sewer, Heating Distribution, Cooling Distribution, Fuel Distribution, Other Site Mechanical Utilities

G40 Site Electrical Utilities: Electrical Distribution, Site Lighting, Site Communications & Security, Other Site Electrical Utilities

G90 Other Site Construction: Service and Pedestrian Tunnels, Other Site Systems & Equipment

Using the Assemblies Approach

The assemblies format allows you to examine the various elements of the building and their impact on total project cost. It is not necessary for the design professional to assign specific materials or detailed dimensions to every element in a design before arriving at an assemblies cost. He or she can select assemblies in various combinations within basic, predetermined limitations. Comparing assemblies costs will lead to the choice that best accommodates the project's budget, building codes, and the owner's special requirements.

In the UNIFORMAT II organization of the assemblies estimate, the same construction materials end up being included as part of many different systems in the estimate. While a unit price estimate would group all concrete items in MasterFormat Division 3, Concrete, the assemblies estimate will include concrete in several categories, including Substructure, Shell, Interiors, and Sitework, depending on the assembly used. Also, items listed in separate divisions in MasterFormat can be combined into one assembly in UNIFORMAT II.

Assemblies consolidate the many different elements that go into a complex system. Assemblies are easy to use, since they reflect the way the building is constructed. This makes it easier to get an overview of what each part of the building costs. Interior partitions are a good case in point. In a unit price estimate, interior partitions may be composed of Division 4 (Masonry), Division 5 (Metals), Division 6 (Wood & Plastics), Division 8 (Doors & Windows), and Division 9 (Finishes). In the assemblies estimate, these are combined in Division C (Interiors). This relocation of familiar items may seem a bit confusing at first, but the increase in estimating speed can be well worth the time it takes to get to know UNIFORMAT II.

Making Trade-offs During Initial Design

Assemblies estimating is normally performed before detailed plans and specifications have been developed, and helps the designer to bring the project within the owner's budget. This may involve making decisions and comparisons on each of the various assemblies. All must be studied in relation to each other and to the project as a whole. Some of the criteria may include:

- Price of each assembly
- Quantity of assembly
- Appearance and quality
- Story height
- Clear span
- Site complications and restrictions
- Zoning limitations
- Thermal characteristics
- Life cycle costs
- Maintenance costs

5

- Energy considerations
- Environmental controls
- Acoustical characteristics
- Fireproofing characteristics
- Owner's special requirements in excess of building code requirements
- Budget restraints
- Type of occupancy

Organizing the Assemblies Estimate

When organizing an assemblies estimate, it is important to gather all available information on the project. Figure 1.2 is a preliminary estimate data sheet that can be used to collect project information. Fill in as much as possible, gathering information from:

- Building code requirements
- Owner's requirements
- Preliminary assumptions
- Site inspection and investigation
- Schedule requirements

Assemblies Estimating Forms

Estimating should be well organized to save time, to make the process as easy as possible, to provide a checklist to ensure that no items are overlooked, and to increase the overall accuracy of the estimate. A standard estimating form is extremely helpful. The sample form shown in Figure 1.3. can be used for an entire building or one or two divisions.

In addition to the preliminary estimate form, special cost tables in books such as *Means Assemblies Cost Data* can be used to:

- Determine costs of individual footings, columns, and other building features.
- Develop cost comparisons for different assemblies, such as floor or roof systems.
- Select and combine materials to develop the system price for an item like a wall or ceiling.
- Calculate the loads for structural, mechanical, and electrical components in order to determine total foundation and superstructure costs.

Square Foot and Cubic Foot Estimates

A rough idea of the project cost is necessary at the earliest stage of design. At this point there are usually no plans; only basic requirements are known. A square foot or cubic foot estimate can provide this information with less than an hour's work using tables such as those found in *Means Assemblies Cost Data*. You need only know the type of building or facility and the proposed number of square feet. Estimates are usually within + or – 15% of the final project cost, providing allowances are made for the following:

- Classification of the job (such as apartment, hotel, or school).
- Type of owner (such as government or industrial).
- Location of project (using City Cost Indexes).
- Relative size of the project (using size modifier).
- Construction period (accounting for future escalation of costs).

All final square foot costs are derived from the thousands of components that make up a building. Means square foot costs include

PRELIMINARY ESTIMATE

PROJECT _____   TOTAL SITE AREA _____

BUILDING TYPE _____   OWNER _____

LOCATION _____   ARCHITECT _____

DATE OF CONSTRUCTION _____   ESTIMATED CONSTRUCTION PERIOD _____

BRIEF DESCRIPTION _____

_____

_____

TYPE OF PLAN _____   TYPE OF CONSTRUCTION _____

QUALITY _____   BUILDING CAPACITY _____

Floor			Wall Areas					
Below Grade Levels			Foundation Walls	L.F.		Ht.		S.F.
Area		S.F.	Frost Walls	L.F.		Ht.		S.F.
Area		S.F.	Exterior Closure			Total		S.F.
Total Area		S.F.	Comment					
Ground Floor			Fenestration			%		S.F.
Area		S.F.				%		S.F.
Area		S.F.	Exterior Wall			%		S.F.
Total Area		S.F.				%		S.F.
Supported Levels			Site Work					
Area		S.F.	Parking			S.F. (For		Cars)
Area		S.F.	Access Roads			L.F. (X		Ft. Wide)
Area		S.F.	Sidewalk			L.F. (X		Ft. Wide)
Area		S.F.	Landscaping			S.F. (		% Unbuilt Site)
Area		S.F.	Building Codes					
Total Area		S.F.	City			County		
Miscellaneous			National			Other		
Area		S.F.	Loading					
Area		S.F.	Roof		psf	Ground Floor		psf
Area		S.F.	Supported Floors		psf	Corridor		psf
Area		S.F.	Balcony		psf	Partition, allow		psf
Total Area		S.F.	Miscellaneous					psf
Net Finished Area		S.F.	Live Load Reduction					
Net Floor Area		S.F.	Wind					
Gross Floor Area		S.F.	Earthquake			Zone		
Roof			Comment					
Total Area		S.F.	Soil Type					
Comments			Bearing Capacity					K.S.F.
			Frost Depth					Ft.
Volume			Frame					
Depth of Floor System			Type			Bay Spacing		
Minimum		In.	Foundation, Standard					
Maximum		In.	Special					
Foundation Wall Height		Ft.	Substructure					
Floor to Floor Height		Ft.	Comment					
Floor to Ceiling Height		Ft.	Superstructure, Vertical			Horizontal		
Subgrade Volume		C.F.	Fireproofing			☐ Columns		Hrs.
Above Grade Volume		C.F.	☐ Girders		Hrs.	☐ Beams		Hrs.
Total Building Volume		C.F.	☐ Floor		Hrs.	☐ None		

Figure 1.2

Estimating Form

ASSEMBLY NUMBER	DESCRIPTION	QTY	UNIT	TOTAL COST UNIT	TOTAL COST TOTAL	COST PER S.F.
A10	Foundations					
A20	Basement Construction					
B10	Superstructure					
B20	Exterior Closure					
B30	Roofing					

Figure 1.3

materials and installation. In general, these costs represent the cost of the building only. Furnishings, equipment, specialties, and site development are not included, and must be added.

Square foot estimates are helpful not only at the conceptual stage when no details are available, but also after the bids are received and a reality check is required. Components of the estimate can be compared to sections of the square foot assemblies estimate. If costs are not within 15 percent of the assemblies estimate, further investigation may be advisable. As soon as details become available in the project design, the square foot approach should be discontinued and the project priced by assemblies and/or unit prices.

Cubic foot estimates are useful when the floor-to-floor heights differ from the typical heights for a building use. For instance, if the typical floor-to-floor height of an office building is 12 feet, and the proposed building is to have 16' floor-to-floor height, a cubic foot estimate will provide a more accurate cost projection.

Project Size Modifiers

The Square Foot Base Size chart in Figure 1.4 shows that the typical high-rise apartment building is 145,000 square feet.

Dividing the proposed building area of 172,000 S.F. by 145,000 S.F. yields a size factor of 1.2.

Enter the Area Conversion Scale in Figure 1.4 with the factor 1.2 on the horizontal axis. Project vertically to intersect the cost modifier curve line on the graph. Now, project horizontally. The figure for the cost multiplier is 0.98.

$$\$12,340,000 \times 0.98 = \$12,093,200$$

The national average cost of the 200-unit apartment building would be $12,093,200.

City Cost Indexes

Costs may also be modified by a City Cost Index, as published in Means annual cost data publications and electronic data. Any cost, when totaled and multiplied by the appropriate City Cost Index, will result in a total cost for an entire building that will be close to the actual cost for construction at a specific location in the United States and Canada. The relative index for material, installation, and total construction costs is listed for the 316 largest U.S. and Canadian cities. To determine the cost in a specific location, multiply the national average cost times the location factor divided by 100. *(Note: Location factors for Canadian cities reflect the cost in Canadian dollars. Chapter 16 provides more discussion on City Cost Indexes.)*

Chapter 2 explains in more detail how square foot estimates are put together and includes examples of how they are used in the design stage of various projects.

Order of Magnitude Estimates

The order of magnitude estimate is the most preliminary of all of the types, with an accuracy of + or –20%. It can be completed with minimal information, and is often referred to as an "educated guess" or "ballpark" figure for planning projects. Order of magnitude estimates are most useful when the project is not yet defined, but the owner and

Square Foot Project Size Modifier

One factor that affects the S.F. cost of a particular building is the size. In general, for buildings built to the same specifications in the same locality, the larger building will have the lower S.F. cost. This is due mainly to the decreasing contribution of the exterior walls plus the economy of scale usually achievable in larger buildings. The Area Conversion Scale shown below will give a factor to convert costs for the typical size building to an adjusted cost for the particular project.

The Square Foot Base Size lists the median costs, most typical project size in our accumulated data and the range in size of the projects.

The Size Factor for your project is determined by dividing your project area in S.F. by the typical project size for the particular Building Type. With this factor, enter the Area Conversion Scale at the appropriate Size Factor and determine the appropriate cost multiplier for your building size.

Example: Determine the cost per S.F. for a 100,000 S.F. Mid-rise apartment building.

$$\frac{\text{Proposed building area} = 100,000 \text{ S.F.}}{\text{Typical size from below} = 50,000 \text{ S.F.}} = 2.00$$

Enter Area Conversion scale at 2.0, intersect curve, read horizontally the appropriate cost multiplier of .94. Size adjusted cost becomes .94 x $69.50 = $65.35 based on national average costs.

Note: For Size Factors less than .50, the Cost Multiplier is 1.1
For Size Factors greater than 3.5, the Cost Multiplier is .90

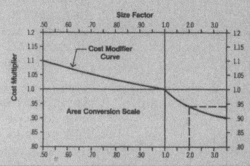

Square Foot Base Size							
Building Type	Median Cost per S.F.	Typical Size Gross S.F.	Typical Range Gross S.F.	Building Type	Median Cost per S.F.	Typical Size Gross S.F.	Typical Range Gross S.F.
Apartments, Low Rise	$ 55.05	21,000	9,700 - 37,200	Jails	$168.00	40,000	5,500 - 145,000
Apartments, Mid Rise	69.50	50,000	32,000 - 100,000	Libraries	99.15	12,000	7,000 - 31,000
Apartments, High Rise	79.70	145,000	95,000 - 600,000	Medical Clinics	94.80	7,200	4,200 - 15,700
Auditoriums	92.05	25,000	7,600 - 39,000	Medical Offices	89.05	6,000	4,000 - 15,000
Auto Sales	56.95	20,000	10,800 - 28,600	Motels	68.25	40,000	15,800 - 120,000
Banks	123.00	4,200	2,500 - 7,500	Nursing Homes	91.55	23,000	15,000 - 37,000
Churches	83.15	17,000	2,000 - 42,000	Offices, Low Rise	74.30	20,000	5,000 - 80,000
Clubs, Country	82.85	6,500	4,500 - 15,000	Offices, Mid Rise	78.05	120,000	20,000 - 300,000
Clubs, Social	80.65	10,000	6,000 - 13,500	Offices, High Rise	99.85	260,000	120,000 - 800,000
Clubs, YMCA	83.10	28,300	12,800 - 39,400	Police Stations	125.00	10,500	4,000 - 19,000
Colleges (Class)	109.00	50,000	15,000 - 150,000	Post Offices	92.00	12,400	6,800 - 30,000
Colleges (Science Lab)	159.00	45,600	16,600 - 80,000	Power Plants	691.00	7,500	1,000 - 20,000
College (Student Union)	121.00	33,400	16,000 - 85,000	Religious Education	76.25	9,000	6,000 - 12,000
Community Center	86.60	9,400	5,300 - 16,700	Research	130.00	19,000	6,300 - 45,000
Court Houses	118.00	32,400	17,800 - 106,000	Restaurants	112.00	4,400	2,800 - 6,000
Dept. Stores	51.45	90,000	44,000 - 122,000	Retail Stores	54.65	7,200	4,000 - 17,600
Dormitories, Low Rise	88.75	25,000	10,000 - 95,000	Schools, Elementary	79.60	41,000	24,500 - 55,000
Dormitories, Mid Rise	116.00	85,000	20,000 - 200,000	Schools, Jr. High	81.10	92,000	52,000 - 119,000
Factories	49.85	26,400	12,900 - 50,000	Schools, Sr. High	81.10	101,000	50,500 - 175,000
Fire Stations	87.05	5,800	4,000 - 8,700	Schools, Vocational	80.75	37,000	20,500 - 82,000
Fraternity Houses	85.65	12,500	8,200 - 14,800	Sports Arenas	67.70	15,000	5,000 - 40,000
Funeral Homes	95.70	10,000	4,000 - 20,000	Supermarkets	54.85	44,000	12,000 - 60,000
Garages, Commercial	60.80	9,300	5,000 - 13,600	Swimming Pools	127.00	20,000	10,000 - 32,000
Garages, Municipal	77.75	8,300	4,500 - 12,600	Telephone Exchange	147.00	4,500	1,200 - 10,600
Garages, Parking	31.90	163,000	76,400 - 225,300	Theaters	81.20	10,500	8,800 - 17,500
Gymnasiums	80.45	19,200	11,600 - 41,000	Town Halls	89.30	10,800	4,800 - 23,400
Hospitals	152.00	55,000	27,200 - 125,000	Warehouses	36.80	25,000	8,000 - 72,000
House (Elderly)	75.25	37,000	21,000 - 66,000	Warehouse & Office	42.50	25,000	8,000 - 72,000
Housing (Public)	69.70	36,000	14,400 - 74,400				
Ice Rinks	77.40	29,000	27,200 - 33,600				

Figure 1.4

designer need an indication of the cost of the project at completion so that a decision can be made to proceed. The only information necessary for this type of estimate is the proposed use and size of the planned structure.

Order of magnitude costs may be defined in relation to usable units that have been designed for a facility. For example, hospital administrators who plan to enlarge a hospital will need to know the projected cost per bed. Tables such as Figure 1.5 indicate usable unit costs for several types of structures. This table also includes square foot and cubic foot costs. If the number of beds in the proposed hospital (or the number of apartments in an apartment building, or parking spaces in a garage) is known, along with primary materials to be used for the structure, an order of magnitude estimate can be generated.

Using Historical Costs

An order of magnitude estimate lists historical costs per usable unit of facility. For example, assume that a developer would like an approximate cost for financial planning on a proposed 20-story apartment building with 200 apartments. Figure 1.6 shows the historical cost per apartment.

A quick look at the table reveals a median historical cost per apartment of $61,700.

$$200 \text{ apartments} \times \$61,700/\text{apartment} = \$12,340,000$$

To carry the estimate one step further, use the Space Planning Guide (Figure 1.7), which shows that the typical gross square footage for a high-rise apartment unit is 860 square feet.

$$200 \text{ apartments} \times 860 \text{ S.F./apartment} = 172,000 \text{ S.F.}$$

Order of Magnitude and Square Foot Estimate Comparison

Order of magnitude and square foot estimates may be used to develop preliminary costs only if certain limitations are taken into account. Always consider the following:

- Neither order of magnitude nor square foot estimates reflect actual floor and roof loading, structural systems, clear spans, or column spacing. All of these items can be important cost factors.
- At the time these estimates are assembled, the materials to be used for the exterior closure system, as well as the extent and type of glass and glazing, are not likely to be defined. Selection can have a major impact on cost.
- Neither of the preliminary estimates makes allowances for the type of heating and cooling systems, building automation systems, the extent of plumbing requirements, and the need for fire protection sprinklers—all important cost considerations.
- Some building square foot costs include sitework, and others do not. If possible, determine whether or not this cost is part of the figures being used for the estimate.

Always carefully evaluate the parameters of an order of magnitude or square foot estimate to ensure its validity. At the conceptual stage when there is little definition of the project, try to match the building type as closely as possible to the available reference data. If you are using Means square foot and cubic foot costs, assume that they are for the building only and do not include the cost of sitework or purchasing land.

K1010 | S.F., C.F. and % of Total Costs

		K1010 | S.F. & C.F. Costs	UNIT	UNIT COSTS			% OF TOTAL			
				1/4	MEDIAN	3/4	1/4	MEDIAN	3/4	
410	0010	**GARAGES, PARKING**	S.F.	22	31.90	56.95				**410**
	0020	Total project costs	C.F.	2.13	2.90	4.22				
	2720	Plumbing	S.F.	.59	1	1.54	2.60%	3.40%	3.90%	
	2900	Electrical		.93	1.41	2.07	4.30%	5.20%	6.30%	
	3100	Total: Mechanical & Electrical	↓	1.24	3.64	4.67	6.50%	9.40%	12.80%	
	3200									
	9000	Per car, total cost	Car	9,600	12,000	15,600				
430	0010	**GYMNASIUMS**	S.F.	63	80.45	103				**430**
	0020	Total project costs	C.F.	3.15	4	5.25				
	1800	Equipment	S.F.	1.33	2.71	5.40	2.10%	3.40%	6.70%	
	2720	Plumbing		4	4.93	5.90	5.40%	7.30%	7.90%	
	2770	Heating, ventilating, air conditioning		4.30	6.55	13.15	9%	11.10%	22.60%	
	2900	Electrical		4.79	6	8.05	6.60%	8.30%	10.70%	
	3100	Total: Mechanical & Electrical	↓	14.80	20.40	27.35	20.60%	26.20%	29.40%	
	3500	See also division 11480								
460	0010	**HOSPITALS**	S.F.	130	152	225				**460**
	0020	Total project costs	C.F.	9.20	11.35	16.35				
	1800	Equipment	S.F.	3.16	5.90	10.20	2.50%	3.80%	6%	
	2720	Plumbing		10.85	14.60	18.80	7.80%	9.40%	11.80%	
	2770	Heating, ventilating, air conditioning		15.35	20.35	27.40	8.40%	14.60%	17%	
	2900	Electrical		13.05	16.95	26.70	9.90%	12%	14.50%	
	3100	Total: Mechanical & Electrical	↓	37.15	49.25	80.70	26.60%	33.10%	39%	
	9000	Per bed or person, total cost	Bed	46,800	99,300	156,600				
	9900	See also division 11700 & 11780								
480	0010	**HOUSING** For the Elderly	S.F.	59.65	75.25	92.45				**480**
	0020	Total project costs	C.F.	4.23	5.90	7.55				
	0100	Site work	S.F.	4.16	6.70	9.45	5.10%	8.20%	12.10%	
	0500	Masonry		1.82	6.80	9.90	2.10%	7.10%	12.20%	
	1800	Equipment		1.43	1.95	3.15	1.90%	3.20%	4.40%	
	2510	Conveying systems		1.45	1.95	2.65	1.80%	2.30%	2.90%	
	2720	Plumbing		4.43	5.80	7.75	8.10%	9.70%	10.90%	
	2730	Heating, ventilating, air conditioning		2.27	3.22	4.81	3.30%	5.60%	7.20%	
	2900	Electrical		4.44	6	7.70	7.30%	8.50%	10.50%	
	3100	Total: Mechanical & Electrical	↓	15.30	17.95	23.75	18.10%	22%	29.10%	
	9000	Per rental unit, total cost	Unit	55,500	64,500	72,100				
	9500	Total: Mechanical & Electrical		12,200	14,200	16,600				
500	0010	**HOUSING** Public (Low Rise)	S.F.	51.85	69.70	90.60				**500**
	0020	Total project costs	C.F.	3.98	5.55	6.90				
	0100	Site work	S.F.	6.40	9.20	14.90	8.40%	11.70%	16.50%	
	1800	Equipment		1.36	2.22	3.64	2.30%	3%	5.10%	
	2720	Plumbing		3.40	4.78	6.05	6.80%	9%	11.60%	
	2730	Heating, ventilating, air conditioning		1.82	3.53	3.86	4.20%	6%	6.40%	
	2900	Electrical		3.03	4.52	6.25	5.10%	6.60%	8.30%	
	3100	Total: Mechanical & Electrical	↓	14.40	18.55	20.75	14.50%	17.60%	26.50%	
	9000	Per apartment, total cost	Apt.	55,000	62,600	78,600				
	9500	Total: Mechanical & Electrical		10,000	13,000	15,000				
510	0010	**ICE SKATING RINKS**	S.F.	44.60	77.40	110				**510**
	0020	Total project costs	C.F.	3.15	3.22	3.71				
	2720	Plumbing	S.F.	1.60	3	3.07	3.10%	5.60%	6.70%	
	2900	Electrical		4.59	7.04	7.45	6.70%	15%	15.80%	
	3100	Total: Mechanical & Electrical	↓	7.70	11.10	13.85	9.90%	25.90%	29.80%	
520	0010	**JAILS**	S.F.	132	168	217				**520**
	0020	Total project costs	C.F.	12.30	15.10	19.25				
	1800	Equipment	S.F.	5.45	15.05	25.60	4%	9.40%	15.20%	
	2720	Plumbing		13.30	16.85	22.25	7%	8.90%	13.80%	
	2770	Heating, ventilating, air conditioning		11.75	15.71	30.35	7.50%	9.40%	17.70%	
	2900	Electrical	↓	14.35	18.45	22.80	8.10%	11.40%	12.40%	

Figure 1.5

K1010 | S.F., C.F. and % of Total Costs

		K1010	S.F. & C.F. Costs	UNIT	UNIT COSTS			% OF TOTAL			
					1/4	MEDIAN	3/4	1/4	MEDIAN	3/4	
010	0010	APARTMENTS Low Rise (1 to 3 story)	RK1010 010	S.F.	43.95	55.05	73.60				010
	0020	Total project cost		C.F.							
	0100	Site work		S.F.	4.02	5.50	8.70	6.80%	11%	14.10%	
	0500	Masonry			.82	2.13	3.50	1.50%	4%	6.50%	
	1500	Finishes			4.62	6.05	7.80	9%	10.70%	12.90%	
	1800	Equipment			1.43	2.17	3.15	2.70%	4.10%	6.20%	
	2720	Plumbing			3.39	4.40	5.55	6.70%	9%	10.10%	
	2770	Heating, ventilating, air conditioning			2.18	2.69	3.95	4.20%	5.80%	7.70%	
	2900	Electrical			2.53	3.30	4.52	5.20%	6.70%	8.40%	
	3100	Total: Mechanical & Electrical			8.75	11.15	14	16%	18.20%	23%	
	9000	Per apartment unit, total cost		Apt.	40,500	61,700	92,200				
	9500	Total: Mechanical & Electrical		"	7,700	12,200	15,900				
020	0010	APARTMENTS Mid Rise (4 to 7 story)		S.F.	57.15	69.50	85				020
	0020	Total project costs		C.F.	4.55	6.30	8.60				
	0100	Site work		S.F.	2.33	4.62	8.30	5.20%	6.70%	9.10%	
	0500	Masonry			3.87	5.35	7.60	5.90%	7.60%	10.60%	
	1500	Finishes			7.35	9.35	11.90	10.50%	13.10%	17.70%	
	1800	Equipment			1.91	2.73	3.50	2.80%	3.50%	4.70%	
	2500	Conveying equipment			1.32	1.63	1.99	2%	2.30%	2.70%	
	2720	Plumbing			3.42	5.45	6.05	6.30%	7.40%	10%	
	2900	Electrical			4.02	5.50	6.40	6.70%	7.50%	8.90%	
	3100	Total: Mechanical & Electrical			11.60	15.70	19.25	18.80%	21.60%	25.10%	
	9000	Per apartment unit, total cost		Apt.	65,900	77,800	128,700				
	9500	Total: Mechanical & Electrical		"	12,400	14,400	18,200				
030	0010	APARTMENTS High Rise (8 to 24 story)		S.F.	66.05	79.70	97.40				030
	0020	Total project costs		C.F.	5.62	7.85	9.55				
	0100	Site work		S.F.	2.02	3.87	5.40	2.60%	4.80%	6.20%	
	0500	Masonry			3.82	6.95	8.85	4.70%	9.60%	11.10%	
	1500	Finishes			7.35	9.15	10.80	9.70%	11.80%	13.70%	
	1800	Equipment			2.13	2.67	3.46	2.80%	3.50%	4.30%	
	2500	Conveying equipment			1.50	2.28	3.25	2.20%	2.80%	3.40%	
	2720	Plumbing			4.87	5.75	7.05	7%	9.10%	10.60%	
	2900	Electrical			4.54	5.75	7.75	6.40%	7.70%	8.90%	
	3100	Total: Mechanical & Electrical			13.60	17.35	20.90	18%	22.30%	24.40%	
	9000	Per apartment unit, total cost		Apt.	68,400	74,200	104,900				
	9500	Total: Mechanical & Electrical		"	14,800	16,900	18,000				
040	0010	AUDITORIUMS		S.F.	67.95	92.05	124				040
	0020	Total project costs		C.F.	4.30	6	9				
	2720	Plumbing		S.F.	4.35	6	7.60	6.30%	7.20%	9.05%	
	2900	Electrical			5.55	7.65	9.55	6.80%	9%	10.60%	
	3100	Total: Mechanical & Electrical			10.60	14.25	24.80	14.40%	18.50%	23.60%	
050	0010	AUTOMOTIVE SALES		S.F.	49.50	56.95	83.85				050
	0020	Total project costs		C.F.	3.30	4.02	5.20				
	2720	Plumbing		S.F.	2.48	4.02	4.39	4.70%	6.50%	7%	
	2770	Heating, ventilating, air conditioning			3.58	5.55	5.99	6.40%	10.30%	10.40%	
	2900	Electrical			3.94	6.13	6.95	7.30%	10%	12.40%	
	3100	Total: Mechanical & Electrical			8.90	12.85	16.05	17%	20.40%	26%	
060	0010	BANKS		S.F.	98.60	123	156				060
	0020	Total project costs		C.F.	7.10	9.65	12.65				
	0100	Site work		S.F.	10.70	18.35	26.85	7.50%	13.90%	17.60%	
	0500	Masonry			5.05	8.55	18.70	2.90%	6.20%	11.40%	
	1500	Finishes			8.50	11.75	15.20	5.50%	7.70%	10.30%	
	1800	Equipment			3.98	8.25	18.45	3.30%	7.70%	12.50%	
	2720	Plumbing			3.14	4.49	6.50	2.80%	3.90%	4.90%	
	2770	Heating, ventilating, air conditioning			6.10	7.95	10.55	5%	7.20%	8.50%	
	2900	Electrical			9.55	12.60	16.40	8.30%	10.30%	12.20%	
	3100	Total: Mechanical & Electrical			22.65	30.50	37	16.60%	20.40%	24.40%	
	3500	See also division 11020 & 11030									

Figure 1.6

Time and Cost of Estimating

Figure 1.8 shows the relative time required to complete various types of estimates and their accuracy. A square foot estimate takes more time to complete than an order of magnitude estimate because in many instances the square footage of the building must be developed or determined. Additional information improves the level of accuracy. The final accuracy required must be weighed against the cost of creating the estimate. The "best" or most appropriate estimate depends on the situation for which it is needed and how much specific design detail is available. It is not unusual to find preliminary estimates that have been priced out using a combination of square foot costs, assemblies costs, and unit prices.

The figures in the table below indicate typical ranges in square feet as a function of the "occupant" unit. This table is best used in the preliminary design stages to help determine the probable size requirement for the total project.

Unit Gross Area Requirements

Building Type	Unit	Gross Area in S.F.		
		1/4	Median	3/4
Apartments	Unit	660	860	1,100
Auditorium & Play Theaters	Seat	18	25	38
Bowling Alleys	Lane		940	
Churches & Synagogues	Seat	20	28	39
Dormitories	Bed	200	230	275
Fraternity & Sorority Houses	Bed	220	315	370
Garages, Parking	Car	325	355	385
Hospitals	Bed	685	850	1,075
Hotels	Rental Unit	475	600	710
Housing for the elderly	Unit	515	635	755
Housing, Public	Unit	700	875	1,030
Ice Skating Rinks	Total	27,000	30,000	36,000
Motels	Rental Unit	360	465	620
Nursing Homes	Bed	290	350	450
Restaurants	Seat	23	29	39
Schools, Elementary	Pupil	65	77	90
Junior High & Middle		85	110	129
Senior High		102	130	145
Vocational		110	135	195
Shooting Ranges	Point		450	
Theaters & Movies	Seat		15	

Figure 1.7

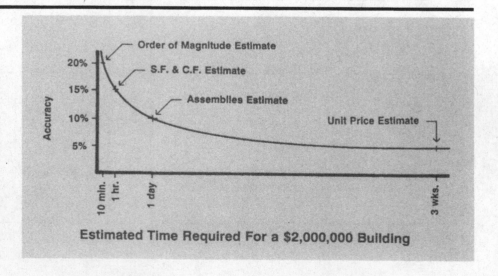

Figure 1.8

Square Foot and Cubic Foot Estimates

Square foot and cubic foot cost data is usually categorized by building type. (Means publishes this information in its annually updated *Assemblies Cost Data*.) Figures 1.5 and 1.6 show typical square foot and cubic foot cost tables based on thousands of buildings constructed within the ten years prior to 2001 and adjusted to January 1, 2001 prices.

Using Square Foot/ Cubic Foot Cost Data

Means square foot and cubic foot costs are shown in ranges of 1/4, median, and 3/4. For the 1/4 figures, three quarters of the projects studied had higher costs reported, and 1/4 had lower costs reported. For the median figures, one-half of the projects had higher costs reported, and one-half had lower. For the 3/4 figures, one-quarter had higher and three-quarters had lower costs reported. If nothing is known about a project except the building's purpose or type, use the median column as a starting point to approximate the building cost.

Figure 1.7, Unit Gross Area Requirements, can be used to approximate the square foot requirements of specified building types. Once an approximate project size has been determined, use the numbers in the tables to derive a representative cost per square foot for the appropriate type of building.

The Area Conversion Scale

The next step is deciding whether to price the building at the 1/4, median, or 3/4 range, or at some other point. For this decision, analyze:

- The owner's quality requirements.
- Unusual design or construction requirements.
- The extent of sitework, installed equipment, and furnishings.
- The extent of air conditioning, heating, and electrical requirements.
- Building configuration.
- Floor-to-floor height.
- Number of floors.

Also consider the building's size. When comparing buildings of similar design and specifications in the same locality, larger buildings tend to have lower square foot costs. This is mainly because of the decreasing contribution of the exterior walls and economy of scale. The Area Conversion Scale in Figure 1.4 provides factors for converting median costs for the model size building to an adjusted cost for a particular size project. The Square Foot Base Size table in the same figure lists the median costs, the typical project size for the accumulated data, and the size range of the projects.

Determine the size factor for the project by dividing the actual area by the typical project size for the building type.

$$\text{Project Size Factor} = \frac{\text{Actual Project Area (S.F.)}}{\text{Typical Project Size (S.F.)}}$$

Enter the Area Conversion Scale in Figure 1.4 on the horizontal axis at the point of the appropriate size factor. Determine the appropriate cost multiplier by projecting a line vertically to intersect the cost modifier curve, and then projecting a line horizontally to read the cost multiplier on the vertical scale. Multiply the project cost by the cost multiplier to calculate the adjusted project total cost.

Some Examples   The following examples show the advantages square foot estimating offers during early design stages to arrive at an approximate estimate of project costs. The step-by-step calculations refer to related square foot and cubic foot costs and tables.

Example One: Three-Story Office Building

A developer must determine the price of a three-story office building with approximately 22,000 square feet of rentable area. The location is a 40,000-square-foot, level site with 160' of frontage in an office park. Surface parking spaces and landscaping will be required.

What is a reasonable size in square feet for this structure? What would the cost of such a structure be, based on the total square feet (and cubic feet)?

Step One: Develop the Building's First Floor, or "Footprint"
Use Figure 2.1 to determine the Gross to Net Ratio for an office

Floor Area Ratios
Table below lists commonly used gross to net area and net to gross area ratios expressed in % for various building types.

Building Type	Gross to Net Ratio	Net to Gross Ratio	Building Type	Gross to Net Ratio	Net to Gross Ratio
Apartment	156	64	School Buildings (campus type)		
Bank	140	72	Administrative	150	67
Church	142	70	Auditorium	142	70
Courthouse	162	61	Biology	161	62
Department Store	123	81	Chemistry	170	59
Garage	118	85	Classroom	152	66
Hospital	183	55	Dining Hall	138	72
Hotel	158	63	Dormitory	154	65
Laboratory	171	58	Engineering	164	61
Library	132	76	Fraternity	160	63
Office	135	75	Gymnasium	142	70
Restaurant	141	70	Science	167	60
Warehouse	108	93	Service	120	83
			Student Union	172	59

The gross area of a building is the total floor area based on outside dimensions.

The net area of a building is the usable floor area for the function intended and excludes such items as stairways, corridors and mechanical rooms.

In the case of a commercial building, it might be considered as the "leasable area."

Figure 2.1

16

building: 135%. The additional 35% is the non-rentable area required for features such as hallways, elevators, stairwells, mechanical and electrical rooms, and restrooms.

22,000 S.F. of Rental Space × 1.35 = 29,700 S.F. Round to: 30,000 S.F.

$$\frac{30,000 \text{ S.F.}}{3 \text{ Stories}} = 10,000 \text{ S.F./floor}$$

Base the estimate on a square building, where the length of one side is the square root of the area. As the square root of 10,000 is 100, the length of one side is 100 feet.

100' × 100' = 10,000 S.F.

Step Two: Sketch the Site
See Figure 2.2 for a sample sketch.

Step Three: Develop Square Foot and Cubic Foot Costs
Square Foot Cost:

Use Figure 2.3: Offices, Low-Rise

Select the median cost, as the building is of simple configuration and modest specifications.

$74.30 S.F. × 30,000 S.F. = $2,229,000

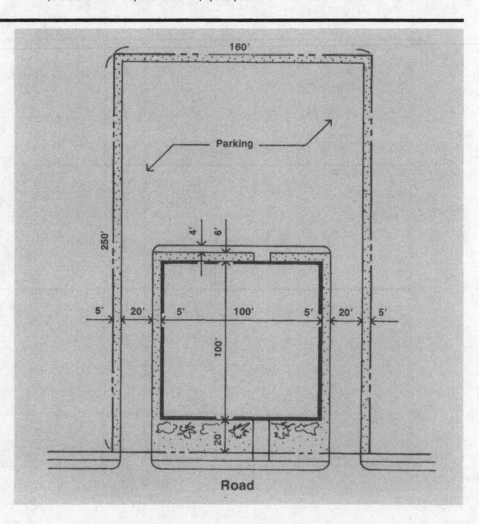

Figure 2.2

K1010 | S.F., C.F. and % of Total Costs

K1010 | S.F. & C.F. Costs

			UNIT	UNIT COSTS			% OF TOTAL			
				1/4	MEDIAN	3/4	1/4	MEDIAN	3/4	
520	3100	Total: Mechanical & Electrical [RK1010-010]	S.F.	36.65	65	76.90	29.20%	31.10%	34.10%	520
530	0010	**LIBRARIES**	S.F.	77.80	99.15	127				530
	0020	Total project costs	C.F.	5.45	6.80	8.85				
	0500	Masonry	S.F.	4.45	9.95	17.10	5.80%	9.50%	11.90%	
	1800	Equipment		1.13	3.03	4.75	1.20%	2.80%	4.50%	
	2720	Plumbing		3.15	4.45	6	3.60%	4.90%	5.70%	
	2770	Heating, ventilating, air conditioning		6.75	11.40	14.85	8%	11%	14.60%	
	2900	Electrical		8	10.25	12.85	8.30%	11%	12.10%	
	3100	Total: Mechanical & Electrical		23.05	31.55	39.50	18.90%	25.30%	27.60%	
550	0010	**MEDICAL CLINICS**	S.F.	76.55	94.80	119				550
	0020	Total project costs	C.F.	5.70	7.40	9.85				
	1800	Equipment	S.F.	2.10	4.41	6.85	1.80%	5.20%	7.40%	
	2720	Plumbing		5.15	7.25	9.70	6.10%	8.40%	10%	
	2770	Heating, ventilating, air conditioning		6.25	8.05	11.85	6.70%	9%	11.30%	
	2900	Electrical		6.50	9.30	12.30	8.10%	10%	12.20%	
	3100	Total: Mechanical & Electrical		20.45	28.80	40.25	22%	27.60%	34.30%	
	3500	See also division 11700								
570	0010	**MEDICAL OFFICES**	S.F.	71.85	89.05	110				570
	0020	Total project costs	C.F.	5.35	7.35	10.05				
	1800	Equipment	S.F.	2.55	4.85	6.80	3%	5.80%	7.20%	
	2720	Plumbing		4.03	6.20	8.45	5.70%	6.80%	8.60%	
	2770	Heating, ventilating, air conditioning		4.87	7.15	9.30	6.20%	8%	9.70%	
	2900	Electrical		5.70	8.30	11.60	7.60%	9.80%	11.40%	
	3100	Total: Mechanical & Electrical		14.20	20.35	30.25	18.50%	22%	24.90%	
590	0010	**MOTELS**	S.F.	46.10	68.25	88				590
	0020	Total project costs	C.F.	4.02	5.65	9.25				
	2720	Plumbing	S.F.	4.67	5.95	7.10	9.40%	10.60%	12.50%	
	2770	Heating, ventilating, air conditioning		2.85	4.24	7.60	5.60%	5.60%	10%	
	2900	Electrical		4.35	5.55	7.25	7.10%	8.20%	10.40%	
	3100	Total: Mechanical & Electrical		14.80	18.50	31.75	18.50%	21%	24.40%	
	5000									
	9000	Per rental unit, total cost	Unit	23,500	44,600	48,200				
	9500	Total: Mechanical & Electrical		4,600	6,900	8,000				
600	0010	**NURSING HOMES**	S.F.	69.25	91.55	112				600
	0020	Total project costs	C.F.	5.55	7.10	9.70				
	1800	Equipment	S.F.	2.32	3.10	5	2.40%	3.70%	6%	
	2720	Plumbing		6.50	8.30	11.50	9.40%	10.70%	14.20%	
	2770	Heating, ventilating, air conditioning		6.45	9.30	11.50	9.30%	11.40%	11.80%	
	2900	Electrical		7.15	8.95	12	9.70%	11%	13%	
	3100	Total: Mechanical & Electrical		17.05	23.85	34.95	26%	29.90%	30.50%	
	3200									
	9000	Per bed or person, total cost	Bed	30,000	36,900	49,100				
610	0010	**OFFICES** Low Rise (1 to 4 story)	S.F.	58.35	74.30	98.95				610
	0020	Total project costs	C.F.	4.22	5.90	8				
	0100	Site work	S.F.	4.40	7.50	11.60	5.30%	9.70%	14%	
	0500	Masonry		2.03	4.75	8.95	2.90%	5.80%	8.70%	
	1800	Equipment		.73	1.32	3.64	1.20%	1.50%	4%	
	2720	Plumbing		2.22	3.36	4.76	3.70%	4.50%	6.10%	
	2770	Heating, ventilating, air conditioning		4.80	6.65	9.80	7.20%	10.50%	11.90%	
	2900	Electrical		4.94	6.85	9.55	7.50%	9.60%	11.10%	
	3100	Total: Mechanical & Electrical		11.65	16.15	23.60	18%	21.80%	26.50%	
620	0010	**OFFICES** Mid Rise (5 to 10 story)	S.F.	64.35	78.05	106				620
	0020	Total project costs	C.F.	4.50	5.70	8.25				
	2720	Plumbing	S.F.	1.95	3.02	4.34	2.80%	3.70%	4.50%	
	2770	Heating, ventilating, air conditioning		4.90	7	11.15	7.60%	9.40%	11%	

Figure 2.3

18

Use the Square Foot Project Size Modifier (see Figure 1.4).

$$\frac{30,000}{20,000} = 1.5 \text{ Size Factor}$$

Cost Multiplier = 0.97

Modified Cost:

| $2,229,000 × 0.97 | = $2,162,130 |

Add the architect's fee.

Use Figure 2.4: 7.75%
.0775 × $2,162,130   = + 167,565
Total Square Foot Cost: $2,329,695

Cubic Foot Cost:

Assume: 9'-0" Ceilings
        12'-0" Floor to Floor

Use Figure 2.3: $5.90/C.F.

3 Floors × 10,000 S.F. × 12' × $5.90/C.F. = $2,124,000

Add the architect's fee.

Use Figure 2.4: 7.75%   + 164,610
Total Cubic Foot Cost $2,288,610

Example Two: Two-Story Office Building, Limited Cost

A client wants to construct a two-story office building, and has limited the project cost to $1,210,000. How big a facility can be built within this budget?

Step One: Establish Square Foot Cost

Consult Figure 2.3: Offices (Low-rise)

Median Cost   = $74.30
Architect's fee (Figure 2.4): 7.9%   + 5.87
                                     $80.17

$1,210,000/$80.17 = 15,093 S.F.

Architectural Fees

Tabulated below are typical percentage fees by project size, for good professional architectural service. Fees may vary from those listed depending upon degree of design difficulty and economic conditions in any particular area.

Rates can be interpolated horizontally and vertically. Various portions of the same project requiring different rates should be adjusted

proportionately. For alterations, add 50% to the fee for the first $500,000 of project cost and add 25% to the fee for project cost over $500,000.

Architectural fees tabulated below include Structural, Mechanical and Electrical Engineering Fees. They do not include the fees for special consultants such as kitchen planning, security, acoustical, interior design, etc.

Building Types	Total Project Size in Thousands of Dollars						
	100	250	500	1,000	5,000	10,000	50,000
Factories, garages, warehouses, repetitive housing	9.0%	8.0%	7.0%	6.2%	5.3%	4.9%	4.5%
Apartments, banks, schools, libraries, offices, municipal buildings	12.2	12.3	9.2	8.0	7.0	6.6	6.2
Churches, hospitals, homes, laboratories, museums, research	15.0	13.6	12.7	11.9	9.5	8.8	8.0
Memorials, monumental work, decorative furnishings	—	16.0	14.5	13.1	10.0	9.0	8.3

Figure 2.4

Step Two: Modify for Size

Use the information in Figure 1.4.

$$\frac{15{,}093 \text{ S.F.}}{20{,}000 \text{ S.F.}} = 0.75$$

Cost Multiplier = 1.05

$$\frac{\$1{,}210{,}000}{\$80.17/\text{S.F.} \times 1.05} = 14{,}375 \text{ S.F.}$$

Figure 2.5 is a sketch of a two-story office building with 14,375 S.F. of floor area.

Example Three: Three-Story Office Building, Limited Height

A developer requires that as large a building as possible be constructed on a site that is limited in size.

What is the ideal dimension for the building footprint?

What would be the total project cost of the building based on that ideal dimension, taking into account certain criteria as outlined below?

Restrictions:	Local building codes
Height:	3 floors above grade (or 45')
Parking	.66 spaces per occupant
Landscaping:	20% of site

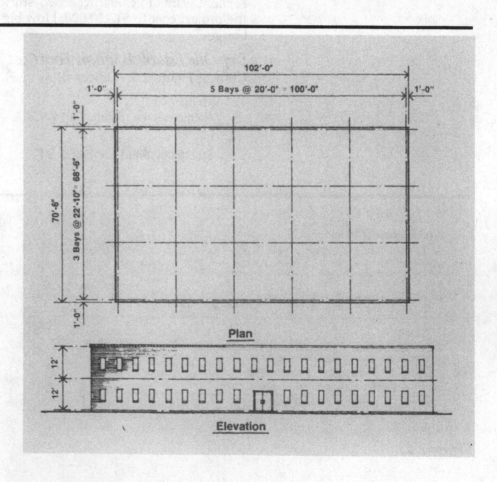

Figure 2.5

Step One: Develop Total Area Available for Building and Parking

Site Area:	40,000 S.F.
Required Landscaping (20%):	– 8,000 S.F.
	32,000 S.F.
Walks, Entrances, etc. (assume 2%):	– 800 S.F.
Available for Building and Parking	31,200 S.F.

Consult Figure 2.6 to determine the occupancy. The minimum occupancy for an office building is 100 S.F. per person. Consult Figure 2.1 for floor area ratios. The floor area ratio for an office building is 135%.

100 S.F. × 1.35 = 135 S.F. of building per person

Parking Spaces: See Figure 2.7 for the parking lot design. The typical range per person is 300–400 S.F. for parking and maneuvering. For this example, use 390 S.F.

.66 spaces per occupant × 390 S.F. = 260 S.F. per occupant

Step Two: Develop the Area Required for Parking and the Building Footprint

Let "U" equal the number of persons per floor

Parking area		Footprint area
(3 floors × 260 S.F. × U)	+	(135 × U) = 31,200 S.F.

Solving for U:

915 U = 31,200 S.F.   U = 34 persons per floor

Parking Area: 3 Floors × 34 persons per floor × 260 S.F. = 26,520 S.F.
Building Footprint: 34 persons × 135 S.F. = 4,590 S.F.

Occupancy Determinations

Description		S.F. Required per Person		
		BOCA	SBC	UBC
Assembly Areas	Fixed Seats	**	6	7
	Movable Seats		15	15
	Concentrated	7		
	Unconcentrated	15		
	Standing Space	3		
Educational	Unclassified			
	Classrooms	20	40	20
	Shop Areas	50	100	50
Institutional	Unclassified		125	
	In-Patient Areas	240		
	Sleeping Areas	120		
Mercantile	Basement	30	30	20
	Ground Floor	30	30	30
	Upper Floors	60	60	50
Office		100	100	100

BOCA = Building Officials & Code Administrators' National Building Code
SBC = Standard Building Code
UBC = Uniform Building Code

** The occupancy load for assembly area with fixed seats shall be determined by the number of fixed seats installed.

Figure 2.6

Assume a building 3 bays long by 2 bays wide with the entrance at the center of the building and approximately 25'-0" column spacing.

Front: 3 × 25'-0" = 75'-0" + 2 × 8" = 76'-4"
Side: 2 × 30'-0" = 60'-0" + 2 × 8" = 61'-4"
Footprint area: 76.33 × 61.33 = 4,681 S.F.
Building area: 3 floors = 4,681 S.F. × 3 = 14,040 S.F.

Sketch the building floor plan and elevation as shown in Figure 2.8.

Summary:

Building Footprint	4,681 S.F.
Parking	26,520 S.F.
Landscape	8,000 S.F.
Walks and Miscellaneous	+ 800 S.F.
Total	40,001 S.F.

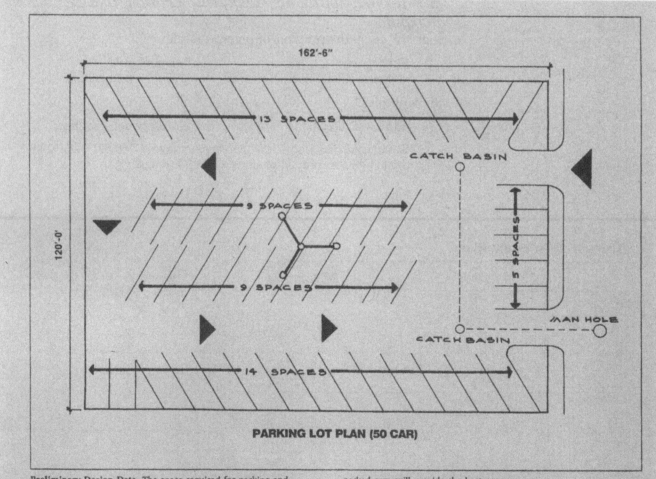

PARKING LOT PLAN (50 CAR)

Preliminary Design Data: The space required for parking and maneuvering is between 300 and 400 S.F. per car, depending upon engineering layout and design.

Ninety degree (90°) parking, with a central driveway and two rows of parked cars, will provide the best economy.

Diagonal parking is easier than 90° for the driver and reduces the necessary driveway width, but requires more total space.

Figure 2.7

Square Foot Cost:

14,040 S.F. × $74.30/S.F.	=	$1,043,172
Use Size Modifier (× 1.05)		1,095,330
Architect's fee (8.0%)		+ 87,625
Total		$1,182,955

Cubic Foot Cost

12' Floor to Floor		
14,040 S.F. × 12' × $5.90/C.F.	=	$ 994,032
Architect's fee (8.0%)		+ 79,523
Total		$1,073,555

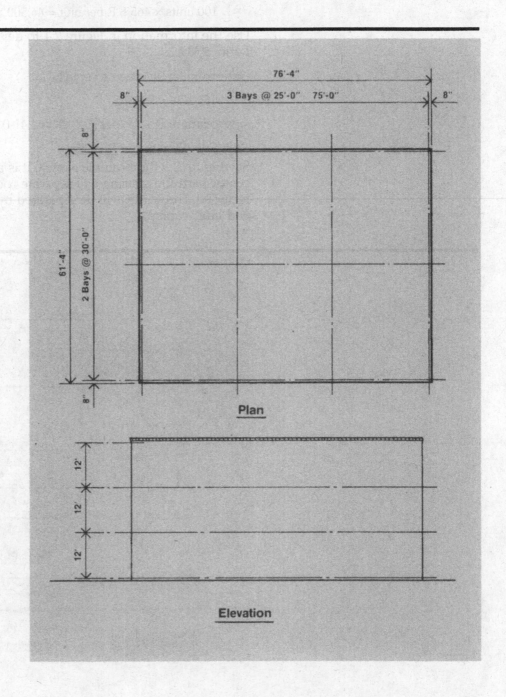

Plan

Elevation

Figure 2.8

Example Four: Two-Story, 100-Unit Motel

A developer needs to know how much it would cost to build a 100-unit motel. What would be a reasonable estimate of the cost based on three types of estimates (square foot, cubic foot, and order of magnitude)?

Assume the following requirements:

- 100 units
- 2 stories
- Includes: restaurant, lobby, and shops

Step One: Determine the Building Area Required for the Project

Use Figure 1.7 to determine the appropriate number of square feet per unit:

100 units × 465 S.F. per unit = 46,500 S.F.

Use the information in Figure 2.1 to develop a Gross to Net Ratio: 158%.

Determine an average room size:

465/1.58 = 294 S.F.

Approximate Room Size: 12'-0" × 24'-0".

Step Two: Sketch the Building

See Figure 2.9 for a sample sketch. It is assumed that 13'-0" center-to-center partition spacing will separate rooms. A 6'-0" corridor is also included. The wings will be separated by an area that includes a pool and landscaping.

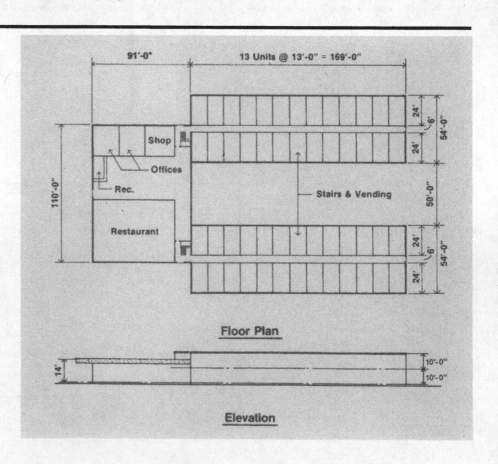

Figure 2.9

Step Three: Determine the Various Areas Involved

Gross Area:	46,500 S.F.
Motel Area from the sketch:	
2 × 169' × 54' × 2 floors	− 36,504 S.F.
Service Area:	9,996 S.F.

From the sketch, determine the service area length by adding the room plus the corridor width on each wing, plus the distance between the wings.

$$(2 × 30') + 50' = 110'$$

$$\frac{9,996 \text{ S.F.}}{110'} = 91'$$

Step Four: Develop a Square Foot Cost

Use Figure 2.3, Motels, to develop a square foot cost: $68.25/S.F.

46,500 S.F. × $68.25/S.F.	= $3,173,625
Use Size Modifier (× .99)	3,141,889
Architect's fee: 7.5%	+ 235,642
Total	$3,377,531

Step Five: Develop a Cubic Foot Cost

Motel	Res. & Office	Stairs

$$[4 × 169' × 54' × 10'] + [110' × 77' × 14'] + [2 × 20' × 10' × 3.331'] = 484,952 \text{ C.F.}$$

484,952 C.F. × $5.65/C.F.	= $2,739,979
Architect's fee: 7.5%	+ 205,498
Total (no sitework included):	$2,945,477

Step Six: Develop an Order of Magnitude Estimate

Use Figure 2.3 to develop an order of magnitude estimate.

100 Units × $44,600 = $4,460,000

Step Seven: Compare the Square Foot and Cubic Foot Estimates

Square Foot Estimate:	$3,377,531
Cubic Foot Estimate:	− 2,945,477
Difference:	$ 432,054 (13%)

Conclusion

Square foot and cubic foot estimates can be developed for buildings with site, occupancy, building code, or construction cost restrictions. However, the following limitations must be recognized:

* If extensive sitework or costly foundation systems are required, the costs must be modified accordingly. This is also true for complex wall systems, mechanical and electrical systems, and structural systems with large bay sizes.
* Make a sketch of the proposed building and its site to make sure that your estimating assumptions are correct.
* When projects have square foot requirements that differ significantly from the model shown in Figure 1.4, use the size modifier to properly adjust the calculated cost.

- If you are using a national average for cost information, such as found in *Means Assemblies Cost Data*, adjust the estimate to your area using the City Cost Index.

If these modifications are realistically applied, the final cost of the project should be within 20% of an order of magnitude estimate, and 15% of the square foot or cubic foot estimate.

Chapter 3

Developing
Structural Loads

At the early stages of a project, you may be asked to provide a preliminary "guesstimate" of project costs. Frequently there is little time to produce this figure, and yet it can be the most significant one in the entire project. If the figure is too high, the project may be deemed to be too expensive and may not proceed. A vague, "off-the-cuff" figure used as a preliminary estimate may lead to an estimate that is too low from the very start. If the figure is low, it will certainly be remembered as "the project budget." An estimate that is too low may cause compromises in quality and quantity of building components throughout the process.

However, it is possible to produce a realistic construction cost estimate in just a few hours that is tailored to the particular project and its scope. Square foot assemblies estimates look at a building's construction, element by element, in a logical and factually-based progression. Costs are developed using a rational and verifiable thought process. The assembly square foot estimating techniques that follow, described in step-by-step examples, will produce such an estimate. This is by far the most desirable approach to use, as it allows you to prepare a detailed preliminary estimate within a reasonable time frame.

Developing Loads

Developing an assemblies square foot estimate begins by calculating the loads that will be applied to the structure's foundation. This chapter explains how to develop the basic structural loads and other design criteria required to prepare a conceptual cost estimate. *(Note: In subsequent chapters, there are additional examples to show how loads are accumulated when various floor systems are used, such as steel or concrete. Other examples demonstrate the use of support systems such as columns, bearing walls, and pile foundations. The procedures shown throughout this book are for assemblies and square foot cost development only and are not intended to replace competent design practice.)*

Assemblies costs are useful for developing structural loads during the first stage of square foot estimating. Four pieces of general information are required in this stage:

- Cost limitations (the owner's budget)
- Building site, size, and height limitations

- Occupancy requirements
- An outline sketch of the building, or drawings of an existing building that is similar to the proposed structure. Plans for a similar building may be updated to reflect current prices and adjusted for material and design differences.

Assembly Versatility

An assemblies estimate of any superstructure must be extremely adaptable if it is to be accurate. A range of assemblies data must be available so the estimator can compare and substitute them, analyzing the merits of each particular combination. Designs and costs must be determined for:

- Foundation
- Columns and load-bearing walls
- Stairs
- Floor assemblies
- Roof assemblies

Each of the components of these subsystems has characteristics that vary by:

- Use
- Cost
- Building code requirements
- Weight to foundation
- Life cycle costs
- Thermal and fireproofing qualities
- Appearance

Soil Bearing Capacity

If soil bearing capacity is limited, particular attention must be paid to the weight of the systems and the bay spacing in the superstructure. It is far more economical to build a structure with many short spans and no piles or caissons than a long-span structure that requires piles or caissons. For that reason, it is a good idea to examine the soil bearing capacity of the site before proceeding.

The soil bearing capacity must be determined before wall or column loads are developed to be certain that the proper foundation is selected for the structure. Without soil bearing capacity information, it is difficult to develop accurate foundation costs.

Soil bearing capacity is the amount of pressure in "kips per square foot" that the soil will support without excessive settlement. The term "kip," a short form of "kilopound," has been used by engineers so often over the years that it has become a standard abbreviation. One kilopound is equivalent to 1,000 pounds (454 kilograms). A kip is a convenient unit of force for use in structural calculations. For the purposes of the assemblies square foot estimates that follow in this book, assume that the soil bearing capacity used is in the range of 3 to 6 kips per square foot (K.S.F.). This range is between stiff clay at the lower end of the scale, and compact sand at the upper end of the scale.

NOTE: *To further clarify the concept of soil pressure, simply think of a 200-pound man standing on the ball of his foot on sand and assume that the contact area of the ball of his foot is six square inches (3" × 2").*

200 lbs./6 Sq. In. = 33.33 lbs./Sq. In.
33.33 lbs./Sq. In. × 144 Sq. In./S.F. = 4800 lbs./S.F. or 4.8 kips/S.F.

The soil bearing capacity must be determined for the project site in order to determine the foundation structure of the proposed building. If this information is not yet available, consult the following sources:

- The local building inspector or material testing laboratories.
- Local contractors, engineers, or architects.
- Borings or boring logs, usually available in building inspectors' offices, for other jobs in close proximity.
- Figure 3. 1.

Load Requirements

Many different types of loads are involved in a structure. To develop floor and roof assemblies, an accurate accounting must be made for:

- *Live loads*: Minimum requirements determined by planned occupancy and building codes. (See Figure 3.2.)
- *Miscellaneous loads*: Ceilings, partitions, or mechanical systems supported by the structure. (See Figure 3.3.)
- *Bay sizes*: Clear spans or column spacing required by planned building occupancy.

Specific floor and roof assemblies can be selected based on cost and other factors. Cost may not be the governing criteria. At the start of an assemblies square foot estimate, it is more important to consider the loads themselves and examine each possible assembly in terms of its application to the structure.

The superimposed loads applied to a floor assembly are a combination of live and miscellaneous loads. The total load of a floor or roof assembly is the superimposed load added to the weight of the floor system.

The following examples show the process of assembling the systems that transmit actual loads to the foundation of a specific building, beginning with a straightforward concrete structure where only the superimposed floor loads are analyzed. They show step-by-step procedures for developing the total load to the foundation.

COST DETERMINATION

1. Determine Soil Bearing Capacity by a known value or using this table as a guide.

Soil Bearing Capacity in Kips per S.F.

Bearing Material	Typical Allowable Bearing Capacity
Hard sound rock	120 KSF
Medium hard rock	80
Hardpan overlaying rock	24
Compact gravel and boulder-gravel; very compact sandy gravel	20
Soft rock	16
Loose gravel; sandy gravel; compact sand; very compact sand-inorganic silt	12
Hard dry consolidated clay	10
Loose coarse to medium sand; medium compact fine sand	8
Compact sand-clay	6
Loose fine sand; medium compact sand-inorganic silts	4
Firm or stiff clay	3
Loose saturated sand-clay; medium soft clay	2

Figure 3.1

Minimum Design Live Loads in Pounds per S.F. for Various Building Codes

Occupancy	Description	Minimum Live Loads, Pounds per S.F.		
		BOCA	IBC	UBC
Access Floors	Office use		50	50
	Computer use		100	100
Armories		150	150	150
Assembly	Fixed seats	60	60	50
	Movable seats	100	100	100
	Platforms or stage floors	125	125	125
Commercial & Industrial	Light manufacturing	125	125	75
	Heavy manufacturing	250	250	125
	Light storage	125	125	125
	Heavy storage	250	250	250
	Stores, retail, first floor	100	100	100
	Stores, retail, upper floors	75	75	100
	Stores, wholesale	125	125	100
Court rooms		100		
Dance halls	Ballrooms	100	100	
Dining rooms	Restaurants	100	100	
Fire escapes	Other than below	100	100	100
	Single family residential	40	40	100
Garages	Passenger cars only	50	50	50
Gymnasiums	Main floors and balconies	100	100	
Hospitals	Operating rooms, laboratories	60	60	
	Private room	40	40	40
	Wards	40	40	40
	Corridors, above first floor	80	80	
Libraries	Reading rooms	60	60	60
	Stack rooms	150	150	125
	Corridors, above first floor	80	80	
Marquees		75	75	
Office Buildings	Offices	50	50	50
	Lobbies	100	100	
	Corridors, above first floor	80	80	
Residential	Multi family private apartments	40	40	40
	Multi family, public rooms	100	100	
	Multi family, corridors	100	100	100
	Dwellings, first floor	40	40	40
	Dwellings, second floor & habitable attics	30	30	
	Dwellings, uninhabitable attics	20	20	
	Hotels, guest rooms	40	40	40
	Hotels, public rooms	100	100	
	Hotels, corridors serving public rooms	100	100	
	Hotels, corridors	40	100	
Schools	Classrooms	40	40	40
	Corridors	80	100	
Sidewalks	Driveways, etc. subject to trucking	250	250	250
Stairs	Exits	100	100	100
Theaters	Aisles, corridors and lobbies	100		
	Orchestra floors	60		
	Balconies	60		
	Stage floors	125	125	
Yards	Terrace, pedestrian	100	100	

BOCA = Building Officials & Code Administration International, Inc. National Building Code
IBC = International Building Code, International Code Council, Inc.
UBC = Uniform Building Code, International Conference of Building Officials

Figure 3.2

Example, Part One: Load

Start by developing the superimposed loads.

- Building: 30,000 S.F.
- Structure: Concrete Flat Plate
- Stories: 3
- Dimensions: 10,000 S.F./floor; 12'-0" Floor to Floor
- Building Occupancy: Office
- Bay Size: 25'-0" × 25'-0"
- Building Code: BOCA

Step One: Analyze the Floor Loads

Minimum Live Load for		
Office Building	50 psf	BOCA Code (Figure 3.2)
Partitions	20 psf	(Figure 3.3)
Ceiling	5 psf	(Figure 3.3)
Superimposed Floor Load	75 psf	

Step Two: Develop the Superimposed Roof Load

Snow Load	20 psf	(Figure 3.4)
Roofing and Insulation	10 psf	(Figure 3.5)
Ceiling	5 psf	(Figure 3.3)
Mechanical	5 psf	(Figure 3.3)
Superimposed Roof Load	40 psf	

Step Three: Sketch the Floor Plan and Elevation

At this point, it is a good idea to make a sketch of the floor plan to show column spacing and bay sizes. Also sketch an elevation to show the number of suspended floors, the roof, and the floor-to-floor relationship. See Figure 3.6.

Example, Part Two: Column Derivation

The next step is to begin the selection of various building systems. Continue by following the steps below. Remember, the superimposed loads in psf, bay sizes, and the building layout have already been developed.

Step One: Determine the Floor Load and Maximum Column Size

- Use the information in Figure 3.7, Floor Construction.
- Enter the table, knowing the bay size is 25' × 25'.

Superimposed Dead Load Ranges

Component	Load Range (PSF)
Ceiling	5-10
Partitions	10-20
Mechanical	4-8

Figure 3.3

Snow Load in Pounds Per Square Foot on the Ground

Figure 3.4

Design Weight per S.F. for Roof Coverings

Type		Description	Weight Per S.F.	Type	Wall Thickness	Description	Weight Per S.F.
Sheathing	Gypsum	1" thick	4	Metal	Aluminum	Corr. & ribbed, .024" to .040"	.4-.8
	Wood	¾" thick	3		Copper	or tin	1.0.
Insulation	per 1"	Loose	.5		Steel	Corrugated, 29 ga. to 12 ga.	.6-5.0
		Poured in place	2	Shingles	Asphalt	Strip shingles	3
		Rigid	1.5		Clay	Tile	9-20
Built-up	Tar & gravel	3 ply felt	5.5		Slate	¼" thick	10
		5 ply felt			Wood		2

Figure 3.5

32

- Go to the line in the bay size listing for a 75 psf superimposed floor load from the example above.
- Total Floor Load (psf): 194 psf (from Figure 3.7)
- Minimum Column Size (in.): 24" (from Figure 3.7)

Step Two: Determine the Roof and Total Loads

- Use Figure 3.7, knowing the bay size is 25' x 25'
- The superimposed roof load from the Load Example above is 40 psf.
- Total Roof Load (psf): 152 psf.

When using concrete floor and column systems for low-rise buildings, it is usually more economical to use the same size concrete column for each floor to minimize formwork changes. In each step up to this point, we have determined the minimum column size for the floor and roof system. In making the final decision on column size for the building, use the largest of the columns, as minimum required column size for the greatest load governs the overall selection process.

The total load as recorded in Steps One and Two is actually the sum of the superimposed load plus the dead load of the structure:

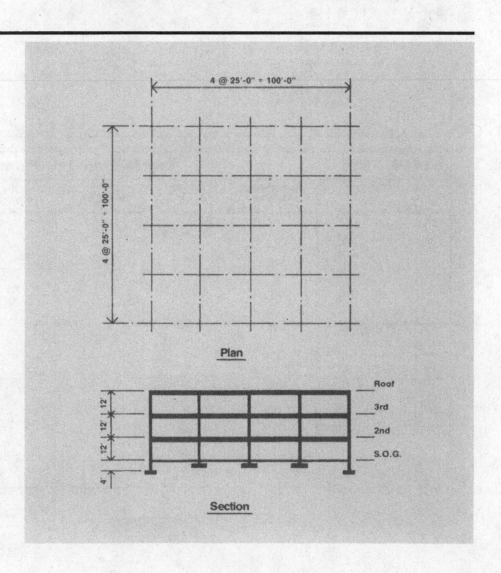

Plan

Section

Figure 3.6

33

B10 Superstructure

B1010 Floor Construction

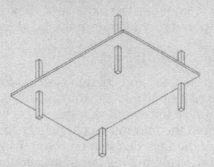

General: Flat Plates: Solid uniform depth concrete two-way slab without drops or interior beams. Primary design limit is shear at columns.

Design and Pricing Assumptions:
Concrete f'c to 4 KSI, placed by concrete pump.
Reinforcement, fy = 60 KSI.
Forms, four use.
Finish, steel trowel.
Curing, spray on membrane.
Based on 4 bay x 4 bay structure.

System Components	QUANTITY	UNIT	COST PER S.F.		
			MAT.	INST.	TOTAL
SYSTEM B1010 223					
15'X15' BAY 40 PSF S. LOAD, 12" MIN. COL.					
Forms in place, flat plate to 15' high, 4 uses	.992	S.F.	.92	3.78	4.70
Edge forms to 6" high on elevated slab, 4 uses	.065	L.F.	.03	.18	.21
Reinforcing in place, elevated slabs #4 to #7	1.706	Lb.	.55	.53	1.08
Concrete ready mix, regular weight, 3000 psi	.459	C.F.	1.21		1.21
Place and vibrate concrete, elevated slab less than 6", pump	.459	C.F.		.49	.49
Finish floor, monolithic steel trowel finish for finish floor	1.000	S.F.		.60	.60
Cure with sprayed membrane curing compound	.010	C.S.F.	.04	.06	.10
TOTAL			2.75	5.64	8.39

B1010 223		Cast in Place Flat Plate						
	BAY SIZE (FT.)	SUPERIMPOSED LOAD (P.S.F.)	MINIMUM COL. SIZE (IN.)	SLAB THICKNESS (IN.)	TOTAL LOAD (P.S.F.)	COST PER S.F.		
						MAT.	INST.	TOTAL
2000	15 x 15	40	12	5-1/2	109	2.75	5.65	8.40
2200	RB1010 -010	75	14	5-1/2	144	2.77	5.65	8.42
2400		125	20	5-1/2	194	2.86	5.70	8.56
2600		175	22	5-1/2	244	2.93	5.75	8.68
3000	15 x 20	40	14	7	127	3.16	5.70	8.86
3400	RB1010 -105	75	16	7-1/2	169	3.36	5.85	9.21
3600		125	22	8-1/2	231	3.70	6	9.70
3800		175	24	8-1/2	281	3.72	6	9.72
4200	20 x 20	40	16	7	127	3.15	5.70	8.85
4400		75	20	7-1/2	175	3.39	5.85	9.24
4600		125	24	8-1/2	231	3.71	6	9.71
5000		175	24	8-1/2	281	3.74	6.05	9.79
5600	20 x 25	40	18	8-1/2	146	3.68	6	9.68
6000		75	20	9	188	3.81	6.05	9.86
6400		125	26	9-1/2	244	4.11	6.25	10.36
6600		175	30	10	300	4.26	6.35	10.61
7000	25 x 25	40	20	9	152	3.80	6.05	9.85
7400		75	24	9-1/2	194	4.04	6.20	10.24
7600		125	30	10	250	4.28	6.40	10.68
8000								

Figure 3.7

25'-0" Square Bay Superimposed Load		75 psf
Floor Dead Load:		
Slab Thickness (from Figure 3.7: .9'-1/2")	.79" × 150 pcf =	119 psf
Total Load (Floor)		194 psf

25'-0" Square Bay Superimposed Load		40 psf
Roof Dead Load:		
Slab Thickness (from Figure 3.7: .9")	.75" × 150 pcf =	112 psf
Total Load (Roof)		152 psf

The total load for floor assemblies using steel beams and concrete decks includes the beams, steel deck (if used in the system), concrete slab, and sprayed-on fireproofing required for the system. Floor assemblies using joists do not include fireproofing, since the assemblies are usually fireproofed by applying a rated ceiling below the joists.

The total load for roof systems of steel joists and beams, or steel joists and joist girders with steel roof decks, includes an allowance for insulation, roofing, and miscellaneous loads, but does not include fireproofing.

It is important to carefully examine each assembly for the makeup of the total load, since roof assemblies have an allowance for insulation, and floor systems used as roofs do not.

Examine each assembly used to determine the makeup of the total load.

Step Three: Develop the Area Supported by an Interior Column

Sketch a cross section of an interior column, allowing adequate space for calculations and load accumulation as shown in Figure 3.8.

Develop the area supported by an interior column as shown in Figure 3.9.

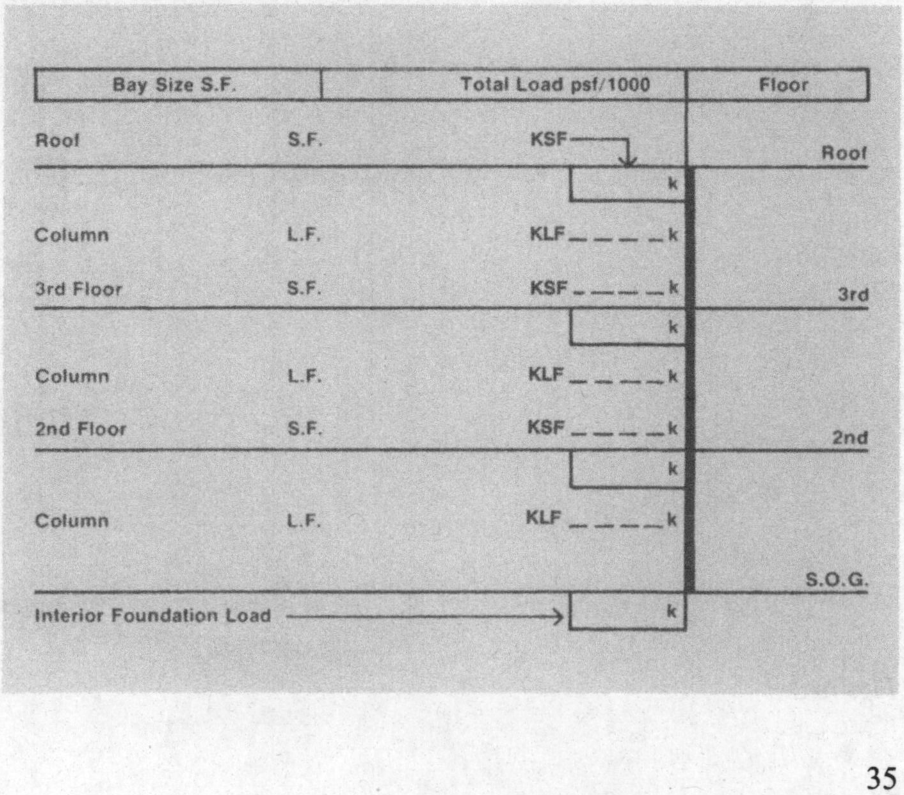

Figure 3.8

Step Four: Determine the Load Imposed by the Roof or Floor

To determine the load for an interior column, first multiply the bay size by the total load in psf. The column load for concrete columns is the column size in feet multiplied by the column height multiplied by 150 pcf (the weight of reinforced concrete in pounds per cubic foot).

To ensure that all roof, floor, and column loads are accounted for, use the method shown in Figure 3.10. With the calculations shown, these loads have been determined:

Minimum column load (95 kips + 7 kips)	=	102 kips
Maximum column load	=	358 kips
Load to the foundation	=	358 kips

Step Five: Determine the Required Concrete Column

As previously determined, the minimum allowable column size shown for a 25'-0" square bay with 75 psf superimposed load using a cast in place flat plate is 24" × 24".

If round columns will be used, select a column with an area equal to or greater than 24" × 24" (576 Sq. In.).

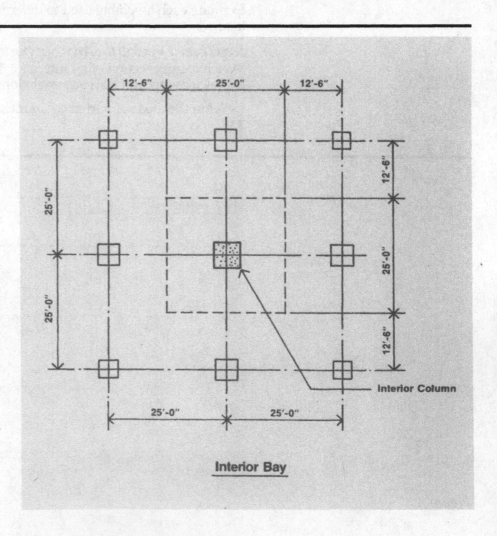

Interior Bay

Figure 3.9

36

In this example, use square columns. Enter the second portion of Figure 3.11 (C.I.P. Columns, Square Tied Minimum Reinforcing) with the column size (minimum column required 24") and note the allowable load in kips (700 kips).

700 kips Allowable > 358 kips required: Column O.K.

If the column load is in excess of the 700 kips allowed, enter the first portion of Figure 3.11 (C.I.P. Columns, Square Tied) under Column Size (24") and proceed down the row to the first 24" column with 12'-0" story height. The allowable load is 900 kips.

The cost of the column is determined in this way:

$$\frac{\text{Cost of 900 kips Column} + \text{Cost of MR Column}}{2} \times \text{Total Length}$$

If the load to be supported is in excess of the load allowable on a 24" square column, it may be necessary to use a larger size column or a concrete column of a higher strength to satisfy conditions. Follow the same procedure to obtain costs.

Conclusion

The above example shows the importance of understanding the loads in a building in order to arrive at an accurate building cost estimate. Remember these four points when developing an assemblies estimate:

- Know the basic characteristics of the local soil. With this information, it is sometimes possible to consider alternatives, such as a lighter building design that will save money over a more costly foundation.
- A careful assemblies analysis ensures an accurate inventory of all loads, greatly reducing the need for redesign at an advanced stage.
- Simple sketches and drawings are helpful in answering the many

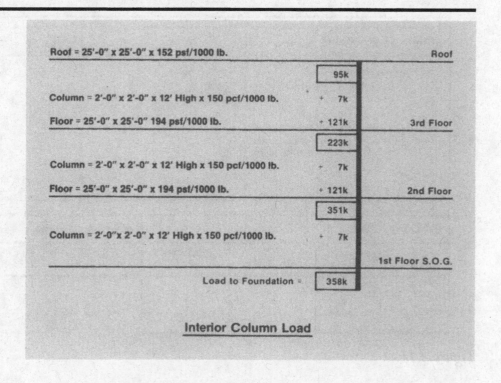

Figure 3.10

B1010 Floor Construction

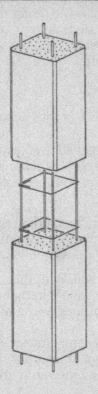

CONCRETE COLUMNS

General: It is desirable for purposes of consistency and simplicity to maintain constant column sizes throughout the building height. To do this, concrete strength may be varied (higher strength concrete at lower stories and lower strength concrete at upper stories), as well as varying the amount of reinforcing.

The first portion of the table provides probable minimum column sizes with related costs and weights per lineal foot of story height for bottom level columns.

The second portion of the table provides costs by column size for top level columns with minimum code reinforcement. Probable maximum loads for these columns are also given.

How to Use Table:

1. Enter the second portion (minimum reinforcing) of the table with the minimum allowable column size from the selected cast in place floor system.

   If the total load on the column does not exceed the allowable working load shown, use the cost per L.F. multiplied by the length of columns required to obtain the column cost.

2. If the total load on the column exceeds the allowable working load shown in the second portion of the table, enter the first portion of the table with the total load on the column and the minimum allowable column size from the selected cast in place floor system.

Select a cost per L.F. for bottom level columns by total load or minimum allowable column size.

Select a cost per L.F. for top level columns using the column size required for bottom level columns from the second portion of the table.

$$\frac{\text{Btm. + Top Col. Costs/L.F.}}{2} = \text{Avg. Col. Cost/L.F}$$

Column Cost = Average Col. Cost/L.F. x Length of Cols. Required.

See reference section in back of book to determine total loads.

Design and Pricing Assumptions:
Normal wt. concrete, f'c = 4 or 6 KSI, placed by pump.
Steel, fy = 60 KSI, spliced every other level.
Minimum design eccentricity of 0.1t.
Assumed load level depth is 8″ (weights prorated to full story basis).
Gravity loads only (no frame or lateral loads included).

Please see the reference section for further design and cost information.

System Components			COST PER V.L.F.		
	QUANTITY	UNIT	MAT.	INST.	TOTAL
SYSTEM B1010 203					
SQUARE COLUMNS, 100K LOAD, 10′ STORY, 10′ SQUARE					
Forms in place, columns, plywood, 10″ x 10″, 4 uses	3.323	SFCA	2	20.50	22.50
Chamfer strip, wood, 3/4″ wide	4.000	L.F.	1.32	2.76	4.08
Reinforcing in place, column ties	1.405	Lb.	1.92	3.54	5.46
Concrete ready mix, regular weight, 4000 psi	.026	C.Y.	1.96		1.96
Placing concrete, incl. vibrating, 12″ sq./round columns, pumped	.026	C.Y.		1.36	1.36
Finish, break ties, patch voids, burlap rub w/grout	3.323	S.F.	.10	2.47	2.57
TOTAL			7.30	30.63	37.93

B1010 203		C.I.P. Column, Square Tied						
	LOAD (KIPS)	STORY HEIGHT (FT.)	COLUMN SIZE (IN.)	COLUMN WEIGHT (P.L.F.)	CONCRETE STRENGTH (PSI)	COST PER V.L.F.		
						MAT.	INST.	TOTAL
0640	100	10	10	96	4000	7.30	31	38.30
0680	RB1010 -112	12	10	97	4000	7.15	30.50	37.65
0700		14	12	142	4000	9.25	37	46.25
0710								

Figure 3.11

B1010 Floor Construction

B1010 203   G.I.P. Column, Square Tied

	LOAD (KIPS)	STORY HEIGHT (FT.)	COLUMN SIZE (IN.)	COLUMN WEIGHT (P.L.F.)	CONCRETE STRENGTH (PSI)	COST PER V.L.F.		
						MAT.	INST.	TOTAL
0740	150	10	10	96	4000	7.30	31	38.30
0780		12	12	142	4000	9.30	37	46.30
0800		14	12	143	4000	9.25	37	46.25
0840	200	10	12	140	4000	9.35	37.50	46.85
0860		12	12	142	4000	9.30	37	46.30
0900		14	14	196	4000	11.70	42.50	54.20
0920	300	10	14	192	4000	12	43	55
0960		12	14	194	4000	11.85	43	54.85
0980		14	16	253	4000	13.40	47	60.40
1020	400	10	16	248	4000	14.60	49	63.60
1060		12	16	251	4000	14.45	49	63.45
1080		14	16	253	4000	14.35	48.50	62.85
1200	500	10	18	315	4000	19.25	58.50	77.75
1250		12	20	394	4000	19.70	61	80.70
1300		14	20	397	4000	19.60	61	80.60
1350	600	10	20	388	4000	23	67.50	90.50
1400		12	20	394	4000	22.50	66.50	89
1600		14	20	397	4000	22.50	66.50	89
1900	700	10	20	388	4000	23.50	69	92.50
2100		12	22	474	4000	23	68.50	91.50
2300		14	22	478	4000	23	68.50	91.50
2600	800	10	22	388	4000	33	86.50	119.50
2900		12	22	474	4000	33	86.50	119.50
3200		14	22	478	4000	32.50	86	118.50
3400	900	10	24	560	4000	32.50	87	119.50
3800		12	24	567	4000	32	86.	118
4000		14	24	571	4000	32	85.50	117.50
4250	1000	10	24	560	4000	39.50	99.50	139
4500		12	26	667	4000	34.50	89.50	124
4750		14	26	673	4000	34.50	89	123.50
5600	100	10	10	96	6000	7.40	31	38.40
5800		12	10	97	6000	7.25	30.50	37.75
6000		14	12	142	6000	9.30	37	46.30
6200	150	10	10	96	6000	7.30	30.50	37.80
6400		12	12	98	6000	9.30	37	46.30
6600		14	12	143	6000	9.30	37	46.30
6800	200	10	12	140	6000	9.75	38	47.75
7000		12	12	142	6000	9.30	37	46.30
7100		14	14	196	6000	11.70	42.50	54.20
7300	300	10	14	192	6000	11.85	43	54.85
7500		12	14	194	6000	11.75	42.50	54.25
7600		14	14	196	6000	11.70	42.50	54.20
7700	400	10	14	192	6000	11.85	43	54.85
7800		12	14	194	6000	11.75	42.50	54.25
7900		14	16	253	6000	14.35	48	62.35
8000	500	10	16	248	6000	14.60	49	63.60
8050		12	16	251	6000	14.45	49	63.45
8100		14	16	253	6000	14.35	48.50	62.85
8200	600	10	18	315	6000	17.60	56.50	74.10
8300		12	18	319	6000	17.40	56	73.40
8400		14	18	321	6000	17.30	55.50	72.80

Figure 3.11 (cont.)

B10 Superstructure

B1010 Floor Construction

B1010 203	C.I.P. Column, Square Tied							
	LOAD (KIPS)	STORY HEIGHT (FT.)	COLUMN SIZE (IN.)	COLUMN WEIGHT (P.L.F.)	CONCRETE STRENGTH (PSI)	COST PER V.L.F.		
						MAT.	INST.	TOTAL
8500	700	10	18	315	6000	19.05	59	78.05
8600		12	18	319	6000	18.80	58.50	77.30
8700		14	18	321	6000	18.65	58	76.65
8800	800	10	20	388	6000	19.90	61.50	81.40
8900		12	20	394	6000	19.70	61.50	81.20
9000		14	20	397	6000	19.60	61	80.60
9100	900	10	20	388	6000	28	77	105
9300		12	20	394	6000	28	76	104
9600		14	20	397	6000	27.50	75.50	103
9800	1000	10	22	469	6000	25	72	97
9840		12	22	474	6000	25	71.50	96.50
9900		14	22	478	6000	25.50	72.50	98

B1010 203	C.I.P. Column, Square Tied-Minimum Reinforcing							
	LOAD (KIPS)	STORY HEIGHT (FT.)	COLUMN SIZE (IN.)	COLUMN WEIGHT (P.L.F.)	CONCRETE STRENGTH (PSI)	COST PER V.L.F.		
						MAT.	INST.	TOTAL
9913	150	10-14	12	135	4000	9.10	37	46.10
9918	300	10-14	16	240	4000	13.30	46.50	59.80
9924	500	10-14	20	375	4000	19.40	60.50	79.90
9930	700	10-14	24	540	4000	27.50	78	105.50
9936	1000	10-14	28	740	4000	35	92.50	127.50
9942	1400	10-14	32	965	4000	46	107	153
9948	1800	10-14	36	1220	4000	55.50	124	179.50
9954	2300	10-14	40	1505	4000	63.50	135	198.50

Figure 3.11 (cont.)

"what if" questions of owners and contractors. The same sketches can serve as a basic outline of the project during conceptual stages.
- Local building code requirements must be addressed. Weather conditions in some locations may dictate particular structural and exterior wall materials. Also, consider the contractor's ability to construct the selected type of structure within the time frame allowed for the project.

Assemblies Estimate: Three-Story Office Building

Occupancy:	Office Building
Live Load:	Floor: 50 psf
Roof:	30 psf (snow load)
Allowable Soil Bearing Capacity:	6 K.S.F.
Footprint:	100' × 100' = 10,000 S.F.
Floors:	3 floors
Floor to Floor Height:	12'-0"
Bay Size:	25'-0" × 25'-0"
Construction:	**Floor:** Open web joists and concrete slab on steel columns and beams. **Roof:** Open web joists and metal deck on steel columns and beams.

Superimposed Load

Floor:

Live Load	50 psf	BOCA Code (Figure 3.2)
Partition	20 psf	(Figure 3.3)
Ceiling	5 psf	(Figure 3.3)
Superimposed Floor Load:	75 psf	

Roof:

Snow Load	30 psf	BOCA Code (Figure 3.2)
Ceiling	5 psf	(Figure 3.3)
Mechanical	5 psf	(Figure 3.3)
Superimposed Roof Load:	40 psf	

Calculate Total (Demand and Live) Loads

Enter the table in Figure 3.12 with 25'-0" × 25'-0" bays and a superimposed load of 75 psf. The result is:

Total Floor Load = 120 psf

Enter the table in Figure 3.13 with 25'-0" × 25'-0" bays and a superimposed load of 40 psf. The result is:

Total Roof Load = 60 psf

Loading Diagram

Compute the loads to an interior column as shown in Figure 3.14. Note that steel columns are usually spliced in two-story sections. The splice has been shown 3'-0" above the third floor.

Column Loads

Develop Column Loads as shown in Figure 3.14.

Note: The three-story office building estimate begins here and resumes for each division at the end of each chapter. This example will always appear shaded and inside a box.

B1010 Floor Construction

B1010 250	Steel Joists, Beams & Slab on Columns

	BAY SIZE (FT.)	SUPERIMPOSED LOAD (P.S.F.)	DEPTH (IN.)	TOTAL LOAD (P.S.F.)	COLUMN ADD	COST PER S.F.		
						MAT.	INST.	TOTAL
3700	20x25	40	44	83		5.15	3.62	8.77
3800					column	.55	.24	.79
3900	20x25	65	26	110		5.60	3.84	9.44
4000					column	.55	.24	.79
4100	20x25	75	26	120		5.55	3.68	9.23
4200					column	.66	.29	.95
4300	20x25	100	26	145		5.90	3.81	9.71
4400					column	.66	.29	.95
4500	20x25	125	29	170		6.60	4.17	10.77
4600					column	.77	.33	1.10
4700	25x25	40	23	84		5.50	3.77	9.27
4800					column	.53	.23	.76
4900	25x25	65	29	110		5.85	3.93	9.78
5000					column	.53	.23	.76
5100	25x25	75	26	120		6.10	3.92	10.02
5200					column	.62	.26	.88
5300	25x25	100	29	145		6.85	4.27	11.12
5400					column	.62	.26	.88
5500	25x25	125	32	170		7.20	4.44	11.64
5600					column	.69	.30	.99
5700	25x30	40	29	84		5.75	3.95	9.70
5800					column	.52	.22	.74
5900	25x30	65	29	110		6	4.09	10.09
6000					column	.52	.22	.74
6050	25x30	75	29	120		6.45	3.73	10.18
6100					column	.57	.25	.82
6150	25x30	100	29	145		7.05	3.95	11
6200					column	.57	.25	.82
6250	25x30	125	32	170		7.55	4.85	12.40
6300					column	.66	.29	.95
6350	30x30	40	29	84		6	3.56	9.56
6400					column	.48	.21	.69
6500	30x30	65	29	110		6.85	3.88	10.73
6600					column	.48	.21	.69
6700	30x30	75	32	120		7	3.94	10.94
6800					column	.55	.24	.79
6900	30x30	100	35	145		7.75	4.25	12
7000					column	.64	.28	.92
7100	30x30	125	35	172		8.50	5.30	13.80
7200					column	.71	.30	1.01
7300	30x35	40	29	85		6.80	3.86	10.66
7400					column	.41	.17	.58
7500	30x35	65	29	111		7.60	4.89	12.49
7600					column	.53	.23	.76
7700	30x35	75	32	121		7.60	4.89	12.49
7800					column	.54	.24	.78
7900	30x35	100	35	148		8.25	4.41	12.66
8000					column	.66	.29	.95
8100	30x35	125	38	173		9.15	4.77	13.92
8200					column	.67	.29	.96
8300	35x35	40	32	85		7	3.93	10.93
8400					column	.47	.20	.67

Figure 3.12

B1020   Roof Construction

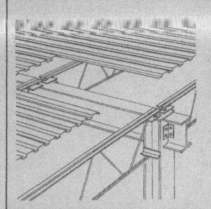

Design and Pricing... nnn nnnn nnn nn nnnn per S.F. for a roof system with steel columns, beams, and deck, using open web steel joists and 1-1/2" galvanized metal deck.

Roof deck is 1-1/2", 22 gauge galvanized steel. Joist cost includes appropriate bridging. Deflection is limited to 1/240 of the span. Fireproofing is not included.

Design and Pricing Assumptions:
Columns are 18' high.
Building is 4 bays long by 4 bays wide.
Joists are 5'-0" O.C. and span the long direction of the bay.
Joists at columns have bottom chords extended and are connected to columns. Column costs are not included but are listed separately per S.F. of floor.

Design Loads	Min.	Max.
Joists & Beams	3 PSF	5 PSF
Deck	2	2
Insulation	3	3
Roofing	6	6
Misc.	6	6
Total Dead Load	20 PSF	22 PSF

System Components			COST PER S.F.		
	QUANTITY	UNIT	MAT.	INST.	TOTAL
SYSTEM B1020 112					
METAL DECK AND JOISTS, 15'X20' BAY, 20 PSF S. LOAD					
Structural steel	.954	Lb.	.64	.23	.87
Open web joists	1.260	Lb.	.64	.33	.97
Metal decking, open, galvanized, 1-1/2" deep, 22 gauge	1.050	S.F.	.72	.42	1.14
TOTAL			2	.98	2.98

B1020 112		Steel Joists, Beams, & Deck on Columns						
	BAY SIZE (FT.)	SUPERIMPOSED LOAD (P.S.F.)	DEPTH (IN.)	TOTAL LOAD (P.S.F.)	COLUMN ADD	COST PER S.F.		
						MAT.	INST.	TOTAL
1100	15x20	20	16	40		2	.98	2.98
1200					columns	1.04	.37	1.41
1300		30	16	50		2.20	1.04	3.24
1400					columns	1.04	.37	1.41
1500		40	18	60		2.25	1.09	3.34
1600					columns	1.04	.37	1.41
1700	20x20	20	16	40		2.15	1.01	3.16
1800					columns	.78	.28	1.06
1900		30	18	50		2.38	1.11	3.49
2000					columns	.78	.28	1.06
2100		40	18	60		2.66	1.23	3.89
2200					columns	.78	.28	1.06
2300	20x25	20	18	40		2.29	1.09	3.38
2400					columns	.62	.23	.85
2500		30	18	50		2.57	1.27	3.84
2600					columns	.62	.23	.85
2700		40	20	60		2.59	1.24	3.83
2800					columns	.83	.29	1.12
2900	25x25	20	18	40		2.67	1.23	3.90
3000					columns	.50	.17	.67
3100		30	22	50		2.89	1.39	4.28
3200					columns	.66	.24	.90
3300		40	20	60		3.11	1.41	4.52

Figure 3.13

Column Loads:

Upper Columns	37.5 kips + 1.2 kips			
	(incl. col. and fireproofing)	=	38.7 kips	
Lower Columns	38.7 + 75 + 1.2 + 75 + 1.2	=	191 kips	

Foundation Loads

Exterior or 1/2 bay columns and corner or 1/4 bay columns assume less load from the superstructure than full bay columns. It would appear that the load reduction for exterior or corner columns should be 1/2 or 1/4 respectively, but usually exterior walls, slab overhangs, and other elements of the structure contribute to the column load on the exterior. Apply the following factors to obtain loads for exterior and corner columns:

- To determine the load on exterior columns, multiply the interior column load by 0.6.
- To determine the load on corner columns, multiple the interior column load by 0.45.

The factors shown above apply to buildings with normal fascia and floor-to-floor height. For special cases, calculate wall loads or use the interior column load for the load on exterior corner columns.

Interior Foundation Load	Figure 3.14	=	191 kips
Exterior Foundation Load	191 kips × .6	=	115 kips
Corner Foundation Load	191 kips × .45	=	86 kips

For spread footings with 6 K.S.F. allowable soil bearing capacity, see Figure 3.15.

Interior Footing	191 kips (200 kips)	6'-0" square × 20" deep
Exterior Footing	115 kips (125 kips)	5'-0" square × 16" deep
Corner Footing	86 kips (100 kips)	4'-6" square × 15" deep

	Bay Size, S.F.		Total Load, kips			
Roof	625 S.F.		0.06 ksi			
					37.5 k	
Column	12 L.F.	x	0.1 k/L.F.	=	1.2 k	Note: Splice column at 3'-0" above finished floor.
3rd Floor	625 S.F.		0.12 ksi	=	75 k	
					113.7 k	
Column	12 L.F.	x	0.1 k/L.F.	=	1.2 k	
2nd Floor	625 S.F.	x	0.12 ksi	=	75 k	
					189.9	
Column	12 L.F.	x	0.1 k/L.F.	=	1.2 k	
1st Floor						
Interior Foundation load				→	191.1 k	Footing

Figure 3.14

A10 Foundations

A1010 Standard Foundations

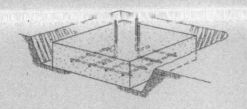

The Spread Footing System includes excavation; backfill; forms (four uses); all reinforcement; 3,000 p.s.i. concrete (chute placed); and screed finish.

Footing systems are priced per individual unit. The Expanded System Listing at the bottom shows footings that range from 3' square x 12" deep, to 18' square x 52" deep. It is assumed that excavation is done by a truck mounted hydraulic excavator with an operator and oiler.

Backfill is with a dozer, and compaction by air tamp. The excavation and backfill equipment is assumed to operate at 30 C.Y. per hour.

Please see the reference section for further design and cost information.

System Components	QUANTITY	UNIT	COST EACH		
			MAT.	INST.	TOTAL
SYSTEM A1010 210					
SPREAD FOOTINGS, LOAD 25K, SOIL CAPACITY 3 KSF, 3' SQ X 12" DEEP					
Bulk excavation	.590	C.Y.		3.44	3.44
Hand trim	9.000	S.F.		5.67	5.67
Compacted backfill	.260	C.Y.		.56	.56
Formwork, 4 uses	12.000	S.F.	7.92	39.96	47.88
Reinforcing, fy = 60,000 psi	.006	Ton	3.48	5.07	8.55
Dowel or anchor bolt templates	6.000	L.F.	5.94	16.56	22.50
Concrete, f'c = 3,000 psi	.330	C.Y.	23.43		23.43
Place concrete, direct chute	.330	C.Y.		5.15	5.15
Screed finish	9.000	S.F.		3.33	3.33
TOTAL			40.77	79.74	120.51

A1010 210	Spread Footings	COST EACH		
		MAT.	INST.	TOTAL
7090	Spread footings, 3000 psi concrete, chute delivered			
7100	Load 25K, soil capacity 3 KSF, 3'-0" sq. x 12" deep	41	79.50	120.50
7150	Load 50K, soil capacity 3 KSF, 4'-6" sq. x 12" deep	84	138	222
7200	Load 50K, soil capacity 6 KSF, 3'-0" sq. x 12" deep	41	79.50	120.50
7250	Load 75K, soil capacity 3 KSF, 5'-6" sq. x 13" deep	131	196	327
7300	Load 75K, soil capacity 6 KSF, 4'-0" sq. x 12" deep	68.50	118	186.50
7350	Load 100K, soil capacity 3 KSF, 6'-0" sq. x 14" deep	164	235	399
7410	Load 100K, soil capacity 6 KSF, 4'-6" sq. x 15" deep	103	163	266
7450	Load 125K, soil capacity 3 KSF, 7'-0" sq. x 17" deep	259	335	594
7500	Load 125K, soil capacity 6 KSF, 5'-0" sq. x 16" deep	131	196	327
7550	Load 150K, soil capacity 3 KSF, 7'-6" sq. x 18" deep	310	395	705
7610	Load 150K, soil capacity 6 KSF, 5'-6" sq. x 18" deep	173	246	419
7650	Load 200K, soil capacity 3 KSF, 8'-6" sq. x 20" deep	440	525	965
7700	Load 200K, soil capacity 6 KSF, 6'-0" sq. x 20" deep	226	305	531
7750	Load 300K, soil capacity 3 KSF, 10'-6" sq. x 25" deep	805	850	1,655
7810	Load 300K, soil capacity 6 KSF, 7'-6" sq. x 25" deep	425	510	935
7850	Load 400K, soil capacity 3 KSF, 12'-6" sq. x 28" deep	1,275	1,275	2,550
7900	Load 400K, soil capacity 6 KSF, 8'-6" sq. x 27" deep	585	665	1,250
7950	Load 500K, soil capacity 3 KSF, 14'-0" sq. x 31" deep	1,750	1,675	3,425
8010	Load 500K, soil capacity 6 KSF, 9'-6" sq. x 30" deep	800	865	1,665

Note: Row 7150/7200 contains box marking "RA1010 -120"

Figure 3.15

Chapter 4

Substructure

Substructure, UNIFORMAT II Division A, consists of Foundations (A10) and Basement Construction (A20). The foundation is the entire substructure below the first floor or frame of a building and includes the footing or other elements on which the building rests. Once structural elements have been chosen and sized, loads to the foundation can be determined. *(See Chapter 3 for developing structural loads.)* This information enables the estimator to select and price the foundation components.

The appropriate foundation for the imposed loads is based on the allowable soil bearing capacity at a given depth below the slab on grade or below grade.

Spread Footings   Spread footings are the most commonly used type of footing for concentrated loads. They are used to convert a column or grade beam load into an allowable area load on the supporting soil. They tend to minimize excavation, and allow for easy examination of soil conditions at the bearing area. The depth of interior footings should be below topsoil or soil materials containing loose fill or vegetation. Exterior footings must be lower than frost penetration. (See Figures 4.1a and 4.1b.)

Exterior or 1/2 bay footings and corner or 1/4 bay footings assume less load from the superstructure than interior or full bay footings. In Figure 4.1a there are:

- 9 Interior Footings (full bay)
- 12 Exterior Footings (1/2 bay)
- 4 Corner Footings (1/4 bay)

As noted in Chapter 3, exterior walls, slab overhangs, and other elements of the structure contribute additional load to the column loads on the exterior of a structure. To account for that additional load, the load on exterior footings is determined by multiplying the interior foundation load by 0.6. To determine the load on corner footings, multiply the interior foundation load by 0.45. (These factors apply to buildings with normal fascia and floor-to-floor height.) For special cases, calculate wall loads or, to be safe, use the interior column load for the load on exterior or corner footings.

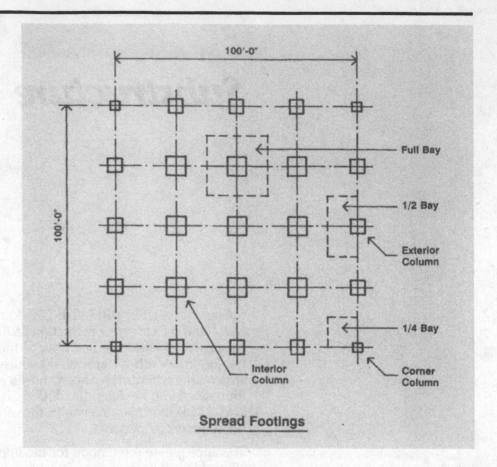

Spread Footings

Figure 4.1a

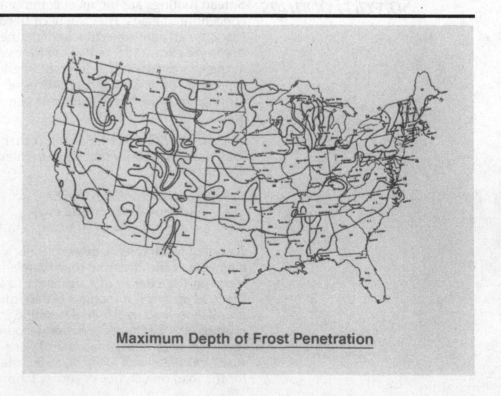

Maximum Depth of Frost Penetration

Figure 4.1b

Example: Spread Footing Selection

Assume a 6 K.S.F. soil bearing capacity. Using the 358K interior column load that was derived in the Total Load example in Chapter 3 (Part Two: Column Derivation, Step 4), the foundation loads and sizes for interior, exterior, and corner columns can be determined as follows:

Interior Column Load	358K	
Exterior Column Load	358K × 0.6	= 215K
Corner Column Load	358K × 0.45	= 161K

Enter the Spread Footings from the table in Figure 4.2.

358K (400K)	8'-6" square × 27" deep
215K (300K)	7'-6" square × 25" deep
161K (200K)	6'-0" square × 20" deep

Concrete Piers

Concrete piers are often required to extend the superstructure column at the ground or basement floor to the spread footing. This may occur due to frost penetration, poor soil conditions, or the existence of basement walls.

Use the information in Figure 3.11 for piers. For concrete columns, assume the pier is the same size as the concrete column. For steel or wood columns, assume the pier is approximately 10" larger than the column size to allow for base plates, anchor bolts, and grouting. Exterior piers usually protrude inside concrete walls because column lines are set back to allow the building fascia to cover the column. Protruding exterior piers interrupt wall formwork and are time-consuming and costly to erect.

Figure 4.3 shows pier sizes for a 24" square concrete column, a 10" steel column, and an 8" square wood column.

Strip Footings

Strip footings are used to convert a lineal wall load to an allowable load for the soil bearing capacity. They may also be employed as a leveling pad to facilitate the placement of formwork for concrete walls. For low-rise, normal-span buildings, reaching the soil bearing capacity is of little consequence. With this in mind, choose a foundation width twice the bearing wall thickness, with a minimum thickness of the strip footing equal to the bearing wall.

Example One: Determining Loads to Strip Footings

This example shows how to develop the load to the footings in units of one thousand pounds per linear foot (K.L.F.). Then select the appropriate system based on the load information.

A two-story building for elderly housing is to be constructed. The only available information about the project follows:

Floor to Ceiling Height: 8'
Span: 25'
Structure: 8" Concrete Plank on 8" Block Wall
Building Occupancy: Elderly Housing
Building Code Requirement: Reinforced Footings

A10 Foundations

A1010 Standard Foundations

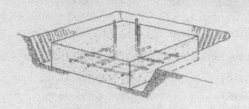

The Spread Footing System includes: excavation; backfill; forms (four uses); all reinforcement; 3,000 p.s.i. concrete (chute placed); and screed finish.

Footing systems are priced per individual unit. The Expanded System Listing at the bottom shows footings that range from 3' square x 12" deep, to 18' square x 52" deep. It is assumed that excavation is done by a truck mounted hydraulic excavator with an operator and oiler.

Backfill is with a dozer, and compaction by air tamp. The excavation and backfill equipment is assumed to operate at 30 C.Y. per hour.

Please see the reference section for further design and cost information.

System Components			COST EACH		
	QUANTITY	UNIT	MAT.	INST.	TOTAL
SYSTEM A1010 210					
SPREAD FOOTINGS, LOAD 25K, SOIL CAPACITY 3 KSF, 3' SQ X 12" DEEP					
Bulk excavation	.590	C.Y.		3.44	3.44
Hand trim	9.000	S.F.		5.67	5.67
Compacted backfill	.260	C.Y.		.56	.56
Formwork, 4 uses	12.000	S.F.	7.92	39.96	47.88
Reinforcing, fy = 60,000 psi	.006	Ton	3.48	5.07	8.55
Dowel or anchor bolt templates	6.000	L.F.	5.94	16.56	22.50
Concrete, f'c = 3,000 psi	.330	C.Y.	23.43		23.43
Place concrete, direct chute	.330	C.Y.		5.15	5.15
Screed finish	9.000	S.F.		3.33	3.33
TOTAL			40.77	79.74	120.51

A1010 210	Spread Footings		COST EACH		
			MAT.	INST.	TOTAL
7090	Spread footings, 3000 psi concrete, chute delivered				
7100	Load 25K, soil capacity 3 KSF, 3'-0" sq. x 12" deep		41	79.50	120.50
7150	Load 50K, soil capacity 3 KSF, 4'-6" sq. x 12" deep	RA1010 -120	84	138	222
7200	Load 50K, soil capacity 6 KSF, 3'-0" sq. x 12" deep		41	79.50	120.50
7250	Load 75K, soil capacity 3 KSF, 5'-6" sq. x 13" deep		131	196	327
7300	Load 75K, soil capacity 6 KSF, 4'-0" sq. x 12" deep		68.50	118	186.50
7350	Load 100K, soil capacity 3 KSF, 6'-0" sq. x 14" deep		164	235	399
7410	Load 100K, soil capacity 6 KSF, 4'-6" sq. x 15" deep		103	163	266
7450	Load 125K, soil capacity 3 KSF, 7'-0" sq. x 17" deep		259	335	594
7500	Load 125K, soil capacity 6 KSF, 5'-0" sq. x 16" deep		131	196	327
7550	Load 150K, soil capacity 3 KSF 7'-6" sq. x 18" deep		310	395	705
7610	Load 150K, soil capacity 6 KSF, 5'-6" sq. x 18" deep		173	246	419
7650	Load 200K, soil capacity 3 KSF, 8'-6" sq. x 20" deep		440	525	965
7700	Load 200K, soil capacity 6 KSF, 6'-0" sq. x 20" deep		226	305	531
7750	Load 300K, soil capacity 3 KSF, 10'-6" sq. x 25" deep		805	850	1,655
7810	Load 300K, soil capacity 6 KSF, 7'-6" sq. x 25" deep		425	510	935
7850	Load 400K, soil capacity 3 KSF, 12'-6" sq. x 28" deep		1,275	1,275	2,550
7900	Load 400K, soil capacity 6 KSF, 8'-6" sq. x 27" deep		585	665	1,250
7950	Load 500K, soil capacity 3 KSF, 14'-0" sq. x 31" deep		1,750	1,675	3,425
8010	Load 500K, soil capacity 6 KSF, 9'-6" sq. x 30" deep		800	865	1,665

Figure 4.2

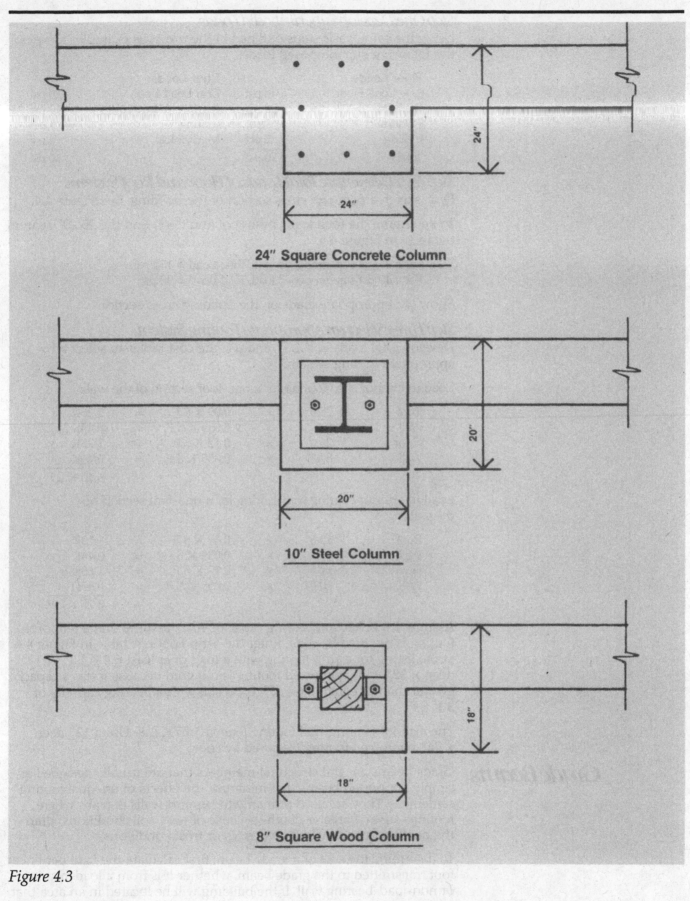

24" Square Concrete Column

10" Steel Column

8" Square Wood Column

Figure 4.3

51

Step One: Develop Superimposed Loads

Using the same basic steps outlined in the previous examples, develop the following superimposed loads:

Floor Loads:		Roof Loads:	
Live Load Floor	40 psf	Live Load Roof	20 psf
Partitions	20 psf	Roofing & Insulation	10 psf
Flooring	5 psf	Ceiling	5 psf
Ceiling	5 psf	Mechanical	5 psf
Totals	70 psf		40 psf

Step Two: Determine Total Loads of Floor and Roof Systems

First, sketch a plan and cross section of the building. See Figure 4.4.

To determine the total loads from roof and floor, find the 25'-0" span in the table in Figure 4.5.

For a 75 psf superimposed load, Total Load = 130 psf.
For a 40 psf superimposed load, Total Load = 90 psf.

Show the appropriate loads on the building cross section.

Step Three: Select an Appropriate Footing System

Develop total loads in K.L.F., and use the cost tables to select an appropriate footing system.

Load to interior strip footing for a one-foot section of the wall:

Roof	25.0'	×	0.09 K.S.F.	=	2.25K
Wall	8.0'	×	0.055 K.S.F.	=	0.44K
Floor	25.0'	×	0.13 K.S.F.	=	3.25K
Wall	8.33'	×	0.055 K.S.F.	=	0.46K
					6.40 K.L.F.

Load to exterior bearing wall footing for a one-foot section of the wall:

Roof	12.5'	×	0.09 K.S.F.	=	1.13K
Wall	8.0'	×	0.055 K.S.F.	=	0.44K
Floor	12.5'	×	0.13 K.S.F.	=	1.63K
Wall	10.01	×	0.055 K.S.F.	=	0.55K
					3.75 K.L.F.

Assume a 3 K.S.F. soil bearing capacity. Also, assume that a reinforced footing is required by code. Enter the Strip Footings table in Figure 4.6 at the listing for a strip footing with a load of at least 6.4 K.S.F. A 12" deep × 32" wide reinforced footing is required because it has a capacity greater than or equal to 6.4 K.L.F. on soil with a bearing capacity of 3 K.S.F.

The exterior bearing wall footing load is 3.75 K.L.F. Use a 12" deep × 24" reinforced footing (required by code).

Grade Beams

Grade beams are stiff structural members that are usually designed as simply-supported beams that minimize the effects of unequal footing settlement. They are used primarily to support walls or slabs where footings are at a greater depth (because of poor soil conditions) than the depth required by building codes or frost conditions.

To determine the cost of a grade beam, first calculate the load per linear foot transmitted to the grade beam, whether it is from a load-bearing or non-load-bearing wall. If the building will be located in an area that

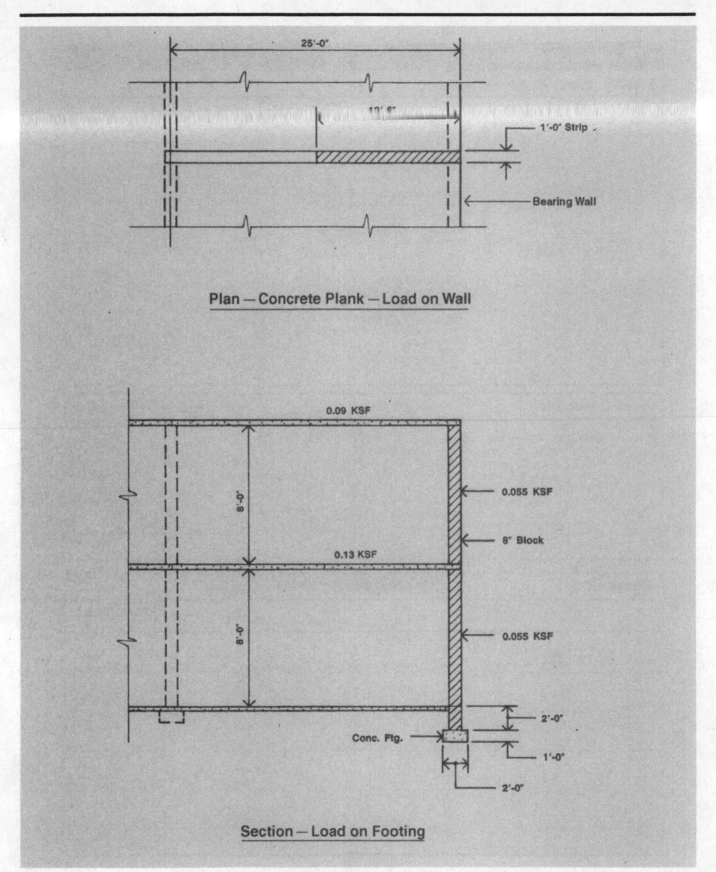

Plan — Concrete Plank — Load on Wall

25'-0"

1'-0" Strip

Bearing Wall

0.09 KSF

0.055 KSF

8" Block

8'-0"

0.13 KSF

8'-0"

0.055 KSF

2'-0"

Conc. Ftg.

1'-0"

2'-0"

Section — Load on Footing

Figure 4.4

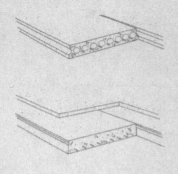

General: Units priced here are for plant produced prestressed members, transported to site and erected.

Normal weight concrete is most frequently used. Lightweight concrete may be used to reduce dead weight.

Structural topping is sometimes used on floors: insulating concrete or rigid insulation on roofs.

Camber and deflection may limit use by depth considerations.

Prices are based upon 10,000 S.F. to 20,000 S.F. projects, and 50 mile to 100 mile transport.

Concrete is f'c = 5 KSI and Steel is fy = 250 or 300 KSI

Note: Deduct from prices 20% for Southern states. Add to prices 10% for Western states.

Description of Table: Enter table at span and load. Most economical sections will generally consist of normal weight concrete without topping. If acceptable, note this price, depth and weight. For topping and/or lightweight concrete, note appropriate data.

Generally used on masonry and concrete bearing or reinforced concrete and steel framed structures.

The solid, 4" slabs are used for light loads and short spans. The 6" to 12" thick hollow core units are used for longer spans and heavier loads. Cores may carry utilities.

Topping is used structurally for loads or rigidity and architecturally to level or slope surface.

Camber and deflection and change in direction of spans must be considered (door openings, etc.), especially untopped.

System Components	QUANTITY	UNIT	COST PER S.F.		
			MAT.	INST.	TOTAL
SYSTEM B1010 229					
10' SPAN, 40 LBS S.F. WORKING LOAD, 2" TOPPING					
Precast prestressed concrete roof/floor slabs 4" thick, grouted	1.000	S.F.	5.40	2.13	7.53
Edge forms to 6" high on elevated slab, 4 uses	.100	L.F.	.04	.28	.32
Welded wire fabric 6 x 6 - W1.4 x W1.4 (10 x 10), 21 lb/csf, 10% lap	.010	C.S.F.	.08	.26	.34
Concrete ready mix, regular weight, 3000 psi	.170	C.F.	.45		.45
Place and vibrate concrete, elevated slab less than 6", pumped	.170	C.F.		.18	.18
Finishing floor, monolithic steel trowel finish for resilient tile	1.000	S.F.		.55	.55
Curing with sprayed membrane curing compound	.010	C.S.F.	.04	.06	.10
TOTAL			6.01	3.46	9.47

B1010 229		Precast Plank with No Topping							
	SPAN (FT.)	SUPERIMPOSED LOAD (P.S.F.)	TOTAL DEPTH (IN.)	DEAD LOAD (P.S.F.)	TOTAL LOAD (P.S.F.)	COST PER S.F.			
						MAT.	INST.	TOTAL	
0720	10	40	4	50	90	5.40	2.13	7.53	
0750	RB1010 -010	75	6	50	125	5.20	1.83	7.03	
0770		100	6	50	150	5.20	1.83	7.03	
0800	15	40	6	50	90	5.20	1.83	7.03	
0820	RB1010 -105	75	6	50	125	5.20	1.83	7.03	
0850		100	6	50	150	5.20	1.83	7.03	
0875	20	40	6	50	90	5.20	1.83	7.03	
0900		75	6	50	125	5.20	1.83	7.03	
0920		100	6	50	150	5.20	1.83	7.03	
0950	25	40	6	50	90	5.20	1.83	7.03	
0970		75	8	55	130	5.80	1.60	7.40	
1000		100	8	55	155	5.80	1.60	7.40	
1200	30	40	8	55	95	5.80	1.60	7.40	
1300		75	8	55	130	5.80	1.60	7.40	
1400		100	10	70	170	6.55	1.41	7.96	
1500	40	40	10	70	110	6.55	1.41	7.96	
1600		75	12	70	145	6.75	1.21	7.96	

Figure 4.5

A10 Foundations

A1010 Standard Foundations

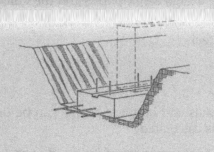

The Strip Footing System includes: excavation; hand trim; all forms needed for footing placement; forms for 2" x 6" keyway (four uses); dowels; and 3,000 p.s.i. concrete.

The footing size required varies for different soils. Soil bearing capacities are listed for 3 KSF and 6 KSF. Depths of the system range from 8" to 24". Widths range from 16" to 96". Smaller strip footings may not require reinforcement.

Please see the reference section for further design and cost information.

System Components	QUANTITY	UNIT	COST PER L.F.		
			MAT.	INST.	TOTAL
SYSTEM A1010 110					
STRIP FOOTING, LOAD 5.1KLF, SOIL CAP. 3 KSF, 24"WIDE X12"DEEP, REINF.					
Trench excavation	.148	C.Y.		.85	.85
Hand trim	2.000	S.F.		1.26	1.26
Compacted backfill	.074	C.Y.		.16	.16
Formwork, 4 uses	2.000	S.F.	1.72	5.70	7.42
Keyway form, 4 uses	1.000	L.F.	.32	.73	1.05
Reinforcing, fy = 60000 psi	3.000	Lb.	.96	1.26	2.22
Dowels	2.000	Ea.	.90	3.70	4.60
Concrete, f'c = 3000 psi	.074	C.Y.	5.25		5.25
Place concrete, direct chute	.074	C.Y.		1.15	1.15
Screed finish	2.000	S.F.		.92	.92
TOTAL			9.15	15.73	24.88

A1010 110	Strip Footings		COST PER L.F.		
			MAT.	INST.	TOTAL
2100	Strip footing, load 2.6KLF, soil capacity 3KSF, 16"wide x 8"deep plain		4.58	9.40	13.98
2300	Load 3.9 KLF, soil capacity, 3 KSF, 24"wide x 8"deep, plain		5.80	10.45	16.25
2500	Load 5.1KLF, soil capacity 3 KSF, 24"wide x 12"deep, reinf.	RA1010 -140	9.15	15.75	24.90
2700	Load 11.1KLF, soil capacity 6 KSF, 24"wide x 12"deep, reinf.		9.15	15.75	24.90
2900	Load 6.8 KLF, soil capacity 3 KSF, 32"wide x 12"deep, reinf.		11.30	17.30	28.60
3100	Load 14.8 KLF, soil capacity 6 KSF, 32"wide x 12"deep, reinf.		11.30	17.30	28.60
3300	Load 9.3 KLF, soil capacity 3 KSF, 40"wide x 12"deep, reinf.		13.35	18.80	32.15
3500	Load 18.4 KLF, soil capacity 6 KSF, 40"wide x 12"deep, reinf.		13.45	18.90	32.35
3700	Load 10.1KLF, soil capacity 3 KSF, 48"wide x 12"deep, reinf.		15.05	20.50	35.55
3900	Load 22.1KLF, soil capacity 6 KSF, 48"wide x 12"deep, reinf.		16	21.50	37.50
4100	Load 11.8KLF, soil capacity 3 KSF, 56"wide x 12"deep, reinf.		17.65	22.50	40.15
4300	Load 25.8KLF, soil capacity 6 KSF, 56"wide x 12"deep, reinf.		19.05	24	43.05
4500	Load 10KLF, soil capacity 3 KSF, 48"wide x 16"deep, reinf.		19.10	23.50	42.60
4700	Load 22KLF, soil capacity 6 KSF, 48"wide, 16"deep, reinf.		19.50	24	43.50
4900	Load 11.6KLF, soil capacity 3 KSF, 56"wide x 16"deep, reinf.		22	32.50	54.50
5100	Load 25.6KLF, soil capacity 6 KSF, 56"wide x 16"deep, reinf.		23	34	57
5300	Load 13.3KLF, soil capacity 3 KSF, 64"wide x 16"deep, reinf.		25	28	53
5500	Load 29.3KLF, soil capacity 6 KSF, 64"wide x 16"deep, reinf.		26.50	30	56.50
5700	Load 15KLF, soil capacity 3 KSF, 72"wide x 20"deep, reinf.		33.50	33.50	67
5900	Load 33KLF, soil capacity 6 KSF, 72"wide x 20"deep, reinf.		35.50	36	71.50
6100	Load 18.3KLF, soil capacity 3 KSF, 88"wide x 24"deep, reinf.		47.50	42.50	90
6300	Load 40.3KLF, soil capacity 6 KSF, 88"wide x 24"deep, reinf.		51.50	47.50	99
6500	Load 20KLF, soil capacity 3 KSF, 96"wide x 24"deep, reinf.		52	45	97
6700	Load 44 KLF, soil capacity 6 KSF, 96" wide x 24" deep, reinf.		54.50	49	103.50

Figure 4.6

has frozen earth through the winter months, frost protection may be necessary. Choose a grade beam with a width no less than that of the supported wall. Also take into consideration the load supported and the minimum depth required. A typical grade beam system cost per linear foot is derived from designing a simply supported beam, such as the one in Figure 4.7, using design assumptions and the following items:

- Excavation
- Reinforcing steel
- Backfill
- Concrete
- Formwork

The entire assemblies cost of grade beams can be determined quickly once the loading is calculated.

Example Two: Determining Grade Beam Load
The following example illustrates how to develop the load on grade beams (in K.L.F.). The calculations for the pier and total load to the spread footing are shown below. (See Figure 4.8.)

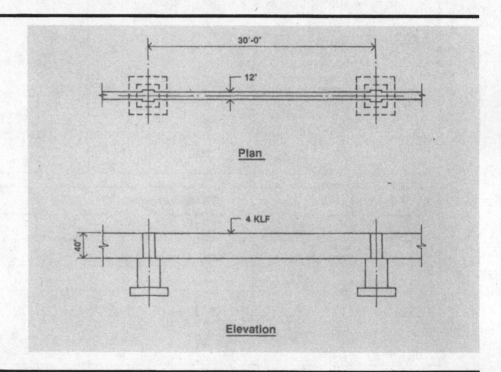

Figure 4.7

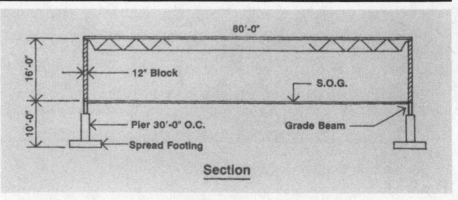

Figure 4.8

56

Construction: Steel Joists on Walls. Use Figure 4.9 for load figures.
Superimposed Load: 40 psf (determined in previous examples)
Frost Requirements: 36"
Roof Total Load: 64 psf (from Figure 4.9)

Roof Load:	$40.0 \times .064$ K.S.F.	=	2.56 K.L.F.
Wall 12" Block:	$16.0' \times .085$ K.S.F.	=	1.36 K.L.F.
Total Load to Grade Beam			3.92 K.L.F.

Grade Beam for Bearing Walls:

Enter the Grade Beams table in Figure 4.10, using a 36" minimum depth and a 30'-0" span. Choose a 40" deep × 12" wide grade beam, 4 K.L.F. capacity, which is greater than 3.92 K.L.F. required.

Load to Pier:

Wall and Roof Load: 30' × 3.92 K.L.F.   =   117.6K

Grade Beam Load:

$$30' \times \left(\frac{40" \text{ beam}}{12"/\text{L.F.}}\right) \times 0.15 \text{ K.C.F.} \qquad = \qquad 15.0K$$

Load to Pier: 132.6K

(Note: Weight of concrete equals 0.15 K.C.F.)

Enter the table in Figure 3.11, C.I.P. Columns—Square Tied, Minimum Reinforcing: 16 square pier maximum load is 300K, which is greater than 132.6K required.

Load to Spread Footing: 3 K.S.F. soil bearing capacity

Load to Pier:		132.6K
Pier: $5' \times 1.33' \times 1.33' \times 0.15$ K.C.F.	=	1.3K
Load to Spread Footing:		133.9K

The Spread Footing System selection (see Figure 4.2):

133.9K (150K), 7'-6" square × 18" deep.

Piles, Pile Caps, and Caissons

The Geotechnical Report details the soil type and bearing capacity at various depths at many locations on the proposed site, and is used by a structural engineer to design the building foundation. The recommendations of the Geotechnical Report, which will specify the foundation system best suited to the project, should be followed.

Piles

Piles are column-like shafts that receive and support superstructure loads, overturning forces, or uplift forces. They receive these loads from isolated column or pier foundations (pile caps), foundation walls, grade beams, or foundation mats. The piles then transfer the loads through shallower poor soil layers to deeper soil that offers adequate support strength and acceptable settlement.

Piles are usually associated with difficult foundation problems and poor soil conditions. They are used when the soil bearing capacity is insufficient to support the weight of the proposed building using simple strip and spread footings. Pile foundations involve extensive structural design and use of specialized materials, labor, and equipment to develop sufficient load bearing capacity. Pile foundations are costly. Different pile types have been developed to suit specific ground

B10 Superstructure

B1020 Roof Construction

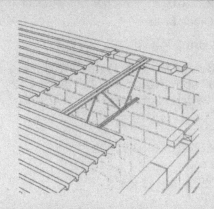

Description: Table below lists cost per S.F. for a roof system using open web steel joists and 1-1/2″ galvanized metal deck. The system is assumed supported on bearing walls or other suitable support. Costs for the supports are not included.

Design and Pricing Assumptions:
Joists are 5′-0″ O.C.
Roof deck is 1-1/2″, 22 gauge galvanized.

System Components		QUANTITY	UNIT	COST PER S.F.		
				MAT.	INST.	TOTAL
SYSTEM B1020 116						
METAL DECK AND JOISTS, 20' SPAN, 20 PSF S. LOAD						
Open web joists, horiz. bridging T.L. lot, to 30' span		1.114	Lb.	.57	.29	.86
Metal decking, open type, galv 1-1/2″ deep		1.050	S.F.	.72	.42	1.14
	TOTAL			1.29	.71	2

B1020 116		Steel Joists & Deck on Bearing Walls						
	BAY SIZE (FT.)	SUPERIMPOSED LOAD (P.S.F.)	DEPTH (IN.)	TOTAL LOAD (P.S.F.)		COST PER S.F.		
						MAT.	INST.	TOTAL
1100	20	20	13-1/2	40		1.29	.71	2
1200		30	15-1/2	50		1.32	.72	2.04
1300		40	15-1/2	60		1.43	.78	2.21
1400	25	20	17-1/2	40		1.43	.79	2.22
1500		30	17-1/2	50		1.56	.85	2.41
1600		40	19-1/2	60		1.59	.86	2.45
1700	30	20	19-1/2	40		1.58	.86	2.44
1800		30	21-1/2	50		1.62	.88	2.50
1900		40	23-1/2	60		1.76	.95	2.71
2000	35	20	23-1/2	40		1.74	.79	2.53
2100		30	25-1/2	50		1.80	.81	2.61
2200		40	25-1/2	60		1.93	.85	2.78
2300	40	20	25-1/2	41		1.96	.86	2.82
2400		30	25-1/2	51		2.09	.92	3.01
2500		40	25-1/2	61		2.17	.95	3.12
2600	45	20	27-1/2	41		2.20	1.28	3.48
2700		30	31-1/2	51		2.34	1.36	3.70
2800		40	31-1/2	61		2.47	1.44	3.91
2900	50	20	29-1/2	42		2.48	1.45	3.93
3000		30	31-1/2	52		2.72	1.59	4.31
3100		40	31-1/2	62		2.90	1.70	4.60
3200	60	20	37-1/2	42		3.18	1.49	4.67
3300		30	37-1/2	52		3.55	1.66	5.21
3400		40	37-1/2	62		3.55	1.66	5.21
3500	70	20	41-1/2	42		3.54	1.65	5.19
3600		30	41-1/2	52		3.79	1.75	5.54
3700		40	41-1/2	64		4.65	2.14	6.79

Figure 4.9

conditions. A full investigation of ground conditions early in the selection process is essential to provide information for professional foundation engineering and an acceptable structure. Again, consult the Geotechnical Report for the specific system to be used.

Piles support loads by means of end bearing and/or skin friction and are designated by their principal method of load transfer to soil.

End bearing piles have shafts that pass through soft strata or thin hard strata. They have tips that either bear on bedrock, or penetrate some distance into a dense, adequate soil (sand or gravel).

Friction piles have shafts that may be entirely embedded in cohesive soil (moist clay), and provide support mainly by adhesion or "skin-friction" between soil and shaft area.

Piles may be installed singly or in clusters or groups. Some building codes require a minimum of three piles per major column load, or two per foundation wall or grade beam. Single pile capacity is limited by the pile's structural strength or the support strength of the soil. The support capacity of a pile cluster is almost always less than the sum of its individual pile capacities because of the overlap of bearing and friction stresses.

Pile Caps

A *pile cap* is a structural member placed on and fastened to the top of all piles in a group. The purpose of a pile cap is to transfer a column or pier load equally to each pile in a supporting cluster.

Caissons

Caissons are drilled or dug-out cylindrical foundation shafts that serve as short, column-like compression members. Caissons transfer superstructure loads through inadequate soils to bedrock or hard strata. They may be either reinforced or unreinforced concrete, and either straight or belled out at the bearing level.

B10 Superstructure

B1020 Roof Construction

| B1020 116 | | | | Steel Joists & Deck on Bearing Walls | | | | |

	BAY SIZE (FT.)	SUPERIMPOSED LOAD (P.S.F.)	DEPTH (IN.)	TOTAL LOAD (P.S.F.)		COST PER S.F.		
						MAT.	INST.	TOTAL
3800	80	20	45-1/2	44		5.15	2.35	7.50
3900		30	45-1/2	54		5.15	2.35	7.50
4000		40	45-1/2	64		5.70	2.61	8.31
4100	90	20	53-1/2	44		4.26	1.86	6.12
4200		30	53-1/2	54		4.53	1.97	6.50
4300		40	53-1/2	65		5.45	2.34	7.79
4400	100	20	57-1/2	44		4.52	1.97	6.49
4500		30	57-1/2	54		5.45	2.34	7.79
4600		40	57-1/2	65		6	2.58	8.58
4700	125	20	69-1/2	44		6.45	2.66	9.11
4800		30	69-1/2	56		7.60	3.09	10.69
4900		40	69-1/2	67		8.70	3.53	12.23

Figure 4.9 (cont.)

A10 Foundations

A1020 Special Foundations

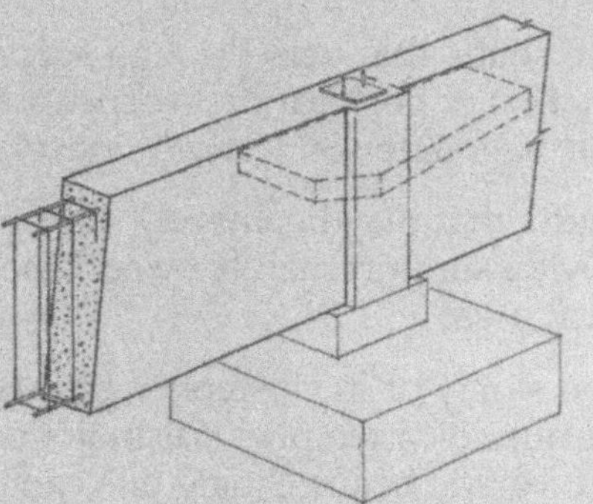

The Grade Beam System includes: excavation with a truck mounted backhoe; hand trim; backfill; forms (four uses); reinforcing steel; and 3,000 p.s.i. concrete placed from chute.

Superimposed loads vary in the listing from 8 Kips per linear foot (KLF) to 50 KLF. In the Expanded System Listing, the span of the beams varies from 15′ to 40′. Depth varies from 28″ to 52″. Width varies from 12″ to 28″.

Please see the reference section for further design and cost information.

System Components	QUANTITY	UNIT	COST PER L.F. MAT.	INST.	TOTAL
SYSTEM A1020 210					
GRADE BEAM, 15′ SPAN, 28″ DEEP, 12″ WIDE, 8 KLF LOAD					
Excavation, trench, hydraulic backhoe, 3/8 CY bucket	.260	C.Y.		1.48	1.48
Trim sides and bottom of trench, regular soil	2.000	S.F.		1.26	1.26
Backfill, by hand, compaction in 6″ layers, using vibrating plate	.170	C.Y.		1.01	1.01
Forms in place, grade beam, 4 uses	4.700	SFCA	2.82	16.59	19.41
Reinforcing in place, beams & girders, #8 to #14	.019	Ton	11.12	12.54	23.66
Concrete ready mix, regular weight, 3000 psi	.090	C.Y.	6.39		6.39
Place and vibrate conc. for grade beam, direct chute	.090	C.Y.		1.12	1.12
TOTAL			20.33	34	54.33

A1020 210	Grade Beams		COST PER L.F. MAT.	INST.	TOTAL
2220	Grade beam, 15′ span, 28″ deep, 12″ wide, 8 KLF load		20.50	34	54.50
2240	14″ wide, 12 KLF load		21	34	55
2260	40″ deep, 12″ wide, 16 KLF load	RA1020 -230	21.50	42	63.50
2280	20 KLF load		24	45.50	69.50
2300	52″ deep, 12″ wide, 30 KLF load		28.50	57	85.50
2320	40 KLF load		34	63.50	97.50
2340	50 KLF load		40	69	109
3360	20′ span, 28″ deep, 12″ wide, 2 KLF load		13.30	26	39.30
3380	16″ wide, 4 KLF load		17.45	29	46.45
3400	40″ deep, 12″ wide, 8 KLF load		21.50	42	63.50
3420	12 KLF load		26.50	47.50	74
3440	14″ wide, 16 KLF load		32	53	85
3460	52″ deep, 12″ wide, 20 KLF load		34.50	64.50	99
3480	14″ wide, 30 KLF load		47.50	76.50	124
3500	20″ wide, 40 KLF load		56	81	137
3520	24″ wide, 50 KLF load		69	92.50	161.50
4540	30′ span, 28″ deep, 12″ wide, 1 KLF load		13.90	26.50	40.40
4560	14″ wide, 2 KLF load		21.50	37	58.50
4580	40″ deep, 12″ wide, 4 KLF load		26	44.50	70.50
4600	18″ wide, 8 KLF load		36	54.50	90.50
4620	52″ deep, 14″ wide, 12 KLF load		45.50	74.50	120
4640	20″ wide, 16 KLF load		56	81.50	137.50
4660	24″ wide, 20 KLF load		69	93	162
4680	36″ wide, 30 KLF load		99	117	216
4700	48″ wide, 40 KLF load		130	142	272
5720	40′ span, 40″ deep, 12″ wide, 1 KLF load		19	39.50	58.50

Figure 4.10

Caissons are used in the following situations:

- For tall structures
- For heavy structures
- For underpinning (extensive use)
- To moderate depth if unsuitable soil exists

Shaft diameters range in size from 20" to 84", with the most typical sizes beginning at 36". If inspection of the bottom is required, the minimum practical diameter is 36". If handwork is required (in addition to mechanical belling), the minimum diameter is 36". The most common shaft diameter is 36" with a 5- or 6-foot bell. The maximum practical bell diameter is 3 times the shaft diameter.

Permanent casings add to the cost and should be avoided if possible for economic reasons. Wet or loose strata are undesirable. The associated installation sometimes involves a mudding operation with bentonite clay slurry to keep the walls of excavation stable. (This cost is not included in the tables.)

Reinforcement may be required, especially for heavy loads. If uplift, bending moment, or lateral loads are present, reinforcement will be necessary. It is standard practice to add a small amount of reinforcement at the top portion of each caisson, even if the above conditions are not present. All reinforcement should extend below the soft strata. Horizontal reinforcement is not required for belled bottoms.

The three basic types of caisson bearings follow:

- **Belled:** Generally used to reduce the bearing pressure on soil, but not recommended for shallow depths or poor soils.

  — Good soils for belling include most clays, hardpan, soft shale, and decomposed rock.

  — Soils not recommended include sand, gravel, silt, and igneous rock. Compact sand and gravel above water table may stand. Water in the bearing strata is undesirable.

  — Soils requiring handwork include hard shale, limestone, and sandstone.

- **Straight Shafted:** The entire length is enlarged to permit safe bearing pressures. They are most economical for light loads on high bearing capacity soil.
- **Socketed or Keyed:** Used for extremely heavy loads. The shaft is sunk into rock for combined friction and bearing support action and is usually reinforced.

Advantages of using caissons include:

- Shafts can pass through soils that piles cannot penetrate.
- There is no soil heaving or displacement during installation.
- There is no vibration during installation.
- They create less noise than pile driving.
- Bearing strata can be visually inspected and tested.

Example Three: Selecting Pilings, Pile Caps, and Caissons
Selecting the type of foundation assembly to use is based on engineering judgment and relative costs.

Select an appropriate piling foundation system, given the following information:

Construction: 8-Story Building

> 30'-0" × 35'-0" bays comp. beam and deck and lightweight concrete slab
> 12'-0" floor to floor

Superimposed Loads:

> Floor: 75 psf
> Roof: 40 psf

Total Load (Figure 4.11):

> Floor: 117 psf
> Roof: 82 psf

Interior Column Load:

Roof	30' × 35'	×	.087 K.S.F.	=	91K
Floors	7 (30' × 35'	×	.117 K.S.F.)	=	860K
Column (estimated) 8 floors		×	12' × .15 K.L.F.	=	14K
Interior Column Load:					965K

See the boring log in Figure 4.12. Firm, fine sand and gravel start at 20' depth and continue to refusal at a depth of 45'-6" below grade.

Good bearing strata at 25'-0" allows the use of pressure-injected piles. End bearing piles or caissons would be driven or placed to the rock strata or 50'-0" + below grade to allow for cutoffs and variations in the elevation of the rock strata.

Option One

> End bearing: Steel H piles, 8 piles required.
>
> | | |
> |---|---|
> | Pile cost per interior column (Figure 4.13) | $10,125 |
> | Pile cap (Figure 4.14) (1243K, 8 piles, 3'-0" spacing) | 1,660 |
> | | $11,785 |

Option Two

> Pressure-injected piles (requires 4'-6" pile spacing),
> 6 piles required.
>
> | | |
> |---|---|
> | Pile cost per interior column (Figure 4.15) | $6,925 |
> | Pile cap, Figure 4.14 (1413K, 6 piles, 4"-6' spacing) | 1,335 |
> | Total | $8,260 |

Option Three

> Caissons (Figure 4.16)
> Caissons 4' × 100' @ 1,200K = $10,800/2 (50' required) = $5,400

The choice appears to be caissons or pressure-injected piles, depending on availability in the area. Add mobilization, demobilization, and pile testing costs to obtain complete prices.

To satisfy the column loading of 956K, the following pile clusters were selected from *Means Assemblies Cost Data*.

C.I.P. concrete piles, 50' long 1200K	15 piles	$27,850
Precast concrete piles, 50' long 1200K	19 piles	34,000
Steel pipe piles, 50' long 1200K	16 piles	26,800
Steel H piles, 50' long 1200K	8 piles	10,125
Step tapered steel pipes, 50' long 1200K	16 piles	12,175
Pressure-injected footings, 25' long 1200K	6 piles	6,925

B1010 Floor Construction

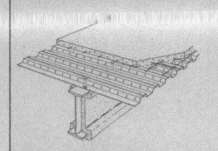

Description: Table below lists costs ($/S.F.) for a floor system using composite steel beams with welded shear studs, composite steel deck, and light weight concrete slab reinforced with W.W.F. Price includes sprayed fiber fireproofing on steel beams.

Design and Pricing Assumptions:
Structural steel is A36, high strength bolted.
Composite steel deck varies from 22 gauge to 16 gauge, galvanized.

Shear Studs are 3/4".
W.W.F., 6 x 6 – W1.4 x W1.4 (10 x 10)
Concrete f'c = 3 KSI, lightweight.
Steel trowel finish and cure.
Fireproofing is sprayed fiber (non-asbestos).

Spandrels are assumed the same as interior beams and girders to allow for exterior wall loads and bracing or moment connections.

System Components	QUANTITY	UNIT	COST PER S.F.		
			MAT.	INST.	TOTAL
SYSTEM B1010 256					
20X25 BAY, 40 PSF S. LOAD, 5-1/2" SLAB, 17-1/2" TOTAL THICKNESS					
Structural steel	4.320	Lb.	2.98	1.29	4.27
Welded shear connectors 3/4" diameter 4-7/8" long	.163	Ea.	.08	.24	.32
Metal decking, non-cellular composite, galv. 3" deep, 22 gauge	1.050	S.F.	1.01	.67	1.68
Sheet metal edge closure form, 12", w/2 bends, 18 ga	.045	L.F.	.09	.07	.16
Welded wire fabric rolls, 6 x 6 - W1.4 x W1.4 (10 x 10), 21 lb/csf	1.000	S.F.	.08	.26	.34
Concrete ready mix, light weight, 3,000 PSI	.333	C.F.	1.42		1.42
Place and vibrate concrete, elevated slab less than 6", pumped	.333	C.F.		.35	.35
Finishing floor, monolithic steel trowel finish for finish floor	1.000	S.F.		.60	.60
Curing with sprayed membrane curing compound	.010	C.S.F.	.04	.06	.10
Shores, erect and strip vertical to 10' high	.020	Ea.		.28	.28
Sprayed mineral fiber/cement for fireproof, 1" thick on beams	.483	S.F.	.21	.37	.58
TOTAL			5.91	4.19	10.10

B1010 256		Composite Beams, Deck & Slab						
	BAY SIZE (FT.)	SUPERIMPOSED LOAD (P.S.F.)	SLAB THICKNESS (IN.)	TOTAL DEPTH (FT. - IN.)	TOTAL LOAD (P.S.F.)	COST PER S.F.		
						MAT.	INST.	TOTAL
2400	20x25	40	5-1/2	1 - 5-1/2	80	5.90	4.19	10.09
2500		75	5-1/2	1 - 9-1/2	115	6.15	4.21	10.36
2750	RB1010-105	125	5-1/2	1 - 9-1/2	167	7.60	4.97	12.57
2900		200	6-1/4	1 - 11-1/2	251	8.55	5.40	13.95
3000	25x25	40	5-1/2	1 - 9-1/2	82	5.90	4	9.90
3100		75	5-1/2	1 - 11-1/2	118	6.60	4.08	10.68
3200		125	5-1/2	2 - 2-1/2	169	6.90	4.42	11.32
3300		200	6-1/4	2 - 6-1/4	252	9.40	5.15	14.55
3400	25x30	40	5-1/2	1 - 11-1/2	83	6.05	3.96	10.01
3600		75	5-1/2	1 - 11-1/2	119	6.50	4.01	10.51
3900		125	5-1/2	1 - 11-1/2	170	7.55	4.54	12.09
4000		200	6-1/4	2 - 6-1/4	252	9.45	5.20	14.65
4200	30x30	40	5-1/2	1 - 11-1/2	81	5.95	4.11	10.06
4400		75	5-1/2	2 - 2-1/2	116	6.50	4.30	10.80
4500		125	5-1/2	2 - 5-1/2	168	7.85	4.84	12.69
4700		200	6-1/4	2 - 9-1/4	252	9.45	5.60	15.05
4900	30x35	40	5-1/2	2 - 2-1/2	82	6.25	4.25	10.50
5100		75	5-1/2	2 - 5-1/2	117	6.85	4.37	11.22
5300		125	5-1/2	2 - 5-1/2	169	8.10	4.93	13.03
5500		200	6-1/4	2 - 9-1/4	254	9.45	5.60	15.05

Figure 4.11

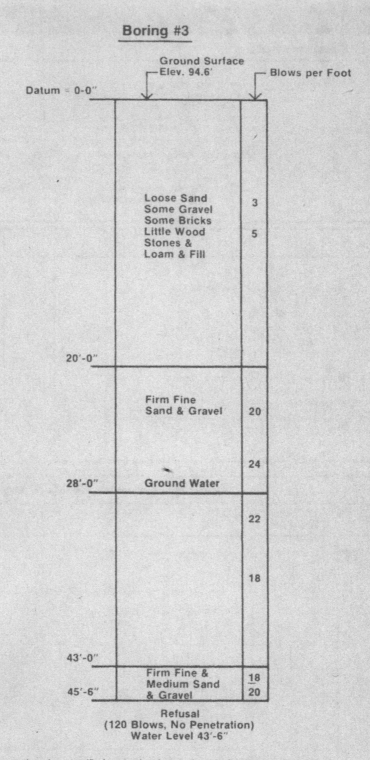

Figure 4.12

64

A1020 Special Foundations

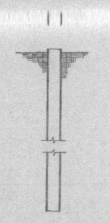

A1020 "H" Pile System includes: steel H sections; heavy duty driving point; splices where applicable and allowance for cutoffs.

The Expanded System Listing shows costs per cluster of piles. Clusters range from one pile to seventeen piles. Loads vary from 50 Kips to 2,000 Kips. All loads for Steel H Pile systems are given in terms of end bearing capacity.

Steel sections range from 10" x 10" to 14" x 14" in the Expanded System Listing. The 14" x 14" steel section is used for all H piles used in applications requiring a working load over 800 Kips.

Please see the reference section for cost of mobilization of the pile driving equipment and other design and cost information.

System Components	QUANTITY	UNIT	COST EACH		
			MAT.	INST.	TOTAL
SYSTEM A1020 140					
STEEL H PILES, 50' LONG, 100K LOAD, END BEARING, 1 PILE					
Steel H piles 10" x 10", 42 #/L.F.	53.000	V.L.F.	612.15	420.29	1,032.44
Heavy duty point, 10"	1.000	Ea.	93.50	119	212.50
Pile cut off, steel pipe or H piles	1.000	Ea.		20.50	20.50
TOTAL			705.65	559.79	1,265.44

A1020 140	Steel H Piles		COST EACH		
			MAT.	INST.	TOTAL
2220	Steel H piles, 50' long, 100K load, end bearing, 1 pile		705	560	1,265
2260	2 pile cluster		1,400	1,125	2,525
2280	200K load, end bearing, 2 pile cluster	RA1020 -100	1,400	1,125	2,525
2300	3 pile cluster		2,125	1,700	3,825
2320	400K load, end bearing, 3 pile cluster		2,125	1,700	3,825
2340	4 pile cluster		2,825	2,250	5,075
2360	6 pile cluster		4,225	3,350	7,575
2380	800K load, end bearing, 5 pile cluster		3,525	2,800	6,325
2400	7 pile cluster		4,950	3,900	8,850
2420	12 pile cluster		8,475	6,725	15,200
2440	1200K load, end bearing, 8 pile cluster		5,650	4,475	10,125
2460	11 pile cluster		7,750	6,150	13,900
2480	17 pile cluster		12,000	9,525	21,525
2500	1600K load, end bearing, 10 pile cluster		8,775	5,875	14,650
2520	14 pile cluster		12,300	8,225	20,525
2540	2000K load, end bearing, 12 pile cluster		10,500	7,050	17,550
2560	18 pile cluster		15,800	10,600	26,400
2580					
3580	100' long, 50K load, end bearing, 1 pile		2,300	1,250	3,550
3600	100K load, end bearing, 1 pile		2,300	1,250	3,550
3620	2 pile cluster		4,625	2,500	7,125
3640	200K load, end bearing, 2 pile cluster		4,625	2,500	7,125
3660	3 pile cluster		6,925	3,700	10,625
3680	400K load, end bearing, 3 pile cluster		6,925	3,700	10,625
3700	4 pile cluster		9,225	4,950	14,175
3720	6 pile cluster		13,900	7,450	21,350

Figure 4.13

A10 Foundations

A1010 Standard Foundations

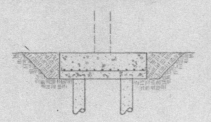

These pile cap systems include excavation with a truck mounted hydraulic excavator/hand trimming, compacted backfill, forms for concrete, templates for dowels or anchor bolts, reinforcing steel and concrete placed and screeded.

Pile embedment is assumed as 6". Design is consistent with the Concrete Reinforcing Steel Institute Handbook f'c = 3000 psi, fy = 60,000.

Please see the reference section for further design and cost information.

System Components	QUANTITY	UNIT	COST EACH		
			MAT.	INST.	TOTAL
SYSTEM A1010 250 **CAP FOR 2 PILES,6'-6"X3'-6"X20",15 TON PILE,8" MIN. COL., 45K COL. LOAD**					
Excavation, bulk, hyd excavator, truck mtd. 30" bucket 1/2 CY	2.890	C.Y.		16.87	16.87
Trim sides and bottom of trench, regular soil	23.000	S.F.		14.49	14.49
Dozer backfill & roller compaction	1.500	C.Y.		3.23	3.23
Forms in place pile cap, square or rectangular, 4 uses	33.000	SFCA	23.10	118.80	141.90
Templates for dowels or anchor bolts	8.000	Ea.	7.92	22.08	30
Reinforcing in place footings, #8 to #14	.025	Ton	13.75	12.38	26.13
Concrete ready mix, regular weight, 3000 psi	1.400	C.Y.	99.40		99.40
Place and vibrate concrete for pile caps, under 5CY, direct chute	1.400	C.Y.		29.09	29.09
Monolithic screed finish	23.000	S.F.		8.51	8.51
TOTAL			144.17	225.45	369.62

A1010 250		Pile Caps						
	NO. PILES	SIZE FT-IN X FT-IN X IN	PILE CAPACITY (TON)	COLUMN SIZE (IN)	COLUMN LOAD (KIPS)	COST EACH		
						MAT.	INST.	TOTAL
5100	2	6-6x3-6x20	15	8	45	144	226	370
5150	RA1020 -330	26	40	8	155	177	274	451
5200		34	80	11	314	239	350	589
5250		37	120	14	473	256	380	636
5300	3	5-6x5-1x23	15	8	75	169	261	430
5350		28	40	10	232	190	294	484
5400		32	80	14	471	218	335	553
5450		38	120	17	709	252	385	637
5500	4	5-6x5-6x18	15	10	103	189	261	450
5550		30	40	11	308	279	375	654
5600		36	80	16	626	325	435	760
5650		38	120	19	945	345	455	800
5700	6	8-6x5-6x18	15	12	156	310	380	690
5750		37	40	14	458	515	555	1,070
5800		40	80	19	936	580	625	1,205
5850		45	120	24	1413	645	690	1,335
5900	8	8-6x7-9x19	15	12	205	465	520	985
5950		36	40	16	610	685	670	1,355
6000		44	80	22	1243	845	815	1,660
6050		47	120	27	1881	915	875	1,790

Figure 4.14

A1020 Special Foundations

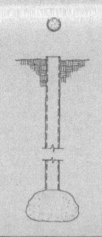

Pressure Injected Piles are usually uncased up to 25' and cased over 25' depending on soil conditions.

These costs include excavation and hauling of excess materials; steel casing over 25'; reinforcement; 3,000 p.s.i. concrete; plus mobilization and demobilization of equipment for a distance of up to fifty miles to and from the job site.

The Expanded System lists cost per cluster of piles. Clusters range from one pile to eight piles. End-bearing loads range from 50 Kips to 1,600 Kips.

Please see the reference section for further design and cost information.

System Components			COST EACH		
	QUANTITY	UNIT	MAT.	INST.	TOTAL
SYSTEM A1020 710 PRESSURE INJECTED FOOTING, END BEARING, 50' LONG, 50K LOAD, 1 PILE					
Pressure injected footings, cased, 30-60 ton cap., 12" diameter	50.000	V.L.F.	547.50	884.50	1,432
Pile cutoff, concrete pile with thin steel shell	1.000	Ea.		10.30	10.30
TOTAL			547.50	894.80	1,442.30

A1020 710	Pressure Injected Footings		COST EACH		
			MAT.	INST.	TOTAL
2200	Pressure injected footing, end bearing, 25' long, 50K load, 1 pile		255	630	885
2400	100K load, 1 pile		255	630	885
2600	2 pile cluster	RA1020 -100	510	1,250	1,760
2800	200K load, 2 pile cluster		510	1,250	1,760
3200	400K load, 4 pile cluster		1,025	2,500	3,525
3400	7 pile cluster		1,775	4,375	6,150
3800	1200K load, 6 pile cluster		2,150	4,775	6,925
4000	1600K load, 7 pile cluster		2,500	5,575	8,075
4200	50' long, 50K load, 1 pile		550	895	1,445
4400	100K load, 1 pile		950	895	1,845
4600	2 pile cluster		1,900	1,775	3,675
4800	200K load, 2 pile cluster		1,900	1,775	3,675
5000	4 pile cluster		3,800	3,575	7,375
5200	400K load, 4 pile cluster		3,800	3,575	7,375
5400	8 pile cluster		7,600	7,175	14,775
5600	800K load, 7 pile cluster		6,650	6,275	12,925
5800	1200K load, 6 pile cluster		6,150	5,350	11,500
6000	1600K load, 7 pile cluster		7,175	6,275	13,450

Figure 4.15

A1020 Special Foundations

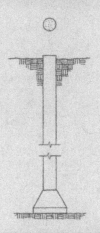

Caisson Systems are listed for three applications: stable ground, wet ground and soft rock. Concrete used is 3,000 p.s.i. placed from chute. Included are a bell at the bottom of the caisson shaft (if applicable) along with required excavation and disposal of excess excavated material up to two miles from job site.

The Expanded System lists cost per caisson. End-bearing loads vary from 200 Kips to 3,200 Kips. The dimensions of the caissons range from 2' x 50' to 7' x 200'.

Please see the reference section for further design and cost information.

System Components	QUANTITY	UNIT	COST EACH		
			MAT.	INST.	TOTAL
SYSTEM A1020 310					
CAISSON, STABLE GROUND, 3000 PSI CONC., 10KSF BRNG, 200K LOAD, 2'X50'					
Caissons, drilled, to 50', 24" shaft diameter, .116 C.Y./L.F.	50.000	V.L.F.	500	1,032.50	1,532.50
Reinforcing in place, columns, #3 to #7	.060	Ton	36.60	70.50	107.10
Caisson bell excavation and concrete, 4' diameter .444 CY	1.000	Ea.	31.50	196	227.50
Load & haul excess excavation, 2 miles	6.240	C.Y.		28.83	28.83
TOTAL			568.10	1,327.83	1,895.93

A1020 310	Caissons	COST EACH		
		MAT.	INST.	TOTAL
2200	Caisson, stable ground, 3000 PSI conc, 10KSF brng, 200K load, 2'x50'	570	1,325	1,895
2400	400K load, 2'-6"x50'-0"	940	2,125	3,065
2600	800K load, 3'-0"x100'-0"	2,575	5,050	7,625
2800	1200K load, 4'-0"x100'-0" RA1020-200	4,450	6,350	10,800
3000	1600K load, 5'-0"x150'-0"	10,100	9,850	19,950
3200	2400K load, 6'-0"x150'-0"	14,600	12,800	27,400
3400	3200K load, 7'-0"x200'-0"	26,400	18,800	45,200
5000	Wet ground, 3000 PSI conc., 10 KSF brng, 200K load, 2'-0"x50'-0"	495	1,850	2,345
5200	400K load, 2'-6"x50'-0"	825	3,175	4,000
5400	800K load, 3'-0"x100'-0"	2,250	8,525	10,775
5600	1200K load, 4'-0"x100'-0"	3,850	12,800	16,650
5800	1600K load, 5'-0"x150'-0"	8,650	27,700	36,350
6000	2400K load, 6'-0"x150'-0"	12,600	34,100	46,700
6200	3200K load, 7'-0"x200'-0"	22,700	54,500	77,200
7800	Soft rock, 3000 PSI conc., 10 KSF brng, 200K load, 2'-0"x50'-0"	495	10,200	10,695
8000	400K load, 2'-6"x50'-0"	825	16,400	17,225
8200	800K load, 3'-0"x100'-0"	2,250	43,100	45,350
8400	1200K load, 4'-0"x100'-0"	3,850	63,500	67,350
8600	1600K load, 5'-0"x150'-0"	8,650	129,500	138,150
8800	2400K load, 6'-0"x150'-0"	12,600	154,000	166,600
9000	3200K load, 7'-0"x200'-0"	22,700	245,500	268,200

Figure 4.16

Steel H piles appear to be the economical choice for end bearing piles. Check Steel H piles 50'-0" long against pressure-injected footings 25'-0" long and caissons 50'-0" long to support a 965K column load.

Both Steel H pile clusters and pressure-injected footing clusters require a pile cap to transfer the column load to each pile in the cluster.

Pile caps are shown in Figure 4.14. Enter the table with a column or pier load, number of piles in the cluster, and the required pile spacing. Pressure-injected footings require 4'-6" pile spacing, while 3'-0" pile spacing is adequate for other pile types.

Summary

Spread and strip footings are the most economical foundation system to use if there is adequate soil bearing capacity at a reasonable depth. Also, these systems can be used with piers, columns, walls, and grade beams.

Pile or caisson assemblies are more costly to install, and more difficult to price since there are many unknown factors involved in their installation.

An assemblies square foot estimate for a foundation system that involves piles or caissons requires additional research. Always collect as much information as possible about subsoil types and the specific systems being used by contacting the following:

- Area building inspector or building department
- Local pile or caisson subcontractors
- Local contractors or engineers

Often, a phone call to the right source provides sufficient information to determine approximate depths to good bearing strata in an area. At best, an assemblies square foot estimate for pile foundation systems is useful only for general budget figures.

Slab on Grade

When estimating a slab on grade, consider its intended use. The applicable building code may prescribe the required floor loading per square foot, which in turn defines the thickness and amount of reinforcement necessary to satisfy the loading conditions. Slabs may be categorized in three distinct groups according to their loading conditions: *non-industrial*, *light industrial*, and *heavy industrial*. Each can be divided further into *reinforced* and *non-reinforced*.

If the specifications or dimensions are predetermined by the project requirements or the estimator's past experience, it may be possible to develop a unit price estimate of the cost of the slab on grade to include such items as:

- Base course
- Topping
- Vapor barrier
- Finish
- Concrete
- Cure
- Reinforcement

However, it is even easier to create an assemblies estimate using assemblies costs developed by the estimator or a published source such as *Means Assemblies Cost Data*. Figure 4.17 shows the assemblies costs with all components listed for reinforced and non-reinforced slabs for

A2020 Basement Walls

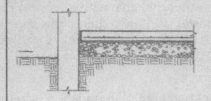

There are four types of Slab on Grade Systems listed: Non-industrial, Light industrial, Industrial and Heavy industrial. Each type is listed two ways: reinforced and non-reinforced. A Slab on Grade system includes three passes with a grader; 6" of compacted gravel fill; polyethylene vapor barrier; 3500 p.s.i. concrete placed by chute; bituminous fibre expansion joint; all necessary edge forms (4 uses); steel trowel finish; and sprayed-on membrane curing compound.

The Expanded System Listing shows costs on a per square foot basis. Thicknesses of the slabs range from 4" to 8". Non-industrial applications are for foot traffic only with negligible abrasion. Light industrial applications are for pneumatic wheels and light abrasion. Industrial applications are for solid rubber wheels and moderate abrasion. Heavy industrial applications are for steel wheels and severe abrasion. All slabs are either shown unreinforced or reinforced with welded wire fabric.

System Components	QUANTITY	UNIT	COST PER S.F.		
			MAT.	INST.	TOTAL
SYSTEM A2020 170					
SLAB ON GRADE, 4" THICK, NON INDUSTRIAL, NON REINFORCED					
Fine grade, 3 passes with grader and roller	.110	S.Y.		.32	.32
Gravel under floor slab, 6" deep, compacted	1.000	S.F.	.17	.19	.36
Polyethylene vapor barrier, standard, .006" thick	1.000	S.F.	.03	.10	.13
Concrete ready mix, regular weight, 3500 psi	.012	C.Y.	.87		.87
Place and vibrate concrete for slab on grade, 4" thick, direct chute	.012	C.Y.		.21	.21
Expansion joint, premolded bituminous fiber, 1/2" x 6"	.100	L.F.	.09	.21	.30
Edge forms in place for slab on grade to 6" high, 4 uses	.030	L.F.	.02	.07	.09
Cure with sprayed membrane curing compound	1.000	S.F.	.04	.06	.10
Finishing floor, monolithic steel trowel	1.000	S.F.		.60	.60
TOTAL			1.22	1.76	2.98

A2020 170	Plain & Reinforced		COST PER S.F.		
			MAT.	INST.	TOTAL
2220	Slab on grade, 4" thick, non industrial, non reinforced		1.22	1.76	2.98
2240	Reinforced		1.30	2.02	3.32
2260	Light industrial, non reinforced	RA1030 -200	1.43	2.15	3.58
2280	Reinforced		1.51	2.41	3.92
2300	Industrial, non reinforced		1.80	3.42	5.22
2320	Reinforced		1.88	3.68	5.56
3340	5" thick, non industrial, non reinforced		1.44	1.80	3.24
3360	Reinforced		1.52	2.06	3.58
3380	Light industrial, non reinforced		1.65	2.19	3.84
3400	Reinforced		1.73	2.45	4.18
3420	Heavy industrial, non reinforced		2.21	3.96	6.17
3440	Reinforced		2.23	4.25	6.48
4460	6" thick, non industrial, non reinforced		1.73	1.77	3.50
4480	Reinforced		1.85	2.11	3.96
4500	Light industrial, non reinforced		1.95	2.16	4.11
4520	Reinforced		2.14	2.62	4.76
4540	Heavy industrial, non reinforced		2.52	4.03	6.55
4560	Reinforced		2.64	4.37	7.01
5580	7" thick, non industrial, non reinforced		1.96	1.83	3.79
5600	Reinforced		2.12	2.20	4.32
5620	Light industrial, non reinforced		2.19	2.22	4.41
5640	Reinforced		2.35	2.59	4.94
5660	Heavy industrial, non reinforced		2.76	3.96	6.72
5680	Reinforced		2.88	4.30	7.18

Figure 4.17

different slab thicknesses. The national average cost of concrete is shown at $69.00/cubic yard. (*Note: Costs can be localized to over 900 locations using the City Cost Indexes in the back of* Means Assemblies Cost Data.) Using the system shown in Figure 4.17 (line #2220) as an example, what is the effect of a delivered concrete price of $85 per cubic yard?

There is 0.012 C.Y./S.F. in a 4" thick slab.

4"/12"/ft/27 C.F./C.Y. = 0.012 C.Y./S.F.

.012 × ($85 - $69) = $.19/S.F. + $2.98/S.F. = $3.17/S.F.

The system price now is $3.17/S.F. based on the new material prices instead of $2.98/S.F., which used $69/C.Y. as its basis.

Slab on Grade Selection and Cost Analysis

Using Figure 4.17, select the appropriate slab thickness and the proper usage. Multiply the system cost per square foot by the total area of the slab to obtain the total cost for slabs on grade. For the area of slab, use the out-to-out dimension. Doing so will provide a built-in allowance for the thickened slab under partitions and at the perimeter, as well as other areas where extra material is needed.

Assembly Cost/S.F. × Area of Slab = Total Cost for Slab on Grade

---

Assemblies Estimate:
Three-Story Office
Building

Column Footings

Loads applied to the foundation and footing size (see "Foundation Loads" in Chapter 3) using a 6 K.S.F. soil capacity:

Interior Columns	191 kips:	6'-0" sq. × 20" deep
Exterior Columns	115 kips:	5'-0" sq. × 16" deep
Corner Columns	86 kips:	4'-6" sq. × 15" deep

Costs:

Interior (Figure 3.15, line #7700)	(200 kips)	9 ea. @ $531	=	$4,779
Exterior (Figure 3.15, line #7500)	(125 kips)	12 ea. @ $327	=	3,924
Corners (Figure 3.15, line #7410)	(100 kips)	4 ea. @ $266	=	1,064
		Total for Column Footings		$9,767

Strip Footings

A strip footing is necessary around the perimeter of the building. To find the correct number of total linear feet of strip footing, deduct the width of both exterior and corner column footings. See Figure 4.18.

Perimeter: 400 L.F.

Exterior Footings: 12 @ 5' each

Corner Footings: 4 @ 4.5' each

400' – (12 × 5') – (4 × 4.5') = 322 L.F.

Costs:

(Figure 4.19, line #2700) 322 L.F. @ $24.90 = $8,018

The load to the strip footing is negligible since the precast concrete fascia is hung from the columns and elevated floor slabs and accounted for in the column loads.

Foundation Wall

The foundation wall needed for the three-story office building is 4' deep × 12" wide.

$$100' \times 4 = 400 \text{ L.F.}$$

Cost:

(Figure 4.20, line #1560) 400 L.F. @ $50.90 = $20,360

Piers

To carry the column loads at the perimeter wall, provide a pier under each perimeter column to provide support directly to the footing instead of the slab on grade (see Figure 4.21). Assume a 20" square pier.

$$16 \times 3.5' = 56 \text{ L.F.}$$

Cost:

(Figure 4.22, line #9924) 56 L.F. @ $79.90 = $4,474

Excavation and Backfill

Excavation and fill (see Figure 4.23) is based on the following assumptions:

- Three feet over-excavation in width is used for a work area at the bottom of excavation. The excavation has an appropriate side slope for the type of material being excavated.
- Backfill from the excavation to the foundation wall is done with excavated material, except for clay (which is assumed to be wasted).
- Costs are shown for material stored on- or off-site. Storage is within a two-mile haul distance.

Variables that affect the cost include:

- Availability and type of equipment.
- Haul distance to spoil area.
- Haul distance to borrow area as the source for fill.
- Protection of the excavation, particularly in deep excavations where sheet piles may be necessary.

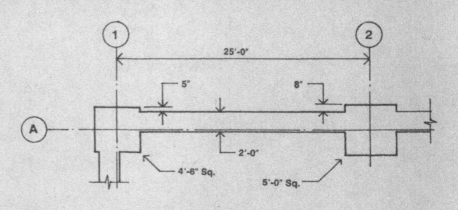

Figure 4.18

A10 Foundations

A1010 Standard Foundations

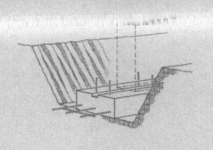

The Strip Footing System includes: excavation; hand trim; all forms needed for footing placement; forms for 2" x 6" keyway (four uses); dowels; and 3,000 p.s.i. concrete.

The footing size required varies for different soils. Soil bearing capacities are listed for 3 KSF and 6 KSF. Depths of the system range from 8" to 24". Widths range from 16" to 96". Smaller strip footings may not require reinforcement.

Please see the reference section for further design and cost information.

System Components	QUANTITY	UNIT	COST PER L.F.		
			MAT.	INST.	TOTAL
SYSTEM A1010 110					
STRIP FOOTING, LOAD 5.1KLF, SOIL CAP. 3 KSF, 24"WIDE X12"DEEP, REINF.					
Trench excavation	.148	C.Y.		.85	.85
Hand trim	2.000	S.F.		1.26	1.26
Compacted backfill	.074	C.Y.		.16	.16
Formwork, 4 uses	2.000	S.F.	1.72	5.70	7.42
Keyway form, 4 uses	1.000	L.F.	.32	.73	1.05
Reinforcing, fy = 60000 psi	3.000	Lb.	.96	1.26	2.22
Dowels	2.000	Ea.	.90	3.70	4.60
Concrete, f'c = 3000 psi	.074	C.Y.	5.25		5.25
Place concrete, direct chute	.074	C.Y.		1.15	1.15
Screed finish	2.000	S.F.		.92	.92
TOTAL			9.15	15.73	24.88

A1010 110	Strip Footings		COST PER L.F.		
			MAT.	INST.	TOTAL
2100	Strip footing, load 2.6KLF, soil capacity 3KSF, 16"wide x 8"deep plain		4.58	9.40	13.98
2300	Load 3.9 KLF, soil capacity, 3 KSF, 24"wide x 8"deep, plain		5.80	10.45	16.25
2500	Load 5.1KLF, soil capacity 3 KSF, 24"wide x 12"deep, reinf.	RA1010 -140	9.15	15.75	24.90
2700	Load 11.1KLF, soil capacity 6 KSF, 24"wide x 12"deep, reinf.		9.15	15.75	24.90
2900	Load 6.8 KLF, soil capacity 3 KSF, 32"wide x 12"deep, reinf.		11.30	17.30	28.60
3100	Load 14.8 KLF, soil capacity 6 KSF, 32"wide x 12"deep, reinf.		11.30	17.30	28.60
3300	Load 9.3 KLF, soil capacity 3 KSF, 40"wide x 12"deep, reinf.		13.35	18.80	32.15
3500	Load 18.4 KLF, soil capacity 6 KSF, 40"wide x 12"deep, reinf.		13.45	18.90	32.35
3700	Load 10.1KLF, soil capacity 3 KSF, 48"wide x 12"deep, reinf.		15.05	20.50	35.55
3900	Load 22.1KLF, soil capacity 6 KSF, 48"wide x 12"deep, reinf.		16	21.50	37.50
4100	Load 11.8KLF, soil capacity 3 KSF, 56"wide x 12"deep, reinf.		17.65	22.50	40.15
4300	Load 25.8KLF, soil capacity 6 KSF, 56"wide x 12"deep, reinf.		19.05	24	43.05
4500	Load 10KLF, soil capacity 3 KSF, 48"wide x 16"deep, reinf.		19.10	23.50	42.60
4700	Load 22KLF, soil capacity 6 KSF, 48"wide, 16"deep, reinf.		19.50	24	43.50
4900	Load 11.6KLF, soil capacity 3 KSF, 56"wide x 16"deep, reinf.		22	32.50	54.50
5100	Load 25.6KLF, soil capacity 6 KSF, 56"wide x 16"deep, reinf.		23	34	57
5300	Load 13.3KLF, soil capacity 3 KSF, 64"wide x 16"deep, reinf.		25	28	53
5500	Load 29.3KLF, soil capacity 6 KSF, 64"wide x 16"deep, reinf.		26.50	30	56.50
5700	Load 15KLF, soil capacity 3 KSF, 72"wide x 20"deep, reinf.		33.50	33.50	67
5900	Load 33KLF, soil capacity 6 KSF, 72"wide x 20"deep, reinf.		35.50	36	71.50
6100	Load 18.3KLF, soil capacity 3 KSF, 88"wide x 24"deep, reinf.		47.50	42.50	90
6300	Load 40.3KLF, soil capacity 6 KSF, 88"wide x 24"deep, reinf.		51.50	47.50	99
6500	Load 20KLF, soil capacity 3 KSF, 96"wide x 24"deep, reinf.		52	45	97
6700	Load 44 KLF, soil capacity 6 KSF, 96" wide x 24" deep, reinf.		54.50	49	103.50

Figure 4.19

A20 Basement Construction

A2020 Basement Walls

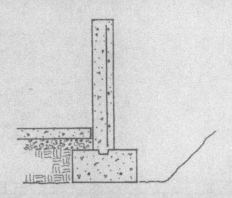

The Foundation Bearing Wall System includes: forms up to 16' high (four uses); 3,000 p.s.i. concrete placed and vibrated; and form removal with breaking form ties and patching walls. The wall systems list walls from 6" to 16" thick and are designed with minimum reinforcement.

Excavation and backfill are not included.

Please see the reference section for further design and cost information.

System Components	QUANTITY	UNIT	COST PER L.F. MAT.	COST PER L.F. INST.	COST PER L.F. TOTAL
SYSTEM A2020 110					
FOUNDATION WALL, CAST IN PLACE, DIRECT CHUTE, 4' HIGH, 6" THICK					
Formwork	8.000	SFCA	5.20	25.68	30.88
Reinforcing	3.300	Lb.	.96	.98	1.94
Unloading & sorting reinforcing		Lb.		.06	.06
Concrete, 3,000 psi	.074	C.Y.	5.25		5.25
Place concrete, direct chute	.074	C.Y.		1.54	1.54
Finish walls, break ties and patch voids, one side	4.000	S.F.	.12	2.48	2.60
TOTAL			11.53	30.74	42.27

A2020 110		Walls, Cast in Place						
	WALL HEIGHT (FT.)	PLACING METHOD	CONCRETE (C.Y./L.F.)	REINFORCING (LBS./L.F.)	WALL THICKNESS (IN.)	COST PER L.F. MAT.	COST PER L.F. INST.	COST PER L.F. TOTAL
1500	4'	direct chute	.074	3.3	6	11.55	30.50	42.05
1520			.099	4.8	8	13.75	31.50	45.25
1540			.123	6.0	10	15.80	32	47.80
1560	RA2020-210		.148	7.2	12	17.90	33	50.90
1580			.173	8.1	14	19.95	33.50	53.45
1600			.197	9.44	16	22	34.50	56.50
1700	4'	pumped	.074	3.3	6	11.55	31.50	43.05
1720			.099	4.8	8	13.75	32.50	46.25
1740			.123	6.0	10	15.80	34	49.80
1760			.148	7.2	12	17.90	34.50	52.40
1780			.173	8.1	14	19.95	35	54.95
1800			.197	9.44	16	22	36	58
3000	6'	direct chute	.111	4.95	6	17.30	46	63.30
3020			.149	7.20	8	20.50	47.50	68
3040			.184	9.00	10	23.50	48.50	72
3060			.222	10.8	12	27	49.50	76.50
3080			.260	12.15	14	30	51	81
3100			.300	14.39	16	33.50	52.50	86

Figure 4.20

For pricing purposes, use Figure 4.23 with the "Footprint" or area of the excavation and the depth required. Determine the type of soil nearest the types listed in the table. Costs are listed per square foot of excavated area. Interpolate for depths and areas other than those shown.

Pricing excavation is not like pricing a ton of steel. When using cost tables, regardless of the source, review the assumptions made in compiling the data and temper the prices with judgment based on the anticipated variables at the project site.

Since there is no basement area in the three-story office building project, bulk excavation, as in Figure 4.23, is not needed. The only excavation necessary is for the perimeter foundation wall.

Waterproofing and Dampproofing

Waterproofing and dampproofing are not included in the three-story office building estimate. However, they can be priced using data such

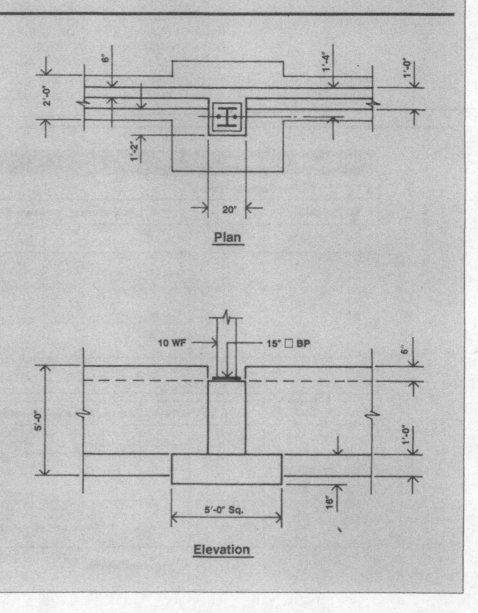

Plan

Elevation

Figure 4.21

as shown in Figure 4.24. Determine the total linear feet of perimeter foundation wall, and then select the type of waterproofing or dampproofing needed. Multiply the cost per lineal foot times the total length of foundation wall. Note that each item in the table is defined for a specific wall height. With some simple math, the costs can be translated into a cost per square foot for wall heights not specifically listed.

Whenever waterproofing is used on a basement wall, it is necessary to provide positive foundation underdrains, or *footing drains*, as they are sometimes called, to carry away water that may accumulate around the wall or strip footing.

Slab on Grade

For this particular application, a 4" thick, non-industrial, reinforced slab on grade is sufficient.

Cost:

(Figure 4.17, line #2240	10,000 S.F. @ $3.32	=	$33,200
Total Cost per Square Foot	$33,200/30,000 S.F.	=	$1.11/S.F.

Refer to Figure 4.25 for the summary of the costs for the slab. Figure 4.25 shows the summary of Division A, Foundation costs for this building. Note that the total cost is calculated by adding the costs of each individual item as well as the cost per square foot, which is the total division cost divided by 30,000 square feet.

B10 Superstructure

B1010 Floor Construction

B1010 203 — C.I.P. Column, Square Tied

	LOAD (KIPS)	STORY HEIGHT (FT.)	COLUMN SIZE (IN.)	COLUMN WEIGHT (P.L.F.)	CONCRETE STRENGTH (PSI)	COST PER V.L.F. MAT.	INST.	TOTAL
8500	700	10	18	315	6000	19.05	59	78.05
8600		12	18	319	6000	18.80	58.50	77.30
8700		14	18	321	6000	18.65	58	76.65
8800	800	10	20	388	6000	19.90	61.50	81.40
8900		12	20	394	6000	19.70	61.50	81.20
9000		14	20	397	6000	19.60	61	80.60
9100	900	10	20	388	6000	28	77	105
9300		12	20	394	6000	28	76	104
9600		14	20	397	6000	27.50	75.50	103
9800	1000	10	22	469	6000	25	72	97
9840		12	22	474	6000	25	71.50	96.50
9900		14	22	478	6000	25.50	72.50	98

B1010 203 — C.I.P. Column, Square Tied-Minimum Reinforcing

	LOAD (KIPS)	STORY HEIGHT (FT.)	COLUMN SIZE (IN.)	COLUMN WEIGHT (P.L.F.)	CONCRETE STRENGTH (PSI)	COST PER V.L.F. MAT.	INST.	TOTAL
9913	150	10-14	12	135	4000	9.10	37	46.10
9918	300	10-14	16	240	4000	13.30	46.50	59.80
9924	500	10-14	20	375	4000	19.40	60.50	79.90
9930	700	10-14	24	540	4000	27.50	78	105.50
9936	1000	10-14	28	740	4000	35	92.50	127.50
9942	1400	10-14	32	965	4000	46	107	153
9948	1800	10-14	36	1220	4000	55.50	124	179.50
9954	2300	10-14	40	1505	4000	63.50	135	198.50

Figure 4.22

A20 Basement Construction

A2010 Basement Excavation

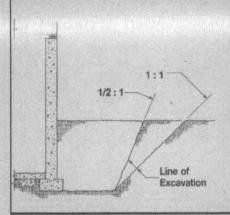

1 : 1

1/2 : 1

Line of Excavation

Primary assumptions are made on excavation is by 2-1/2 C.Y. wheel mounted front end loader and one-third by 1-1/2 C.Y. hydraulic excavator.

Two-mile round trip haul by 12 C.Y. tandem trucks is included for excavation wasted and storage of suitable fill from excavated soil. For excavation in clay, all is wasted and the cost of suitable backfill with two-mile haul is included.

Sand and gravel assumes 15% swell and compaction; common earth assumes 25% swell and 15% compaction; clay assumes 40% swell and 15% compaction (non-clay).

In general, the following items are accounted for in the costs in the table below.

1. Excavation for building or other structure to depth and extent indicated.
2. Backfill compacted in place.
3. Haul of excavated waste.
4. Replacement of unsuitable material with bank run gravel.

Note: Additional excavation and fill beyond this line of general excavation for the building (as required for isolated spread footings, strip footings, etc.) are included in the cost of the appropriate component systems.

System Components			COST PER S.F.		
	QUANTITY	UNIT	MAT.	INST.	TOTAL
SYSTEM A2010 110					
EXCAVATE & FILL, 1000 S.F., 4' DEEP, SAND, ON SITE STORAGE					
Excavating bulk shovel, 1.5 C.Y. bucket, 160 cy/hr	.108	C.Y.		.14	.14
Excavation, front end loader, 2-1/2 C.Y.	.216	C.Y.		.28	.28
Haul earth, 12 C.Y. dump truck	.170	C.Y.		.78	.78
Backfill, dozer bulk push 300', including compaction	.203	C.Y.		.44	.44
TOTAL				1.64	1.64

A2010 110	Building Excavation & Backfill	COST PER S.F.		
		MAT.	INST.	TOTAL
2220	Excav & fill, 1000 S.F. 4' sand, gravel, or common earth, on site storage		1.64	1.64
2240	Off site storage		3.50	3.50
2260	Clay excavation, bank run gravel borrow for backfill	2.51	3.12	5.63
2280	8' deep, sand, gravel, or common earth, on site storage [RG1010-010]		3.81	3.81
2300	Off site storage		8.95	8.95
2320	Clay excavation, bank run gravel borrow for backfill	6.05	7.15	13.20
2340	16' deep, sand, gravel, or common earth, on site storage [RG1030-400]		10.35	10.35
2350	Off site storage		22	22
2360	Clay excavation, bank run gravel borrow for backfill	16.80	18.30	35.10
3380	4000 S.F., 4' deep, sand, gravel, or common earth, on site storage [RG1010-020]		1.28	1.28
3400	Off site storage		2.13	2.13
3420	Clay excavation, bank run gravel borrow for backfill	1.17	1.97	3.14
3440	8' deep, sand, gravel, or common earth, on site storage		2.77	2.77
3460	Off site storage		5.05	5.05
3480	Clay excavation, bank run gravel borrow for backfill	2.77	4.32	7.09
3500	16' deep, sand, gravel, or common earth, on site storage		6.65	6.65
3520	Off site storage		13.80	13.80
3540	Clay, excavation, bank run gravel borrow for backfill	7.40	10.25	17.65
4560	10,000 S.F., 4' deep, sand gravel, or common earth, on site storage		1.17	1.17
4580	Off site storage		1.68	1.68
4600	Clay excavation, bank run gravel borrow for backfill	.72	1.59	2.31
4620	8' deep, sand, gravel, or common earth, on site storage		2.45	2.45
4640	Off site storage		3.80	3.80
4660	Clay excavation, bank run gravel borrow for backfill	1.67	3.39	5.06
4680	16' deep, sand, gravel, or common earth, on site storage		5.50	5.50
4700	Off site storage		9.70	9.70
4720	Clay excavation, bank run gravel borrow for backfill	4.44	7.70	12.14
5740	30,000 S.F., 4' deep, sand, gravel, or common earth, on site storage		1.08	1.08

Figure 4.23

A1010 Standard Foundations

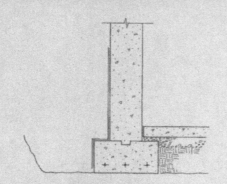

General: Apply foundation wall dampproofing over clean concrete giving particular attention to the joint between the wall and the footing. Use care in backfilling to prevent damage to the dampproofing.

Costs for four types of dampproofing are listed below.

System Components	QUANTITY	UNIT	COST PER L.F.		
			MAT.	INST.	TOTAL
SYSTEM A1010 320					
FOUNDATION DAMPPROOFING, BITUMINOUS, 1 COAT, 4' HIGH					
Bituminous asphalt dampproofing brushed on below grade, 1 coat	4.000	S.F.	.28	2.12	2.40
Labor for protection of dampproofing during backfilling	4.000	S.F.		.84	.84
TOTAL			.28	2.96	3.24

A1010 320	Foundation Dampproofing	COST PER L.F.		
		MAT.	INST.	TOTAL
1000	Foundation dampproofing, bituminous, 1 coat, 4' high	.28	2.96	3.24
1400	8' high	.56	5.90	6.46
1800	12' high	.84	9.15	9.99
2000	2 coats, 4' high	.44	3.68	4.12
2400	8' high	.88	7.35	8.23
2800	12' high	1.32	11.30	12.62
3000	Asphalt with fibers, 1/16" thick, 4' high	.68	3.68	4.36
3400	8' high	1.36	7.35	8.71
3800	12' high	2.04	11.30	13.34
4000	1/8" thick, 4' high	1.24	4.36	5.60
4400	8' high	2.48	8.70	11.18
4800	12' high	3.72	13.35	17.07
5000	Asphalt coated board and mastic, 1/4" thick, 4' high	2.48	4	6.48
5400	8' high	4.96	8	12.96
5800	12' high	7.45	12.30	19.75
6000	1/2" thick, 4' high	3.72	5.60	9.32
6400	8' high	7.45	11.15	18.60
6800	12' high	11.15	17	28.15
7000	Cementitious coating, on walls, 1/8" thick coating, 4' high	6.60	4.68	11.28
7400	8' high	13.20	9.40	22.60
7800	12' high	19.80	14	33.80
8000	Cementitious/metallic slurry, 2 coat, 1/4" thick, 2' high	38.50	134	172.50
8400	4' high	77	268	345
8800	6' high	116	400	516

Figure 4.24

| ASSEMBLY NUMBER | DESCRIPTION | QTY | UNIT | TOTAL COST | | COST PER S.F. |
				UNIT	TOTAL	
DIVISION A	SUBSTRUCTURE					
A1010-210	Column Footings, 6 KSF					
A1010-210-7700	Interior, 190K (200K)	9	EACH	$ 531.00	$ 4,779	
A1010-210-7500	Exterior, 114K (125K)	12	EACH	$ 327.00	$ 3,924	
A1010-210-7410	Corner, 86K (100K)	4	EACH	$ 266.00	$ 1,064	
A1010-110-2700	Strip Footings, 12" x 24"	322	L.F.	$ 24.90	$ 8,018	
A1010-240-1560	Foundation Wall, 4' x 12" Direct Chute	400	L.F.	$ 50.90	$ 20,360	
B1010-203-9924	Piers: 20" Square, 16 x 3.5'	56	V.L.F.	$ 79.90	$ 4,474	
A1030-120-2240	Slab on Grade, 4" thick, Non-industrial, Reinforced	10,000	S.F.	$ 3.32	$ 33,200	
	SUBTOTAL DIVISION A				$ 75,819	$ 2.53

Figure 4.25

Chapter 5

Superstructure

Superstructure is found in Division B of UNIFORMAT II, "Shell" (section B10). The two major components of the superstructure are Floor Construction (B1010) and Roof Construction (B1020). *(Note that exterior bearing walls are included in section B2010, Exterior Walls.)*

There are a number of alternatives to be considered when selecting a structural system, each of which can have a significant impact on cost. To analyze the merits of each combination, data must be available for comparison and substitution of systems. Designs must be determined and costs provided for:

- Elevated floor systems
- Roof framing systems
- Columns

Elevated Floor Assemblies

Determine the design live loads for elevated floor assemblies based on the building use and applicable building code. Once the loads have been determined, select the type of assembly from the appropriate assemblies table. Variables that affect selection include bay size, system weight, system depth, material compatibility, cost, and aesthetics. Depending on the assemblies selected, columns may or may not be included in the cost shown. Make sure you understand any design assumptions or conditions shown on the assemblies table before determining a price. With cost information available, you can compare the following possibilities:

- Alternate materials using the same dimensions
- Alternate dimensions using the same materials

Being able to make quick comparisons like these is one of the major benefits of assemblies estimating. Not only can the designer make informed choices based on cost, but the owner can see more clearly how different systems and materials can affect the final cost of the project.

Superimposed loads and anticipated bay spacing can be established based on design criteria and a comparison of the relative costs of the possible systems. Figure 5.1 provides a quick comparison of assemblies, helping to determine which are most economical without having to spend time gleaning information from a large number of individual assemblies.

Comparative Costs ($/SF) of Floor Systems/Type, (Table Number), Bay Size, & Load

Bay Size	Cast-In-Place Concrete						Precast Concrete			Structural Steel					Wood	
	1 Way BM & Slab B1010 219	2 Way BM & Slab B1010 220	Flat Slab B1010 222	Flat Plate B1010 223	Joist Slab B1010 226	Waffle Slab B1010 227	Beams & Hollow Core Slabs B1010 236	Beams & Hollow Core Slabs Topped B1010 238	Beams & Double Tees Topped B1010 239	Bar Joists on Cols. & Brg. Walls B1010 248	Bar Joists & Beams on cols. B1010 250	Composite Beams & C.I.P. Slab B1010 252	Composite Deck & Slab, W Shapes B1010 254	Composite Beam & Dk., Lt. Wt. Slab B1010 256	Wood Beams & Joists B1010 264	Laminated Wood Beams & Joists B1010 265
Superimposed Load = 40 PSF																
15 x 15	10.41	9.80	9.02	8.40	10.23	—	—	—	—	—	—	—	—	—	7.96	7.98
15 x 20	10.39	10.17	9.28	8.86	10.29	—	—	—	—	6.23	7.21	—	10.62	—	10.47	8.12
20 x 20	10.49	10.67	9.60	8.85	10.26	12.55	14.65	16.55	—	6.34	7.69	—	11.27	—	10.37	8.33
20 x 25	10.62	11.21	10.21	9.68	10.41	12.65	14.27	15.83	—	7.25	8.77	12.25	11.41	10.09	—	—
25 x 25	10.69	11.48	10.49	9.85	10.21	12.90	15.07	17.15	—	7.63	9.27	12.55	12.55	9.90	—	—
25 x 30	10.91	11.92	11.08	—	10.75	13.05	14.52	16.40	14.69	7.93	9.70	13.55	12.60	10.01	—	—
30 x 30	12.00	12.83	11.58	—	11.07	13.65	14.97	17.39	15.72	7.69	9.56	13.50	13.75	10.06	—	—
30 x 35	12.43	13.75	12.25	—	11.48	13.80	15.03	17.65		8.59	10.66	14.00	14.80	10.50	—	—
35 x 35	13.28	14.30	12.50	—	11.57	14.35	15.88	18.09		8.79	10.93	14.45	15.05	11.03	—	—
35 x 40	13.54	15.10	—	—	11.10	14.65	15.99	18.44	14.00	—	—	15.60	15.70	12.02	—	—
40 x 40	—	—	—	—	12.36	15.00	16.79	19.29	15.91	—	—	—	—	—	—	—
40 x 45	—	—	—	—	12.82	15.40	—	—	—	—	—	—	—	—	—	—
40 x 50	—	—	—	—	—	—	—	—	15.04	—	—	—	—	—	—	—
Superimposed Load = 75 PSF																
15 x 15	10.47	9.99	9.14	8.42	10.30	—	—	—	—	—	—	—	—	—	9.79	9.90
15 x 20	10.76	10.95	9.56	9.21	10.52	—	—	—	—	6.74	8.13	—	11.44	—	12.60	10.55
20 x 20	11.28	11.53	9.94	9.24	10.85	12.75	15.35	17.25	—	7.13	8.77	—	12.45	—	12.64	10.47
20 x 25	11.57	12.43	10.73	9.86	10.93	12.95	14.27	16.70	—	8.21	9.23	13.60	13.70	10.36	—	—
25 x 25	11.50	12.39	10.85	10.24	10.85	13.20	15.62	17.15	—	8.10	10.02	14.30	14.20	10.68	—	—
25 x 30	11.66	12.97	11.60	—	11.15	13.35	15.38	17.81	14.69	8.18	10.18	15.05	14.80	10.51	—	—
30 x 30	12.87	13.90	12.15	—	11.45	13.85	15.57	18.39	15.72	8.80	10.94	14.95	15.70	10.80	—	—
30 x 35	13.02	14.40	12.85	—	11.63	13.80	15.83	17.40	—	10.21	12.49	15.90	16.80	11.22	—	—
35 x 35	14.40	14.85	13.20	—	12.10	14.55	16.58	19.10	—	10.88	13.35	16.35	17.10	12.26	—	—
35 x 40	14.55	15.65	—	—	12.48	15.00	—	—	14.95	—	—	17.30	17.80	12.84	—	—
40 x 40	—	—	—	—	12.67	15.45	—	—	16.55	—	—	—	—	—	—	—
40 x 45	—	—	—	—	12.92	15.90	—	—	—	—	—	—	—	—	—	—
40 x 50	—	—	—	—	—	—	—	—	15.29	—	—	—	—	—	—	—
Superimposed Load = 125 PSF																
15 x 15	10.62	10.38	9.38	8.56	10.43	—	—	—	—	—	—	—	—	—	14.58	14.25
15 x 20	11.16	11.83	9.92	9.70	11.01	—	—	—	—	7.85	9.58	—	12.95	—	17.36	15.55
20 x 20	11.89	11.95	10.54	9.71	11.04	12.90	—	—	—	8.76	9.94	—	13.85	—	21.05	15.40
20 x 25	12.28	12.77	11.31	10.36	11.53	13.15	—	—	—	8.80	10.77	15.10	15.85	12.57	—	—
25 x 25	13.25	13.40	11.43	10.68	12.07	13.50	—	—	—	9.43	11.64	16.20	16.75	11.32	—	—
25 x 30	13.35	14.00	11.95	—	12.00	13.60	—	—	—	10.12	12.40	17.50	16.55	12.09	—	—
30 x 30	13.65	14.60	12.55	—	12.11	14.05	—	—	—	11.16	13.80	17.35	18.15	12.69	—	—
30 x 35	14.55	15.70	13.30	—	12.21	14.30	—	—	—	11.28	13.92	19.30	19.30	13.03	—	—
35 x 35	15.25	16.30	13.55	—	12.16	14.85	—	—	—	12.70	14.46	18.80	20.10	14.25	—	—
35 x 40	15.40	16.35	—	—	12.45	15.55	—	—	—	—	—	20.20	20.65	14.55	—	—
40 x 40	—	—	—	—	13.23	15.70	—	—	—	—	—	—	—	—	—	—
40 x 45	—	—	—	—	13.40	16.10	—	—	—	—	—	—	—	—	—	—
40 x 50	—	—	—	—	—	—	—	—	—	—	—	—	—	—	—	—
Superimposed Load = 200 PSF																
15 x 15	11.06	11.08	9.69	—	10.80	—	—	—	—	—	—	—	—	—	25.85	22.15
15 x 20	12.05	12.71	10.16	—	11.33	—	—	—	—	—	—	—	15.40	—	23.55	20.10
20 x 20	12.90	12.90	10.70	—	11.53	13.40	—	—	—	—	—	—	16.25	—	—	21.15
20 x 25	13.20	13.87	11.65	—	12.13	13.55	—	—	—	—	—	18.30	17.45	13.95	—	—
25 x 25	14.38	15.05	11.80	—	12.52	13.80	—	—	—	—	—	19.00	18.90	14.55	—	—
25 x 30	14.49	15.15	12.50	—	12.69	14.55	—	—	—	—	—	20.05	19.95	14.65	—	—
30 x 30	15.20	15.55	13.20	—	12.75	15.00	—	—	—	—	—	20.95	23.75	15.05	—	—
30 x 35	15.30	16.50	—	—	13.09	15.45	—	—	—	—	—	22.70	23.05	15.05	—	—
35 x 35	16.45	17.10	—	—	13.16	15.70	—	—	—	—	—	23.10	25.10	16.30	—	—
35 x 40	16.55	17.45	—	—	13.35	16.60	—	—	—	—	—	24.80	25.65	17.40	—	—
40 x 40	—	—	—	—	—	—	—	—	—	—	—	—	—	—	—	—
40 x 45	—	—	—	—	—	—	—	—	—	—	—	—	—	—	—	—
40 x 50	—	—	—	—	—	—	—	—	—	—	—	—	—	—	—	—

Figure 5.1

Whatever assembly is chosen, compatible materials must be used throughout. For instance, if steel is used for the floor and roof framing systems, steel columns should also be used. Concrete columns should be used with concrete-framed buildings. The same holds true for wood framing and wood columns, although in some instances steel pipe columns may be used with wood framing.

Example: Roof and Floor Assembly Selection

The examples that follow demonstrate some of the many uses for the tables in *Means Assemblies Cost Data*.

With the size of the building, bay size, and number of floors determined in Chapter 4, the next step is to select a floor and roof assembly. Use the example and loads illustrated in Figure 3.6.

Floor:

Live Load Office Building	50 psf	BOCA Code
Partitions	20	
Ceiling	5	(Assume)
	75 psf	Total Superimposed Load

Roof:

Snow Load	30 psf	BOCA Code
Ceiling	5	
Mechanical	5	
	40 psf	Total Superimposed Load

Design Criteria:

25' × 25' Bay Spacing
3 stories with 2 elevated floors and a roof

Step One: Select the Floor Assembly
See Figure 5.1:

Open Web (Bar) Joists and Beams on Columns: $10.02/S.F.
Concrete Flat Plate: $10.24/S.F.

For this example, assume a concrete frame building with flat plate construction. Using the concrete flat plate, additional ceiling construction can be eliminated in some areas for economic and/or architectural reasons.

Step Two: Price the Floor Assembly
With the reference from Figure 5.1, go to the specific table for the floor assembly chosen.

See Figure 3.7:

25' × 25' bay; 75 psf Superimposed Load
Total Load = 194 psf
System Cost = $10.24/S.F.

Step Three: Note Minimum Required Column Size
See Figure 3.7:

25' × 25' bay; 75 psf Superimposed Load Column
Size: 24" square (for concrete floor systems only)

Step Four: Select a Roof Assembly
See Figure 3.7:

> 25' × 25' bay; 40 psf Superimposed Load
> Total Load = 152 psf
> System Cost = $9.85/S.F.

Step Five: Summarize Column Loads
The loads shown in Figure 3.10 were developed for this structure.

The loads are as follows:

Minimum Column Load	=	102 kips
Maximum Column Load	=	358 kips
Load to the Foundation	=	358 kips

Step Six: Determine the Concrete Column Cost
See Figure 3.11.

> 24" square; 358 kips Working Load
> First check the "Minimum Reinforcing" section of Figure 3.11.
> Line # 9930: 24" square column with minimum reinforcing will support
> a total load of 700 kips (358 kips required).
> Column Cost = $105.50/V.L.F.

If the load accumulated had been in excess of 700 kips, then the first portion of the table in Figure 3.11 would have been used.

Example: Proceed down the "Column Size" section to the first 24" square column at a 12' story height.

> Total Load = 900 kips
> Column Cost = $118.00/V.L.F.

To determine the average column cost per vertical linear foot, proceed as follows:

$$\frac{\$105.50/\text{V.L.F.} + \$118.00/\text{V.L.F.}}{2} = \$111.75/\text{V.L.F.}$$

The average cost is more accurate for estimating, because the concrete columns at the upper floors are smaller and therefore not as costly as the concrete columns at the ground level in this example.

Step Seven: Determine the Superstructure Cost

Roof: 100' × 100' × $9.85/S.F.	=	$ 98,500
Elevated Floors: 2 × 100' × 100' × $10.24/S.F.	=	204,800
Columns: 25 each × 36' high × $111.75/V.L.F.	=	100,575
Total Cost		$403,875
Total Cost per Square Foot	=	$13.46/S.F.

Types of Floor Assemblies
In using *Means Assemblies Cost Data* it is important to consider certain design assumptions. Also be aware that some of the assemblies have similar names, but differences in components. Following is a list of assemblies with their recommended uses. *(For fuller descriptions and graphics, refer to the assembly pages in* Means Assemblies Cost Data.*)*

Cast-in-Place Slabs, One Way: May be used in conjunction with precast concrete beams, steel framing, or bearing walls.

Cast-in-Place Beams and Slabs, One Way: Effective system for heavy or concentrated loads, with spans under 20'. These systems have relatively deep floors and complex framing.

Cast-in-Place Beams and Slabs, Two Way: Efficient for heavy or concentrated loads under 30' spans and fairly square bays.

Cast-in-Place Flat Plate with Drop Panels: Efficient for heavier loads and longer spans. Requires smaller columns than flat plate. Good for parking areas, storage, or industrial applications. Also suitable for all types of buildings that have suspended ceilings.

Cast-in-Place Flat Plate: Economically efficient for light loads, modest spans, and minimum depth of construction. Yields flat ceilings that may be used for finish with paint.

Cast-in-Place Multispan Joist Slab: Effective for reducing dead weight of a floor system for long spans.

Cast-in-Place Waffle Slab: Effective for reducing dead weight over long spans and heavy loads. Depth is uniform. May be used with or without suspended ceilings.

Precast Planks: May be used for floors with or without topping. Planks without topping require good alignment of panels, heavy padding, and carpeting for residential or office use. Used in conjunction with concrete beams, steel beams, or bearing walls. Results in fast erection time.

Precast, Double "T" Beams: Used for floors with topping and roofs without topping. Also commonly used with concrete beams or bearing walls for long spans and for wall panels.

Precast Beams and Planks, No Topping: Used for various bay sizes and superimposed loads for precast beams and precast planks, untopped. Ceilings not necessary. Fast erection time.

Precast Beams and Planks, with Topping: Used for various bay sizes and superimposed loads for precast beams and precast planks, with 2" concrete topping. Ceilings not necessary. Fast erection time.

Precast Beams and Precast Double "Ts," With Topping: Used for multiple bay sizes and superimposed loads for precast beams and precast double "T's" with 2" concrete topping. Effective for long spans and fast erection.

Steel Framing: Used for multiple bay sizes and beam girder spacing and orientation, and various superimposed loads. System includes sprayed-on fireproofing. For use with cast-in-place slabs, one-way; precast planks; and metal deck with concrete fill.

Light Gauge Steel: Light gauge "C" or punched joists with plywood subfloor. Can be used with bearing walls or steel stud wall systems. Clean, simple erection.

Steel Joists on Walls: Used for various spans for open web joists, slab form, and 2-1/2" concrete slab supported by walls or other suitable bearing system. Inexpensive and lightweight, it is a popular system. Must be fireproofed with fire-rated, suspended, or gypsum ceiling.

Steel Joists and Beams on Columns with Exterior Bearing Walls: Used for varied bay sizes and superimposed loads for columns, beams, open-web joists, and 2-1/2" concrete slabs. Exterior walls are assumed as bearing walls; the cost of these walls is not included. Inexpensive and lightweight. Must be fireproofed with fire-rated suspended or gypsum ceiling.

Steel Joists and Beams on Columns: Used for varied bay sizes and superimposed loads for columns, beams, open-web joists, and 2-1/2" concrete slabs. Inexpensive and lightweight, it is a popular system. Must be fireproofed with fire-rated, suspended, or gypsum ceiling. Columns are included in the cost for use with one suspended floor and roof.

Composite Beam and Cast-in-Place Slabs: Used for varied bay sizes and superimposed loads for steel beams with shear connectors and formed reinforced concrete slab. Efficient when loads are heavy, and spans are moderately long. Beams are fireproofed with sprayed-on fireproofing.

Steel Beams, Composite Deck, and Concrete Slabs: Used for varied bay sizes and superimposed loads for steel beams, composite steel deck, and concrete slab. Efficient when loads are heavy, and spans are moderately long. Beams and steel deck are fireproofed with sprayed-on fireproofing.

Composite Beams and Deck, Lightweight Concrete Slabs: Varied bay sizes and superimposed loads for composite steel beams with shear connectors, composite steel deck, and lightweight concrete slabs. Popular, relatively inexpensive steel and concrete framing system. Sprayed-on fireproofing for the beams only is included, as the lightweight concrete deck is considered fireproof by most building codes.

Metal Deck and Concrete Fill: Varied spans and superimposed loading for composite steel deck, cellular and non-cellular, and concrete fill of varied thickness. Used with steel framing system.

Wood Joists: Costs are listed per square foot for wood joists and plywood subfloor. Joists are spaced 12", 16", and 24" on-center. Sizes range from 2" x 6" to 4" x 10".

Wood Beam and Joist: Costs are listed per square foot for wood beams, wood joists, and plywood subfloor for various bay sizes and superimposed loading.

Laminated Floor Beams: Costs are listed per square foot for laminated wood beams, wood joists, and plywood subfloor for various bay sizes and superimposed loading.

Wood Deck: Costs are listed per square foot for wood deck of varied species and thicknesses with allowable load and span.

Considerations in Selecting Foundation and Framing Systems

Using the information from Chapter 3 and this chapter, many comparisons can be made to better understand the methods and choices available. For example, the following cost comparisons can be made between:

- Full load and factored load.
- Large-area buildings versus small-area buildings.
- Various foundation costs using load reduction factors.
- Interior versus exterior footings for small and large buildings.

Footing Load Reduction Comparisons
Example One: Small Building

See the footing plan for the small building, Figure 5.2.

        Footings: 1 Interior
                  4 Exterior
                 4 Corner

Assuming all footings are at full bay loading (75 kips load and 3 K.S.F. soil pressure), the footing cost is:

See Figure 3.15, line #7250:

9 ea. @ $327   = $2,943 or

$$\frac{\$2,943}{2,500 \text{ S.F.}} \quad = \quad \$1.18/\text{S.F.}$$

25' × 25' Bays

Now assume interior and exterior footings with load reduction:

Footing Loads: Interior =   75 kips
                Exterior =   45 kips (75 kips × .6)
                Corner   =   34 kips (75 kips × .45)
Soil Pressure = 3 K.S.F.

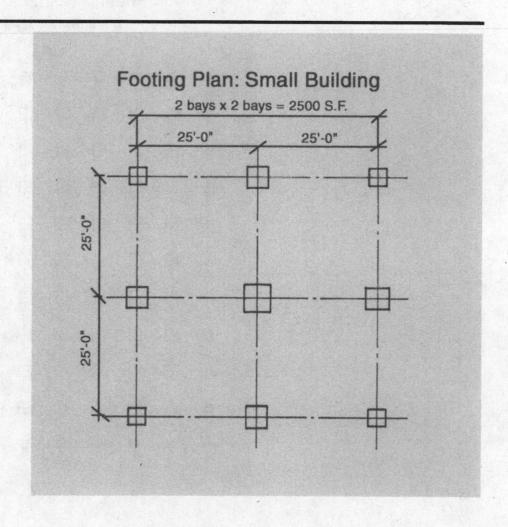

Figure 5.2

Revised Costs:

Interior:	(Figure 3.15, line #7250)	1 ea. @ $327	=	$ 327
Exterior:	(Figure 3.15, line #7150)	4 ea. @ $222.00	=	888
Corner:	(Figure 3.15)			
	(Interpolated between			
	lines #7100 & #7150)	4 ea. @ $157.50	=	630
Total				$1,845
Total Cost per Square Foot			=	$0.74/S.F.

Compare the previous example with a building 10 bays × 10 bays and 62,500 S.F.

Example Two: Large Building
See Figure 5.3.

Footings: 81 Interior
           36 Exterior
           4 Corner

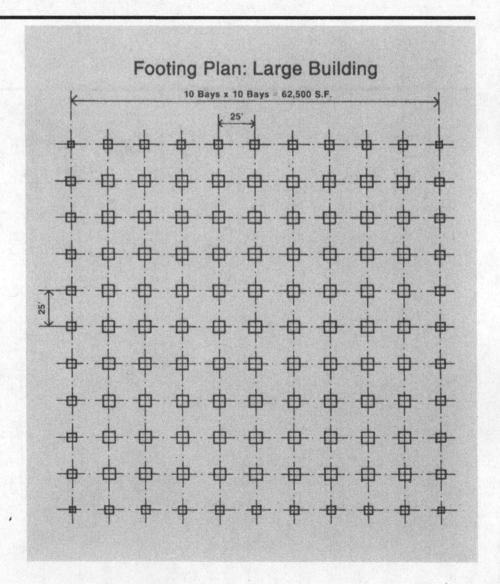

Footing Plan: Large Building

10 Bays x 10 Bays = 62,500 S.F.

25'

25'

Figure 5.3

Assuming all interior footings (75 K.S.F. load and 3 K.S.F. soil pressure), the cost is:

See Figure 3.15, line #7250:

121 ea. @ $327 = $39,567

$$\frac{\$39,567}{\text{█1,█00 █.█}} = \$0.63/\text{S.F.}$$

25' × 25' Bays

Now assume interior and exterior footings with load reduction:

Footing Loads: Interior = 75 kips
Exterior = 45 kips
Corner = 34 kips

Soil Pressure: 3 K.S.F.

Costs:
Interior: (Figure 3.15, line #7250) 81 ea. @ $327 = $26,487
Exterior (Figure 3.15, line #7150) 36 ea. @ $222 = 7,992
Corner (Figure 3.15, Interpolate between lines #7100 and #7150) 4 ea. @ $157.50 = 630

Total $35,109
Total Cost per Square Foot: = $0.56/S.F.

In summary, using full bay loading in lieu of 1/2 and 1/4 bay loading would increase the estimated cost of a 2 × 2 bay building by approximately 59%. For a 10 × 10 bay building, the increase would be 13%.

Small versus Large Concrete Building

Example One: Small Building

Follow the same procedure for a three-story concrete building with 25' square bays and flat plate construction.

Loads: Floors = 242.5 kips
Roof = 95.0 kips
337.5 kips

2 Bay × 2 Bay
2,500 S.F. × 3 floors = 7,500 S.F.

Footing Loads: Interior = 338 kips
Exterior = 110 kips
Corner = 152 kips

Interior (Figure 3.15, line #7850) 1 ea. @ $2,550 = $2,550
Exterior (Figure 3.15, line #7650) 4 ea. @ $965 = 3,860
Corner (Figure 3.15, line #7550) 4 ea. @ $705 = 2,820
Total $9,230
Cost per Square Foot = $1.23/S.F.

All Footings Full Bay Loading
Footing Load = 338 kips
(Figure 3.15, line #7850) 9 ea. @ $2,550 = $22,950
Cost per Square Foot = $3.06/S.F.

Example Two: Large Building

10 Bay × 10 Bay
62,500 S.F. × 3 floors = 187,500 S.F.
Exterior & Interior Footing Loading

```
Footing Loads:   Interior  =  338 kips
                 Exterior  =  110 kips
                 Corner    =  152 kips

Interior  (Figure 3.15, line #7850)  81 ea. @ $2,550  =    $206,550
Exterior  (Figure 3.15, line #7650)  36 ea. @ $965    =      34,740
Corner    (Figure 3.15, line #7550)   4 ea.  @ $705   =       2,820
Total Cost                                                 $244,110
Cost per Square Foot                                  =     $1.30/S.F.
```

All Footings Full Bay Loading
Footing Load = 338 kips

```
          (Figure 3.15, line #7850)  120 ea. @ $2,550  =  $306,000
Cost per Square Foot                                   =    $1.63/S.F.
```

Conclusions Based on a $60/S.F. Building Cost

The estimator must pay the most attention to areas where the work will provide significant cost results, as there never seems to be enough time to look at everything. The previous exercise is time-consuming, but a review of the results reveals where there are potential cost benefits.

On small-perimeter, light steel buildings with a large percentage of exterior footings with all footings fully loaded versus interior, exterior, and corner loading, the cost difference is $0.44/S.F. or 0.7%. It's worthwhile to compute the loads to each footing, then price them accordingly. The cost savings is worth the time spent.

On large-perimeter, light steel buildings with a large percentage of interior footings calculated with total load versus interior, exterior, and corner loading, the cost difference is $0.07/S.F. or 0.1%. The cost saved by sizing each footing is not substantial enough to merit the time spent.

On small-perimeter, heavy concrete buildings with a large percentage of exterior footings with all columns fully loaded versus interior, exterior, and corner loading, the cost difference is $1.83/S.F. or 3.0%. The cost saved is worth the time to size each footing and cost each one out accordingly.

On large-perimeter, heavy concrete buildings with a large percentage of interior footings calculated with fully loaded versus interior, exterior, and corner loading, the cost difference is $0.33/S.F. or 0.5%. The cost saved may or may not be worth the effort.

Large versus Small Bays, Concrete Building

A larger or deeper steel or wood beam or girder is necessary to span a 30' clear span than a 15' clear span. Similarly, a deeper concrete slab is required to span a 30' square column spacing versus a 15' square column spacing. Therefore, it can be assumed that with elevated floor systems of the same components, larger column spacings will increase both the system cost and the ultimate square foot cost.

For example, using Figure 3.7, Cast-in-Place Flat Plates, with a total load of 75 psf and a bay size of 15' × 15', the cost is $8.42 per square foot.

Using the same superimposed load of 75 psf and a bay size of 25' × 25', the cost is now $10.24 per square foot.

```
25' Square Bay:    $10.24/S.F.
15' Square Bay:     $8.42/S.F.
Difference          $1.82/S.F.
```

Using a flat plate concrete slab, and increasing the column spacing or bay size from 15' square to 25' square, the cost of the floor system is increased by $1.82 per square foot and affects the final square foot cost of the building. However, depending on the architect or owner's requirements, it may be desirable to choose the wider bay spacing.

This study of column and bay spacing would not be complete without considering the cost of columns or supports for the slab. For the previous Cast-in Place, Flat Plate System with a superimposed load of 75 psf on a 15' versus 25' square bay, the following analysis may be performed, using Figure 3.7:

Bay Size	Superimposed Load psf	Minimum Col. Size	Slab Thickness	Total Load psf
15' x 15'	75	14"	5-1/2"	144
25' x 25'	75	24"	9-1/2"	194

To support the same 75 psf, but increase the bay spacing from 15' square to 25' square, the slab thickness would have to increase from 5-1/2" to 9-1/2", adding 50 psf to the dead weight of the structure. Likewise, the minimum allowable concrete column size would increase from 14" square to 24".

To understand the effect of columns on the cost differential, proceed as follows:

Load to columns:	15' Square Bay	=	32.4 kips
	25' Square Bay	=	121.3 kips

Using the "Cast-in-Place Columns—Square Tied" table in Figure 3.11, utilize the "Minimum Reinforcement Table" for the minimum column size shown, and assume the columns are 12' high. The cost for columns can be calculated as follows:

14" Square Column
12" Square	=	$46.10/L.F.
Interpolating: 14" Square	=	$52.95/L.F.
16" Square	=	$59.80/L.F.

15' Square Bay, 14" Square Column: $52.95/L.F. × 12' = $635.40 ea.
25' Square Bay, 24" Square Column: $105.50/L.F. × 12' = $1,266 ea.

The column cost per square foot of bay area is:

15' Bay: $2.82/S.F.
25' Bay: $2.03/S.F.

Using the costs per square foot found previously, the cost differential is now:

25' Square Bay:	$10.24 + $2.03	=	$12.27/S.F.
15' Square Bay:	$8.42 + $2.82	=	$11.24/S.F.
Difference			$ 1.03/S.F.

Partial Summary: Increasing the bay size from 15' square to 25' square with the same load increases the elevated framing portion of the building cost by $1.03/S.F. or approximately 8%. To further evaluate the bay spacings, compare the foundations or column footings used to support the different bay sizes.

As mentioned, the footing loads are:

15' Square Bay   =   32.4 kips
25' Square Bay   =   121.3 kips

Assume a 3 K.S.F. soil pressure design condition.

Costs:

15' Square Bay: (Figure 3.15, line #7150)   = $222 ea.
Cost per Square Foot = $0.99/S.F.
25' Square Bay: (Figure 3.15, line #7450)   = $594 ea.
Cost per Square Foot = $0.95/S.F.

To complete the analysis, the cost differential is now:

	Floor		Column		Footing		
25' Square Bay	$10.24	+	$2.03	+	$0.95	=	$13.22/S.F.
15' Square Bay	$8.42	+	$2.82	+	$0.99	=	$12.23/S.F.
Difference							$0.99/S.F.

It costs $0.99/S.F. more, an 8% increase, to change the bay size from 15' square to 25' square for this concrete building. The wider bay spacing may be highly desirable and worth the additional cost.

Small versus Large Bays, Steel Building Bays

For comparison purposes, establish a square foot cost for a structural steel framing system using interior columns, beams, open web joists, and a concrete slab, using Figure 5.4. Exterior walls will be bearing walls. Column costs are included in the systems. Use the same 75 psf superimposed load.

30' Square Bay (Figure 5.4, line #2070)   = $8.80/S.F.
20' Square Bay (Figure 5.4, line # 1400)   = $7.13/S.F.
Difference $1.67/S.F.

Loading: 20' Square Bay   =   47.6 kips
                   30' Square Bay   =   108 kips

Again, assuming 3 K.S.F. allowable soil pressure, the footing costs are:

20' Square Bay (Figure 3.15, line #7150)   = $222 ea.
Cost per Square Foot   = $0.56/S.F.
30' Square Bay (Figure 3.15, line #7450)   = $594 ea.
Cost per Square Foot   = $0.66/S.F.

Total cost for floor and column plus foundation, to increase bay size:

30' Square Bay $8.80 + $0.66   = $9.46/S.F.
20' Square Bay: $7.13 + $0.56   = $7.69/S.F.
Difference $1.77/S.F.

It costs approximately 23% more to increase the bay size from 20' square to 30' square for a steel and bar joist structural system.

B1010 Floor Construction

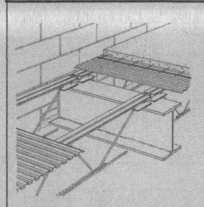

Table below lists costs for a floor system on exterior bearing walls and interior columns and beams using open web steel joists, galvanized steel slab form, 2-1/2" concrete slab reinforced with welded wire fabric.

Design and Pricing Assumptions:
Structural Steel is A36.
Concrete f'c = 3 KSI placed by pump.
WWF 6 x 6 – W1.4 x W1.4 (10 x 10)
Columns are 12' high.
Building is 4 bays long by 4 bays wide.
Joists are 2' O.C. ± and span the long direction of the bay.

Joists at columns have bottom chords extended and are connected to columns.

Slab form is 28 gauge galvanized.
Column costs in table are for columns to support 1 floor plus roof loading in a 2-story building; however, column costs are from ground floor to 2nd floor only. Joist costs include appropriate bridging.
Deflection is limited to 1/360 of the span.
Screeds and steel trowel finish.

Design Loads	Min.	Max.
S.S & Joists	4.4 PSF	11.5 PSF
Slab Form	1.0	1.0
2-1/2" Concrete	27.0	27.0
Ceiling	3.0	3.0
Misc.	7.6	5.5
	43.0 PSF	48.0 PSF

System Components			COST PER S.F.		
	QUANTITY	UNIT	MAT.	INST.	TOTAL
SYSTEM B1010 248					
15'X20' BAY, W.F. STEEL, STEEL JOISTS, SLAB FORM, CONCRETE SLAB					
Structural steel	1.248	Lb.	.63	.28	.91
Open web joists	3.140	Lb.	1.60	.82	2.42
Slab form, galvanized steel 9/16" deep, 28 gauge	1.020	S.F.	.58	.51	1.09
Welded wire fabric 6x6 - W1.4 x W1.4 (10 x 10), 21 lb/CSF roll, 10% lap	1.000	S.F.	.08	.26	.34
Concrete ready mix, regular weight, 3000 psi	.210	C.F.	.55		.55
Place and vibrate concrete, elevated slab less than 6", pumped	.210	C.F.		.22	.22
Finishing floor, monolithic steel trowel finish for finish floor	1.000	S.F.		.60	.60
Curing with sprayed membrane curing compound	.010	C.S.F.	.04	.06	.10
TOTAL			3.48	2.75	6.23

B1010 248		Steel Joists on Beam & Wall						
	BAY SIZE (FT.)	SUPERIMPOSED LOAD (P.S.F.)	DEPTH (IN.)	TOTAL LOAD (P.S.F.)	COLUMN ADD	COST PER S.F.		
						MAT.	INST.	TOTAL
1200	15x20 RB1010 -105	40	17	83		3.48	2.75	6.23
1210					columns	.21	.09	.30
1220	15x20	65	19	108		3.57	2.80	6.37
1230					columns	.21	.09	.30
1250	15x20	75	19	119		3.82	2.92	6.74
1260					columns	.25	.11	.36
1270	15x20	100	19	144		4.36	3.32	7.68
1280					columns	.25	.11	.36
1300	15x20	125	19	170		4.48	3.37	7.85
1310					columns	.33	.14	.47
1350	20x20	40	19	83		3.56	2.78	6.34
1360					columns	.19	.08	.27
1370	20x20	65	23	109		3.96	2.98	6.94
1380					columns	.25	.11	.36
1400	20x20	75	23	119		4.09	3.04	7.13
1410					columns	.25	.11	.36
1420	20x20	100	23	144		4.22	3.09	7.31
1430					columns	.25	.11	.36
1450	20x20	125	23	170		5.10	3.66	8.76
1460					columns	.30	.13	.43

Figure 5.4

B1010 Floor Construction

B1010 248	Steel Joists on Beam & Wall

	BAY SIZE (FT.)	SUPERIMPOSED LOAD (P.S.F.)	DEPTH (IN.)	TOTAL LOAD (P.S.F.)	COLUMN ADD	COST PER S.F.		
						MAT.	INST.	TOTAL
1500	20x25	40	23	84		4.08	3.17	7.25
1510					columns	.20	.09	.29
1520	20x25	65	26	110		4.40	3.33	7.73
1530					columns	.20	.09	.29
1550	20x25	75	26	120		4.71	3.50	8.21
1560					columns	.24	.10	.34
1570	20x25	100	26	145		4.57	3.27	7.84
1580					columns	.24	.10	.34
1670	20x25	125	29	170		5.20	3.60	8.80
1680					columns	.28	.12	.40
1720	25x25	40	23	84		4.35	3.28	7.63
1730					columns	.19	.08	.27
1750	25x25	65	29	110		4.58	3.40	7.98
1760					columns	.19	.08	.27
1770	25x25	75	26	120		4.76	3.34	8.10
1780					columns	.22	.09	.31
1800	25x25	100	29	145		5.40	3.66	9.06
1810					columns	.22	.09	.31
1820	25x25	125	29	170		5.65	3.78	9.43
1830					columns	.25	.11	.36
1870	25x30	40	29	84		4.51	3.42	7.93
1880					columns	.19	.08	.27
1900	25x30	65	29	110		4.80	3.02	7.82
1910					columns	.19	.08	.27
1920	25x30	75	29	120		5.05	3.13	8.18
1930					columns	.21	.09	.30
1950	25x30	100	29	145		5.50	3.31	8.81
1960					columns	.21	.09	.30
1970	25x30	125	32	170		5.95	4.17	10.12
1980					columns	.23	.10	.33
2020	30x30	40	29	84		4.69	3	7.69
2030					columns	.17	.08	.25
2050	30x30	65	29	110		5.30	3.24	8.54
2060					columns	.17	.08	.25
2070	30x30	75	32	120		5.50	3.30	8.80
2080					columns	.20	.08	.28
2100	30x30	100	35	145		6.05	3.53	9.58
2110					columns	.23	.10	.33
2120	30x30	125	35	172		6.65	4.51	11.16
2130					columns	.28	.12	.40
2170	30x35	40	29	85		5.35	3.24	8.59
2180					columns	.17	.07	.24
2200	30x35	65	29	111		6	4.22	10.22
2210					columns	.19	.08	.27
2220	30x35	75	32	121		6	4.21	10.21
2230					columns	.19	.08	.27
2250	30x35	100	35	148		6.50	3.68	10.18
2260					columns	.24	.10	.34
2270	30x35	125	38	173		7.30	3.98	11.28
2280					columns	.24	.11	.35
2320	35x35	40	32	85		5.50	3.29	8.79
2330					columns	.17	.08	.25

Figure 5.4 (cont.)

Assemblies Estimate: Three-Story Office Building

The total superimposed loads from the "Superimposed Load" section are:

Floor: 75 psf
Roof: 40 psf
25' × 25' Bay Spacing
3 stories with 2 elevated floors and a roof

Step One: Select the Floor System

See Figure 5.1:

Open Web Joists and Beams on Columns: $10.02/S.F.
Cast-in-Place Concrete Flat Plate: $10.24/S.F.

The designer and owner agree that the steel framing system is more desirable because of cost, quick erection time, availability, and because structural elements will not be exposed to view. *(Note: A concrete structure is fireproof. Steel and open web joist construction requires fireproofing. Therefore, the cost comparison at this point is not complete.)*

Step Two: Price the Floor System

Choose the appropriate table for an open web joist and beams on columns floor assembly.

(Figure 3.12, line #5100) 25' × 25' Bay; 75 psf Superimposed Load
Total Load   =   120 psf
System Cost   = $10.02/S.F.

Step Three: Select a Roof System

Once the structural floor assembly has been designed and costs have been established, select a roof assembly. The roof framing assembly normally has the same bay spacing and may or may not be of the same construction.

Now select a roof system and price it.

See Figure 3.13, line #3300:

25' × 25' Bay; 40 psf Superimposed Load
Total Load   =   60 psf
System Cost   = $4.52/S.F.

Step Four: Summarize Loads

The loads (depicted in Figure 3.14) were developed for this structure. They are as follows:

Minimum Column Load   =   39 kips
Maximum Column Load   =   191 kips
Load to the Foundation   =   191 kips

Step Five: Price the Columns

The column cost tables are arranged so that the user can enter and make cost determinations when any one of the following situations exists:

- The exact type, size, weight, and length of the columns being estimated have been predetermined.
- The type of column desired and the total load concentrated to the columns are known.
- The bay size has been determined, floor and roof systems (not including columns) have been selected, and the total load concentrated on the columns has been calculated.

Select the column type and determine the cost.

See Figure 5.5:

> 200 kips; 10' unsupported height.
> Use a "W" shape.

(*Note: The unsupported height of the column is the height of the column between areas of lateral bracing such as at beams and floor plates.*)

> Column Cost: $43.45/V.L.F. (1st & 2nd floor) for a W 10 × 45

See Figure 5.5:

> 50 kips; 10' unsupported height.
> Use a "W" shape.
> Column Cost: $19.90/V.L.F. (3rd floor) for a W 5 × 16

Column fireproofing: Figure 5.6 is a cost table for the fireproofing of vertical steel columns. It allows the estimator to choose from several encasement systems. The average size of the columns to be fireproofed is 8". Two-hour protection is required. Drywall is selected as the fireproofing material, because it is compatible with the partitions that will be painted.

Cost:

> See Figure 5.6, line #3450: 859 L.F. @ $16.45/L.F. = $14,131

Beam Fireproofing

Figure 5.7 lists costs for fireproofing steel beams using concrete, gypsum, gypsum and perlite plaster, and sprayed fiber (mineral fiber or cementitious). Multiply the cost per linear foot of the desired system by the total L.F. of beams. If many different size beams are used in the project, use an average size.

Step Six: Determine Total Superstructure Cost

Roof: 100' × 100' × $4.52/S.F.	=	$ 45,200
Suspended Floors: 2 × 100' × 100' × $10.02/S.F.	=	200,400
Columns: 25 ea. × 27.33' high × $43.45/V.L.F.	=	29,687
Columns: 25 ea. × 9.00' high × $19.90/V.L.F.	=	4,478
Column Fireproofing 859 L.F. @ $16.45/L.F.	=	14,131
Total	=	$ 293,896

Figure 5.8 is a summary of superstructure costs.

B10 Superstructure

B1010 Floor Construction

(A) Wide Flange

(B) Pipe

(C) Pipe, Concrete Filled

(D) Square Tube

(E) Square Tube
Concrete Filled

(F) Rectangular Tube

(G) Rectangular Tube, Concrete Filled

General: The following pages provide data for seven types of steel columns: wide flange, round pipe, round pipe concrete filled, square tube, square tube concrete filled, rectangular tube and rectangular tube concrete filled.

Design Assumptions: Loads are concentric; wide flange and round pipe bearing capacity is for 36 KSI steel. Square and rectangular tubing bearing capacity is for 46 KSI steel.

The effective length factor K=1.1 is used for determining column values in the tables. K=1.1 is within a frequently used range for pinned connections with cross bracing.

How To Use Tables:
a. Steel columns usually extend through two or more stories to minimize splices. Determine floors with splices.
b. Enter Table No. below with load to column at the splice. Use the unsupported height.
c. Determine the column type desired by price or design.

Cost:
a. Multiply number of columns at the desired level by the total height of the column by the cost/VLF.
b. Repeat the above for all tiers.

Please see the reference section for further design and cost information.

B1010 208					Steel Columns			
	LOAD (KIPS)	UNSUPPORTED HEIGHT (FT.)	WEIGHT (P.L.F.)	SIZE (IN.)	TYPE	COST PER V.L.F.		
						MAT.	INST.	TOTAL
1000	25	10	13	4	A	10.55	6.95	17.50
1020	RB1010 -130		7.58	3	B	8.70	6.95	15.65
1040			15	3-1/2	C	10.60	6.95	17.55
1060			6.87	3	D	4.82	6.95	11.77
1080			15	3	E	6.30	6.95	13.25
1100			8.15	4x3	F	6.30	6.95	13.25
1120			20	4x3	G	8.30	6.95	15.25
1200		16	16	5	A	12	5.20	17.20
1220			10.79	4	B	11.50	5.20	16.70
1240			36	5-1/2	C	15.95	5.20	21.15
1260			11.97	5	D	7.80	5.20	13
1280			36	5	E	10.55	5.20	15.75
1300			11.97	6x4	F	8.60	5.20	13.80
1320			64	8x6	G	16.90	5.20	22.10
1400		20	20	6	A	14.20	5.20	19.40
1420			14.62	5	B	14.70	5.20	19.90
1440			49	6-5/8	C	21.50	5.20	26.70
1460			11.97	5	D	7.35	5.20	12.55
1480			49	6	E	12.30	5.20	17.50
1500			14.53	7x5	F	9.85	5.20	15.05
1520			64	8x6	G	16.05	5.20	21.25
1600	50	10	16	5	A	12.95	6.95	19.90
1620			14.62	5	B	16.80	6.95	23.75
1640			24	4-1/2	C	12.65	6.95	19.60
1660			12.21	4	D	8.55	6.95	15.50
1680			25	4	E	8.80	6.95	15.75
1700			11.97	6x4	F	9.25	6.95	16.20
1720			28	6x3	G	11.05	6.95	18

Figure 5.5

B1010 Floor Construction

B1010 208				Steel Columns				
	LOAD (KIPS)	UNSUPPORTED HEIGHT (FT.)	WEIGHT (P.L.F.)	SIZE (IN.)	TYPE	COST PER V.L.F.		
						MAT.	INST.	TOTAL
4600	200	10	45	10	A	36.50	6.95	43.45
4620			40.48	10	B	50.50	6.95	57.45
4640			81	8-5/8	C	36.50	6.95	43.45
4660			31.84	8	D	22.50	6.95	29.45
4680			82	8	E	19.30	6.95	26.25
4700			37.69	10x6	F	29	6.95	35.95
4720			70	8x6	G	26	6.95	32.95
4800		16	49	10	A	37	5.20	42.20
4820			49.56	12	B	57	5.20	62.20
4840			123	10-3/4	C	48.50	5.20	53.70
4860			37.60	8	D	24.50	5.20	29.70
4880			90	8	E	25.50	5.20	30.70
4900			42.79	12x6	F	30.50	5.20	35.70
4920			85	10x6	G	28	5.20	33.20
5000		20	58	12	A	41	5.20	46.20
5020			49.56	12	B	54	5.20	59.20
5040			123	10-3/4	C	46	5.20	51.20
5060			40.35	10	D	25	5.20	30.20
5080			90	8	E	24.50	5.20	29.70
5100			47.90	12x8	F	32.50	5.20	37.70
5120			93	10x6	G	34	5.20	39.20
5200	300	10	61	14	A	49.50	6.95	56.45
5220			65.42	12	B	81.50	6.95	88.45
5240			169	12-3/4	C	64	6.95	70.95
5260			47.90	10	D	33.50	6.95	40.45
5280			90	8	E	27.50	6.95	34.45
5300			47.90	12x8	F	37	6.95	43.95
5320			86	10x6	G	39	6.95	45.95
5400		16	72	12	A	54	5.20	59.20
5420			65.42	12	B	75.50	5.20	80.70
5440			169	12-3/4	C	59.50	5.20	64.70
5460			58.10	12	D	37.50	5.20	42.70
5480			135	10	E	33	5.20	38.20
5500			58.10	14x10	F	41.50	5.20	46.70
5600		20	79	12	A	56	5.20	61.20
5620			65.42	12	B	71.50	5.20	76.70
5640			169	12-3/4	C	56.50	5.20	61.70
5660			58.10	12	D	35.50	5.20	40.70
5680			135	10	E	31	5.20	36.20
5700			58.10	14x10	F	39.50	5.20	44.70
5800	400	10	79	12	A	64	6.95	70.95
5840			178	12-3/4	C	84	6.95	90.95
5860			68.31	14	D	48	6.95	54.95
5880			135	10	E	35.50	6.95	42.45
5900			62.46	14x10	F	48.50	6.95	55.45
6000		16	87	12	A	65.50	5.20	70.70
6040			178	12-3/4	C	77.50	5.20	82.70
6060			68.31	14	D	44.50	5.20	49.70
6080			145	10	E	42	5.20	47.20
6100			76.07	14x10	F	54.50	5.20	59.70

Figure 5.5 (cont.)

B1010 Floor Construction

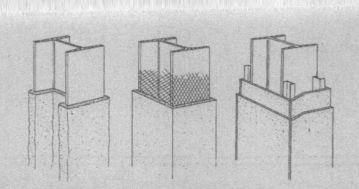

Costs below are costs per V.L.F. for fireproofing by material, column size, thickness and fire rating. Weights listed are for the fireproofing material only.

System Components		QUANTITY	UNIT	COST PER V.L.F.		
				MAT.	INST.	TOTAL
SYSTEM B1010 274						
CONCRETE FIREPROOFING, 8" STEEL COLUMN, 1" THICK, 1 HR. FIRE RATING						
Forms in place, columns, plywood, 4 uses		3.330	SFCA	2.16	18.32	20.48
Welded wire fabric, 2 x 2 #14 galv. 21 lb./C.S.F., column wrap		2.700	S.F.	.45	3.70	4.15
Concrete ready mix, regular weight, 3000 psi		.621	C.F.	1.63		1.63
Place and vibrate concrete, 12" sq./round columns, pumped		.621	C.F.		1.21	1.21
	TOTAL			4.24	23.23	27.47

B1010 274		Steel Column Fireproofing						
	ENCASEMENT SYSTEM	COLUMN SIZE (IN.)	THICKNESS (IN.)	FIRE RATING (HRS.)	WEIGHT (P.L.F.)	COST PER V.L.F.		
						MAT.	INST.	TOTAL
3000	Concrete	8	1	1	110	4.24	23.50	27.74
3050			1-1/2	2	133	4.88	26	30.88
3100			2	3	145	5.50	28	33.50
3150		10	1	1	145	5.50	29	34.50
3200			1-1/2	2	168	6.15	31	37.15
3250			2	3	196	6.85	33	39.85
3300		14	1	1	258	6.95	34	40.95
3350			1-1/2	2	294	7.60	36	43.60
3400			2	3	325	8.35	38	46.35
3450	Gypsum board	8	1/2	2	8	2.35	14.10	16.45
3500	1/2" fire rated	10	1/2	2	11	2.50	14.70	17.20
3550	1 layer	14	1/2	2	18	2.58	15.05	17.63
3600	Gypsum board	8	1	3	14	3.43	18	21.43
3650	1/2" fire rated	10	1	3	17	3.71	19.10	22.81
3700	2 layers	14	1	3	22	3.86	19.65	23.51
3750	Gypsum board	8	1-1/2	3	23	4.71	22.50	27.21
3800	1/2" fire rated	10	1-1/2	3	27	5.35	25	30.35
3850	3 layers	14	1-1/2	3	35	5.95	27.50	33.45
3900	Sprayed fiber	8	1-1/2	2	6.3	2.98	5.20	8.18
3950	Direct application		2	3	8.3	4.11	7.20	11.31
4000			2-1/2	4	10.4	5.30	9.30	14.60
4050		10	1-1/2	2	7.9	3.60	6.30	9.90
4100			2	3	10.5	4.94	8.65	13.59
4150			2-1/2	4	13.1	6.35	11.10	17.45

Figure 5.6

B1010 Floor Construction

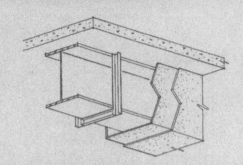

The table below lists fireproofing costs for steel beams by type, beam size, thickness and fire rating. Weights listed are for the fireproofing material only.

System Components	QUANTITY	UNIT	COST PER L.F.		
			MAT.	INST.	TOTAL
SYSTEM B1010 271					
FIREPROOFING, 5/8" F.R. GYP. BOARD, 12"X 4" BEAM, 2" THICK, 2 HR. F.R.					
Corner bead for drywall, 1-1/4" x 1-1/4", galvanized	.020	C.L.F.	.20	2.08	2.28
L bead for drywall, galvanized	.020	C.L.F.	.33	2.44	2.77
Furring, beams & columns, 3/4" galv. channels, 24" O.C.	2.330	S.F.	.26	4.24	4.50
Drywall on beam, no finish, 2 layers at 5/8" thick	3.000	S.F.	2.19	7.29	9.48
Drywall, taping and finishing joints, add	3.000	S.F.	.12	1.08	1.20
TOTAL			3.10	17.13	20.23

B1010 271		Steel Beam Fireproofing						
	ENCASEMENT SYSTEM	BEAM SIZE (IN.)	THICKNESS (IN.)	FIRE RATING (HRS.)	WEIGHT (P.L.F.)	COST PER L.F.		
						MAT.	INST.	TOTAL
0400	Concrete	12x4	1	1	77	4.04	18	22.04
0450	3000 PSI		1-1/2	2	93	4.68	19.65	24.33
0500			2	3	121	5.20	21.50	26.70
0550		14x5	1	1	100	5.20	22.50	27.70
0600			1-1/2	2	122	5.85	25	30.85
0650			2	3	142	6.55	27.50	34.05
0700		16x7	1	1	147	6.25	23.50	29.75
0750			1-1/2	2	169	6.75	24.50	31.25
0800			2	3	195	7.45	27	34.45
0850		18x7-1/2	1	1	172	7.10	27.50	34.60
0900			1-1/2	2	196	7.85	30	37.85
0950			2	3	225	8.70	32.50	41.20
1000		24x9	1	1	264	9.80	33	42.80
1050			1-1/2	2	295	10.60	35	45.60
1100			2	3	328	11.40	37	48.40
1150		30x10-1/2	1	1	366	12.70	41.50	54.20
1200			1-1/2	2	404	13.70	44.50	58.20
1250			2	3	449	14.35	47	61.35
1300	5/8" fire rated	12x4	2	2	15	3.10	17.15	20.25
1350	Gypsum board		2-5/8	3	24	4.46	21	25.46
1400		14x5	2	2	17	3.33	17.85	21.18
1450			2-5/8	3	27	4.92	22.50	27.42
1500		16x7	2	2	20	3.75	19.95	23.70
1550			2-5/8	3	31	5.15	19.25	24.40
1600	5/8" fire rated	18x7-1/2	2	2	22	4.02	21.50	25.52
1650	Gypsum board		2-5/8	3	34	5.90	27	32.90

Figure 5.7

ASSEMBLY NUMBER	DESCRIPTION	QTY	UNIT	TOTAL COST		COST PER S.F.
				UNIT	TOTAL	
DIVISION B	SHELL					
B10	SUPERSTRUCTURE					
B1010-208-1600	Columns, 38K (50K), 25 x 9'	225	V.L.F.	$ 19.90	$ 4,478	
B1010-208-4600	Columns, 191K (200K), 25 x 27.33'	683	V.L.F.	$ 43.45	$ 29,687	
B1010-212-3450	Column Fireproofing, 8" Average, Gyp. Board	859	L.F.	$ 16.45	$ 14,131	
B1010-250-5100	Floor Framing	20,000	S.F.	$ 10.02	$ 200,400	
B1020-112-3300	Roof Framing	10,000	S.F.	$ 4.52	$ 45,200	
	Subtotal, Division B10, Superstructure				$ 293,896	$ 9.80
B20	EXTERIOR CLOSURE					
B30	ROOFING					

Figure 5.8

Chapter 6

Exterior Closure

Exterior closure, UNIFORMAT II Division B20, consists of exterior walls, windows, and doors. This "building skin," or envelope, is the most visible part of a building and usually makes up a significant share of its cost. These exterior systems provide security against intruders and protection from natural elements such as wind, rain, snow, and sun. Some systems, such as bearing and shear walls, may also be designed to function as part of the structural component. Materials most commonly used in structural closure walls are aluminum, concrete (cast-in-place or precast), masonry, steel, glass, and wood.

Assemblies prices for exterior walls vary widely for the same size building depending on the materials selected. Alternatives can be quickly compared using the illustrated assemblies in *Means Assemblies Cost Data.* The tables provide total costs for each assembly with additions and deductions for modifications. For example, for curtain wall systems, a selective system worksheet allows you to prepare a quick estimate to compare the costs of a variety of curtain wall materials, including single glazing, double glazing, ceramic panels, metal panels, etc.

In selecting the exterior closure of a building, be alert to the possibility of the use of several materials. Many buildings have mechanical systems installed in a penthouse on the roof, which require exterior closure and may be of a different material than that used for the rest of the building. Often, this change of material is overlooked in the estimating process.

The Effects of Energy Requirements

Because energy consumption is such a major component in a building's operating costs, it is critical during the initial design and budgeting to compare the energy conservation features of various exterior closure systems. Often the least expensive material is the most costly in terms of energy costs over the life of a building. Another material with a better energy profile may be more costly initially to purchase and install.

The term "R value" has been adopted as the measure of thermal resistance. When heat flows through several materials, as in Figure 6.1, the individual R values can be added together. Using R values also allows the calculation of the Overall Coefficient of Thermal Transmittance (U) of a complete assembly. To determine U values, add up the R values and find the reciprocal ($U = 1/R$).

For example, in Figure 6.1, the first system has a total R value of 4.61. By taking the reciprocal, the U value is 0.22. After adding the insulation, the R value is increased to 14.87, and the U value is reduced to 0.067. The R and U values are inversely proportional: as the R value increases, the U value decreases.

The following example illustrates how the thermal efficiency of a wall can be improved substantially by the addition of low-cost insulation. For Figure 6.1, the cost per square foot of the wall system before adding rigid insulation is as follows:

Brick Veneer/C.M.U. Backup	(Figure 6.2, line #1200)	=	$21.05/S.F.
Plaster	(Figure 6.3, line #0920)	=	2.16/S.F.
Total Cost per Square Foot		=	$23.21/S.F.

After adding the insulation, the cost becomes:

Brick Veneer/C.M.U. Backup/Plaster	=	$23.21/S.F.
Insulation (similar to Figure 6.4, line #1835)	=	0.65/S.F.
Total Cost per Square Foot	=	$23.86/S.F.

If the exterior wall has 11,500 S.F., the cost comparison is as follows:

11,500 S.F. @ $23.86/S.F.	=	$274,390
11,500 S.F. @ $23.21/S.F.	=	$266,915
Cost Difference		$ 7,475

It costs 3% more to increase the wall U value from 0.22 to 0.067, an increase of 228% in the thermal resistance. In analyzing the various options, also add the cost difference for heating and cooling the building with and without the added insulation. The initial cost may be

14" Masonry Cavity Wall with 1" Plaster for the Exterior Closure

Construction	Resistance R.
1. Outside surface (15 mph wind)	0.17
2. Face brick (4 in.)	0.44
3. Air space (2 in., 50° mean temp, 10° diff)	1.02
4. Concrete block (8 in., lightweight)	2.12
5. Plaster (1 in., sand aggregate)	0.18
6. Inside surface (still air)	0.68
Total resistance	**4.61**
U= 1/R=1/4.61=	0.22

Replace item 3 with 2" smooth rigid polystyrene insulation and item 5 with 3/4" furring and 1/2" drywall

Total resistance		4.61
Deduct	3. Airspace	1.02
	5. Plaster	0.18
		1.20
Difference 4.61 - 1.20 =		3.41
Add rigid polystyrene insulation		10.00
3/4" airspace		1.01
1/2" gypsum board		0.45
Total resistance 14.87		
U=1/R=1/14.87=0.067		

Figure 6.1

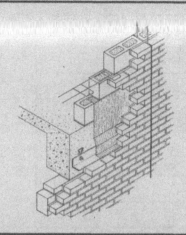

Exterior brick face composite walls are defined in the following terms: type of face brick and backup masonry, thickness of backup masonry and insulation. A special section is included on triple wythe construction at the back. Seven types of face brick are shown with various thicknesses of seven types of backup. All systems include a brick shelf, ties to the backup and necessary dampproofing, flashing, and control joints every 20′.

System Components

System Components	QUANTITY	UNIT	COST PER S.F. MAT.	COST PER S.F. INST.	COST PER S.F. TOTAL
SYSTEM B2010 130					
COMPOSITE WALL, STANDARD BRICK FACE, 6″ C.M.U. BACKUP, PERLITE FILL					
Face brick veneer, standard, running bond	1.000	S.F.	3.16	7.55	10.71
Wash brick	1.000	S.F.	.02	.65	.67
Concrete block backup, 6″ thick	1.000	S.F.	1.44	4.16	5.60
Wall ties	.300	Ea.	.04	.11	.15
Perlite insulation, poured	1.000	S.F.	.31	.22	.53
Flashing, aluminum	.100	S.F.	.09	.29	.38
Shelf angle	1.000	Lb.	.52	.67	1.19
Control joint	.050	L.F.	.10	.05	.15
Backer rod	.100	L.F.		.08	.08
Sealant	.100	L.F.	.02	.20	.22
Collar joint	1.000	S.F.	.33	.42	.75
TOTAL			6.03	14.40	20.43

B2010 130		Brick Face Composite Wall - Double Wythe				COST PER S.F. MAT.	COST PER S.F. INST.	COST PER S.F. TOTAL
	FACE BRICK	BACKUP MASONRY	BACKUP THICKNESS (IN.)	BACKUP CORE FILL		MAT.	INST.	TOTAL
1000	Standard	common brick	4	none		6.45	17.30	23.75
1040		SCR brick	6	none		8.70	15.40	24.10
1080		conc. block	4	none		5.20	13.90	19.10
1120			6	perlite		6.05	14.40	20.45
1160				styrofoam		6.60	14.20	20.80
1200			8	perlite		6.30	14.75	21.05
1240				styrofoam		6.70	14.45	21.15
1280		L.W. block	4	none		5.35	13.80	19.15
1320			6	perlite		6.10	14.30	20.40
1360				styrofoam		6.70	14.10	20.80
1400			8	perlite		6.60	14.65	21.25
1440				styrofoam		7.05	14.35	21.40
1520		glazed block	4	none		11.20	14.85	26.05
1560			6	perlite		11.90	15.30	27.20
1600				styrofoam		12.45	15.10	27.55
1640			8	perlite		12.40	15.65	28.05
1680				styrofoam		12.80	15.40	28.20

Figure 6.2

C1010 Partitions

C1010 144	Plaster Partition Components	COST PER S.F.		
		MAT.	INST.	TOTAL
0060	Metal studs, 16" O.C., including track, non load bearing, 25 gage, 1-5/8"	.18	.74	.92
0080	2-1/2"	.18	.74	.92
0100	3-1/4"	.21	.76	.97
0120	3-5/8"	.21	.76	.97
0140	4"	.26	.77	1.03
0160	6"	.33	.78	1.11
0180	Load bearing, 20 gage, 2-1/2"	.48	.95	1.43
0200	3-5/8"	.57	.96	1.53
0220	4"	.60	.99	1.59
0240	6"	.76	1	1.76
0260	16 gage 2-1/2"	.56	1.08	1.64
0280	3-5/8"	.68	1.11	1.79
0300	4"	.71	1.12	1.83
0320	6"	.90	1.14	2.04
0340	Wood studs, including blocking, shoe and double plate, 2"x4", 12" O.C.	.58	.91	1.49
0360	16" O.C.	.46	.73	1.19
0380	24" O.C.	.35	.59	.94
0400	2"x6", 12" O.C.	.85	1.04	1.89
0420	16" O.C.	.68	.81	1.49
0440	24" O.C.	.52	.64	1.16
0460	Furring one face only, steel channels, 3/4", 12" O.C.	.18	1.43	1.61
0480	16" O.C.	.16	1.27	1.43
0500	24" O.C.	.11	.96	1.07
0520	1-1/2", 12" O.C.	.27	1.60	1.87
0540	16" O.C.	.24	1.40	1.64
0560	24"O.C.	.16	1.10	1.26
0580	Wood strips 1"x3", on wood., 12" O.C.	.21	.66	.87
0600	16"O.C.	.16	.50	.66
0620	On masonry, 12" O.C.	.21	.74	.95
0640	16" O.C.	.16	.56	.72
0660	On concrete, 12" O.C.	.21	1.40	1.61
0680	16" O.C.	.16	1.05	1.21
0700	Gypsum lath. plain or perforated, nailed to studs, 3/8" thick	.43	.44	.87
0720	1/2" thick	.43	.47	.90
0740	Clipped to studs, 3/8" thick	.44	.50	.94
0760	1/2" thick	.44	.53	.97
0780	Metal lath, diamond painted, nailed to wood studs, 2.5 lb.	.18	.44	.62
0800	3.4 lb.	.28	.47	.75
0820	Screwed to steel studs, 2.5 lb.	.18	.47	.65
0840	3.4 lb.	.28	.50	.78
0860	Rib painted, wired to steel, 2.75 lb	.28	.50	.78
0880	3.4 lb	.40	.53	.93
0900	4.0 lb	.41	.58	.99
0910				
0920	Gypsum plaster, 2 coats	.42	1.74	2.16
0940	3 coats	.58	2.09	2.67
0960	Perlite or vermiculite plaster, 2 coats	.40	1.98	2.38
0980	3 coats	.66	2.49	3.15
1000	Stucco, 3 coats, 1" thick, on wood framing	.45	3.50	3.95
1020	On masonry	.25	2.72	2.97
1100	Metal base galvanized and painted 2-1/2" high	.49	1.40	1.89

Figure 6.3

B3010 Roof Coverings

B3010 320	Roof Deck Rigid Insulation	COST PER S.F.		
		MAT.	INST.	TOTAL
0100	Fiberboard low density, 1/2" thick, R1.39			
0150	1" thick R2.78	.36	.44	.80
0300	1 1/2" thick R4.17	.54	.44	.98
0350	2" thick R5.56	.73	.44	1.17
0370	Fiberboard high density, 1/2" thick R1.3	.21	.35	.56
0380	1" thick R2.5	.39	.44	.83
0390	1 1/2" thick R3.8	.63	.44	1.07
0410	Fiberglass, 3/4" thick R2.78	.50	.35	.85
0450	15/16" thick R3.70	.65	.35	1
0500	1-1/16" thick R4.17	.81	.35	1.16
0550	1-5/16" thick R5.26	1.12	.35	1.47
0600	2-1/16" thick R8.33	1.20	.44	1.64
0650	2 7/16" thick R10	1.36	.44	1.80
1000	Foamglass, 1 1/2" thick R4.55	1.56	.44	2
1100	3" thick R9.09	3.23	.50	3.73
1200	Tapered for drainage	1	.59	1.59
1260	Perlite, 1/2" thick R1.32	.29	.34	.63
1300	3/4" thick R2.08	.35	.44	.79
1350	1" thick R2.78	.30	.44	.74
1400	1 1/2" thick R4.17	.42	.44	.86
1450	2" thick R5.56	.59	.50	1.09
1510	Polyisocyanurate 2#/CF density, 1" thick R7.14	.34	.25	.59
1550	1 1/2" thick R10.87	.37	.28	.65
1600	2" thick R14.29	.48	.32	.80
1650	2 1/2" thick R16.67	.54	.34	.88
1700	3" thick R21.74	.76	.35	1.11
1750	3 1/2" thick R25	.79	.35	1.14
1800	Tapered for drainage	.41	.25	.66
1810	Expanded polystyrene, 1#/CF density, 3/4" thick R2.89	.20	.24	.44
1820	2" thick R7.69	.37	.28	.65
1825	Extruded Polystyrene			
1830	Extruded polystyrene, 15 PSI compressive strength, 1" thick R5	.24	.24	.48
1835	2" thick R10	.37	.28	.65
1840	3" thick R15	.83	.35	1.18
1900	25 PSI compressive strength, 1" thick R5	.39	.24	.63
1950	2" thick R10	.76	.28	1.04
2000	3" thick R15	1.14	.35	1.49
2050	4" thick R20	1.19	.35	1.54
2150	Tapered for drainage	.43	.24	.67
2550	40 PSI compressive strength, 1" thick R5	.37	.24	.61
2600	2" thick R10	.74	.28	1.02
2650	3" thick R15	1.08	.35	1.43
2700	4" thick R20	1.44	.35	1.79
2750	Tapered for drainage	.54	.25	.79
2810	60 PSI compressive strength, 1" thick R5	.45	.24	.69
2850	2" thick R10	.80	.29	1.09
2900	Tapered for drainage	.65	.25	.90
4000	Composites with 1-1/2" polyisocyanurate			
4010	1" fiberboard	.90	.44	1.34
4020	1" perlite	.94	.42	1.36
4030	7/16" oriented strand board	1.08	.44	1.52

Figure 6.4

insignificant when compared to the additional operating costs saved over the useful life of the building.

Exterior Wall Systems

The selection of exterior wall systems usually depends on the owner's aesthetic and energy requirements, and zoning ordinances. Some facilities limit architectural elements to those standards used for all buildings in a complex. Building codes and regulations, natural light requirements, and economics (initial cost and life cycle cost) are other factors in the selection process.

Exterior Wall Area

The ratio of exterior wall area to the door and window area can change significantly with the size of the building and the window configuration. Consider the example in Figure 6.5. In a small building like Building A (with individual windows), the percentage of windows to wall area is fairly insignificant. The ratio becomes quite significant in larger wall areas, such as the wall design used for Building B:

Building "A"	Exterior Wall	92%
	Windows	6%
	Doors	2%
Building "B"	Exterior Wall	58%
	Window Wall	41.5%
	Doors	0.5%

Small Footprint versus Large Footprint

When compared to the building footprint area, the exterior closure can reveal some interesting cost discoveries. As the area of the building increases, the proportional contribution of the cost of the exterior wall to total project cost decreases. For instance, compare two medical office buildings with wood truss framing, 10' story height, and a rectangular shape. Building C has an area of 4,000 S.F. Building D has an area of 16,000 S.F.

Building C: The combined cost of the wall construction, windows and doors (Figure 6.6) is $61,000.

The total cost of work in all divisions is $368,000.

$$\frac{\$61,000}{\$368,000} = 16.6\%$$

Building D: The combined cost of the wall construction, windows and doors (Figure 6.7) is $136,900.

The total cost of work in all divisions is $1,265,950.

$$\frac{\$136,900}{\$1,265,950} = 10.8\%$$

Short Perimeter versus Long Perimeter

The proportional cost of the exterior wall will change as the shape of the building is changed. For instance, a square building will have a minimum perimeter, and a building with an irregular shape with wings and more than four corners will have a longer perimeter. Assuming that the cost of the remaining components of the building will not change, the proportional cost of the exterior wall to total cost will change.

For example, the 4,000 S.F. building in the example above was calculated with a perimeter of 280 L.F. Since the total cost of this exterior wall construction is given (Figure 6.6) as $61,000, the cost per linear foot of wall construction is:

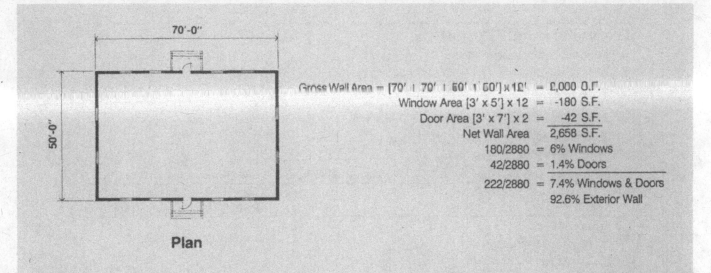

Plan

Gross Wall Area = [70' + 70' + 50' + 50'] x 12' = 2,880 S.F.
Window Area [3' x 5'] x 12 = -180 S.F.
Door Area [3' x 7'] x 2 = -42 S.F.
Net Wall Area = 2,658 S.F.
180/2880 = 6% Windows
42/2880 = 1.4% Doors
222/2880 = 7.4% Windows & Doors
92.6% Exterior Wall

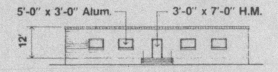

5'-0" x 3'-0" Alum.   3'-0" x 7'-0" H.M.

12'

Elevation Building "A"

Check for Penthouse
Requirements

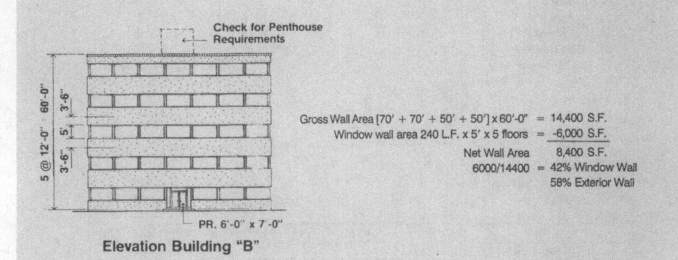

Gross Wall Area [70' + 70' + 50' + 50'] x 60'-0" = 14,400 S.F.
Window wall area 240 L.F. x 5' x 5 floors = -6,000 S.F.

Net Wall Area 8,400 S.F.
6000/14400 = 42% Window Wall
58% Exterior Wall

5 @ 12'-0" = 60'-0"

3'-6"

5'

3'-6"

PR. 6'-0" x 7'-0"

Elevation Building "B"

Figure 6.5

Preliminary Cost Report

Project Name: SF Estimating Methods

Model Type: Medical Office, 1 Story, Brick Veneer / Wood Truss
Stories (Ea.): 1
Story Height (L.F.): 10
Floor Area (S.F.): 4000
Basement: Not Included

Location: National Average Costs
Data Release: 2001
Wage Rate: Union

Costs are derived from a building model with basic components. Scope differences and local market conditions can cause costs to vary significantly.

			$Cost/ Per S.F.	$ Total Cost	% Of Sub-Total
1.0 Foundations					7.0%
1.1	Footings & Foundations		5.30	21,200.00	
1.9	Excavation and Backfill		1.17	4,675.00	
2.0 Substructure					3.6%
2.1	Slab on Grade		3.33	13,300.00	
3.0 Superstructure					5.3%
3.7	Roof		4.88	19,500.00	
4.0 Exterior Closure					16.6%
4.1	Walls		8.13	32,500.00	
4.6	Doors		1.85	7,400.00	
4.7	Windows & Glazed Walls		5.28	21,100.00	
5.0 Roofing					3.2%
5.1	Roof Coverings		1.38	5,500.00	
5.7	Insulation		1.00	4,000.00	
5.8	Openings & Specialties		0.54	2,150.00	
6.0 Interior Construction					27.9%
6.1	Partitions		7.33	29,300.00	
6.4	Interior Doors		7.13	28,500.00	
6.5	Wall Finishes		2.65	10,600.00	
6.6	Floor Finishes		4.67	18,700.00	
6.7	Ceiling Finishes		2.45	9,800.00	
6.9	Interior Surface/Exterior Wall		1.48	5,925.00	
8.0 Mechanical					26.3%
8.1	Plumbing		12.13	48,500.00	
8.2	Fire Protection		2.27	9,075.00	
8.4	Cooling		9.80	39,200.00	
9.0 Electrical					10.1%
9.1	Service & Distribution		1.69	6,775.00	
9.2	Lighting & Power		4.90	19,600.00	
9.4	Special Electrical		2.67	10,700.00	
	Sub-Total		92.00	368,000.00	100%
GENERAL CONDITIONS (Overhead & Profit)		25%	23.00	92,000.00	
ARCHITECTURAL FEES		9%	10.35	41,400.00	
USER FEES		0%	0.00	0.00	
TOTAL BUILDING COST			125.35	501,400.00	

Note: The estimate in this figure is organized using UniFormat rather than UNIFORMAT II.

Figure 6.6

Preliminary Cost Report

Project Name: SF Estimating Methods

Model Type: Medical Office, 1 Story, Brick Veneer / Wood Truss

Stories (Ea.): 1

Story Height (L.F.): 10

Floor Area (S.F.): 16000

Basement: Not Included

Location: National Average Costs

Data Release: 2001

Wage Rate: Union

Costs are derived from a building model with basic components. Scope differences and local market conditions can cause costs to vary significantly.

		$Cost/ Per S.F.	$ Total Cost	% Of Sub-Total
1.0 Foundations				4.8%
1.1	Footings & Foundations	2.65	42,400.00	
1.9	Excavation and Backfill	1.17	18,700.00	
2.0 Substructure				4.2%
2.1	Slab on Grade	3.31	53,000.00	
3.0 Superstructure				6.2%
3.7	Roof	4.88	78,000.00	
4.0 Exterior Closure				10.8%
4.1	Walls	4.06	65,000.00	
4.6	Doors	1.85	29,600.00	
4.7	Windows & Glazed Walls	2.64	42,300.00	
5.0 Roofing				3.2%
5.1	Roof Coverings	1.25	20,000.00	
5.7	Insulation	1.00	16,000.00	
5.8	Openings & Specialties	0.27	4,275.00	
6.0 Interior Construction				31.5%
6.1	Partitions	7.31	117,000.00	
6.4	Interior Doors	7.13	114,000.00	
6.5	Wall Finishes	2.64	42,200.00	
6.6	Floor Finishes	4.69	75,000.00	
6.7	Ceiling Finishes	2.45	39,200.00	
6.9	Interior Surface/Exterior Wall	0.74	11,800.00	
8.0 Mechanical				29.2%
8.1	Plumbing	11.03	176,500.00	
8.2	Fire Protection	2.27	36,300.00	
8.4	Cooling	9.78	156,500.00	
9.0 Electrical				10.1%
9.1	Service & Distribution	0.42	6,775.00	
9.2	Lighting & Power	4.91	78,500.00	
9.4	Special Electrical	2.68	42,900.00	
Sub-Total		79.12	1,265,950.00	**100%**
GENERAL CONDITIONS (Overhead & Profit)	25%	19.78	316,500.00	
ARCHITECTURAL FEES	9%	8.91	142,500.00	
USER FEES	0%	0.00	0.00	
TOTAL BUILDING COST		107.81	1,724,950.00	

Note: The estimate in this figure is organized using UniFormat rather than UNIFORMAT II.

Figure 6.7

$$\frac{\$61,000}{280 \text{ L.F.}} = \$217.85 \text{ per L.F.}$$

The exterior wall contributed 16.6% of the total building cost. This building is shown in Figure 6.8 as Building C.

A square building, such as Building E in Figure 6.8, has an area of 4,000 S.F. and a perimeter of 253 L.F. At $217.85 per linear foot of wall construction, the total cost of the exterior wall would be $55,116. The total cost of work in all divisions would then be $362,116, and the percent contribution of the exterior wall to the whole would be:

$$\frac{\$55,116}{\$362,116} = 15.22\%$$

A long, narrow L-shaped building, Building F in Figure 6.8, has an area of 4,000 S.F. and a perimeter of 440 L.F. At $217.85 per linear foot of wall construction, the total cost of the exterior wall would be $95,854. The total cost of work in all divisions would be $402,854, and the percent contribution of the exterior wall to the whole would be:

$$\frac{\$95,854}{\$402,854} = 23.79\%$$

Assemblies Estimate: Three-Story Office Building

The exterior closure of the sample 30,000 S.F. three-story office building in Chapter 3 will include:

- Insulated concrete masonry units
- Precast concrete wall panels
- Aluminum tube frame windows and door units
- Insulating glass
- Precast terrazzo window sills

Costs:

Insulated Lightweight Concrete Masonry Units (Penthouse)			
(Figure 6.9, line #3310)	493 S.F. @ $6.78/S.F.	=	$ 3,343
Precast Concrete Wall Panels			
(Figure 6.10, line #5750)	9,338 S.F. @ $14.99/S.F.	=	139,977
Aluminum and Glass Doors			
(Figure 6.11, line #6350)	2 pr. @ $3,025/pr.	=	6,050
Aluminum Flush Tube Windows with Thermal Break			
(Figure 6.12, line #2000)	5,750 S.F. @ $21.25/S.F.	=	122,188
Insulating Glass			
(Figure 6.13, line #1000)	5,750 S.F. @ $14.20/S.F.	=	81,650
Precast Terrazzo Window Sills			
(Figure 6.14, line #1880)	771 L.F. @ $13.04/L.F.	=	10,054
Total Cost			$363,262
Cost per Square Foot		=	$12.11/S.F.

Figures 6.15 is an elevation of the three-story office building. All four elevations are similar, except that the side elevations have windows instead of doors. The wall section, Figure 6.16, shows the construction components and relationships of the various elements of the exterior wall.

Figure 6.17 summarizes the square foot costs for the exterior closure for the sample project.

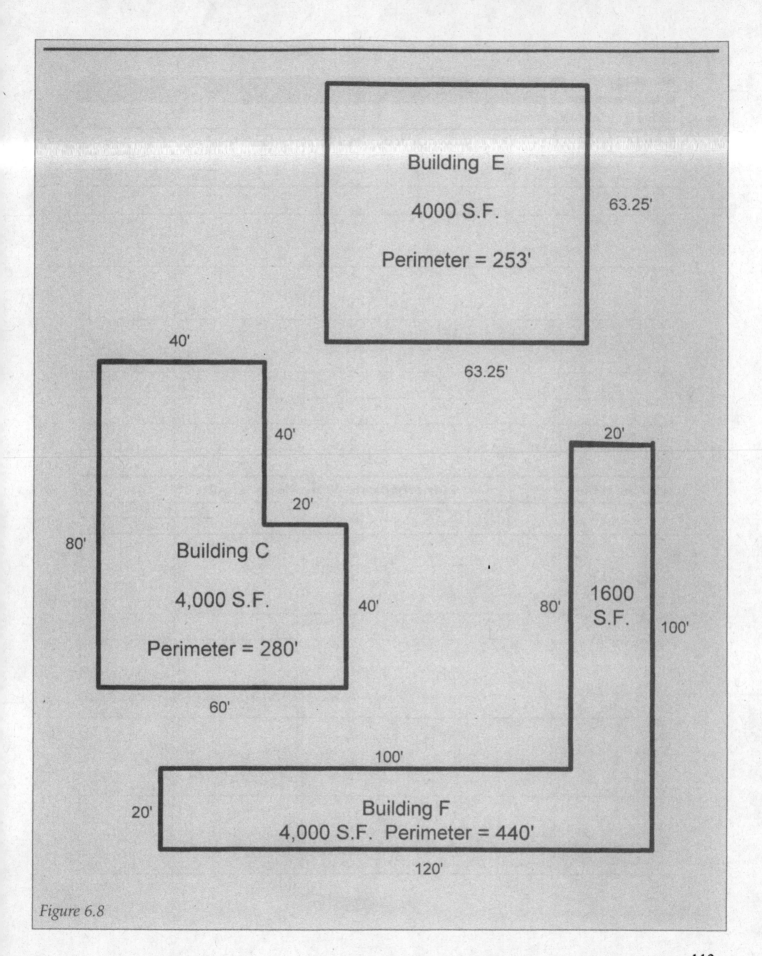

Building E

4000 S.F.

Perimeter = 253'

63.25'

63.25'

40'

40'

20'

20'

80'

Building C

4,000 S.F.

Perimeter = 280'

40'

80'

1600
S.F.

100'

60'

100'

20'

Building F
4,000 S.F. Perimeter = 440'

120'

Figure 6.8

B20 Exterior Enclosure

B2010 Exterior Walls

B2010 109 — Concrete Block Wall - Regular Weight

	TYPE	SIZE (IN.)	STRENGTH (P.S.I.)	CORE FILL		MAT.	INST.	TOTAL
					COST PER S.F.			
2000	75% solid	4x8x16	2,000	none		1.13	4.07	5.20
2050			4,500	none		1.24	4.07	5.31
2100	75% solid	6x8x16	2,000	perlite		1.66	4.48	6.14
2140				none		1.50	4.37	5.87
2150			4,500	perlite		2	4.48	6.48
2190				none		1.84	4.37	6.21
2200		8x8x16	2,000	perlite		2.44	4.81	7.25
2240				none		2.21	4.67	6.88
2250			4,500	perlite		2.71	4.81	7.52
2290				none		2.48	4.67	7.15
2300		12x8x16	2,000	perlite		2.97	6.25	9.22
2340				none		2.60	6	8.60
2350			4,500	perlite		4.02	6.25	10.27
2390				none		3.65	6	9.65
2500	Solid	4x8x16	2,000	none		1.49	4.16	5.65
2550			4,500	none		1.24	4.11	5.35
2600		6x8x16	2,000	none		1.73	4.48	6.21
2650			4,500	none		1.84	4.42	6.26
2700		8x8x16	2,000	none		2.55	4.80	7.35
2750			4,500	none		2.48	4.73	7.21
2800		12x8x16	2,000	none		3.77	6.20	9.97
2850			4,500	none		3.65	6.10	9.75

B2010 109 — Concrete Block Wall - Lightweight

	TYPE	SIZE (IN.)	WEIGHT (P.C.F.)	CORE FILL		MAT.	INST.	TOTAL
					COST PER S.F.			
3100	Hollow	8x4x16	105	perlite		1.46	4.60	6.06
3110				styrofoam		1.89	4.33	6.22
3140				none		1.01	4.33	5.34
3150			85	perlite		3.47	4.50	7.97
3160				styrofoam		3.90	4.23	8.13
3190				none		3.02	4.23	7.25
3200		4x8x16	105	none		1.25	3.93	5.18
3250			85	none		1.35	3.85	5.20
3300		6x8x16	105	perlite		2	4.43	6.43
3310				styrofoam		2.57	4.21	6.78
3340				none		1.69	4.21	5.90
3350			85	perlite		2.03	4.33	6.36
3360				styrofoam		2.60	4.11	6.71
3390				none		1.72	4.11	5.83
3400		8x8x16	105	perlite		2.51	4.77	7.28
3410				styrofoam		2.94	4.50	7.44
3440				none		2.06	4.50	6.56
3450			85	perlite		2.31	4.66	6.97
3460				styrofoam		2.74	4.39	7.13
3490				none		1.86	4.39	6.25
3500		12x8x16	105	perlite		3.92	6.30	10.22
3510				styrofoam		4.28	5.75	10.03
3540				none		3.18	5.75	8.93
3550			85	perlite		3.29	6.15	9.44
3560				styrofoam		3.65	5.60	9.25
3590				none		2.55	5.60	8.15

Figure 6.9

114

B20 Exterior Enclosure

B2010 Exterior Walls

B2010 103 — Flat Precast Concrete

	THICKNESS (IN.)	PANEL SIZE (FT.)	FINISHES	RIGID INSULATION (IN)	TYPE	COST PER S.F.		
						MAT.	INST.	TOTAL
4600	7	4x8	white face	2	low rise	18.15	3.80	21.95
4650		8x8				14.15	3.01	17.16
4700		10x10				12.75	2.75	15.50
4750		20x10				11.75	2.55	14.30
4800	8	4x8	white face	none	low rise	17.65	3.47	21.12
4850		8x8				13.60	2.67	16.27
4900		10x10				12.25	2.41	14.66
4950		20x10				11.30	2.21	13.51
5000	8	4x8	white face	2	low rise	18.55	3.88	22.43
5050		8x8				14.50	3.08	17.58
5100		10x10				13.10	2.82	15.92
5150		20x10				12.15	2.62	14.77

B2010 103 — Fluted Window or Mullion Precast Concrete

	THICKNESS (IN.)	PANEL SIZE (FT.)	FINISHES	RIGID INSULATION (IN)	TYPE	COST PER S.F.		
						MAT.	INST.	TOTAL
5200	4	4x8	smooth gray	none	high rise	11.90	12.45	24.35
5250		8x8				8.50	8.90	17.40
5300		10x10				12.50	3.25	15.75
5350		20x10				10.95	2.84	13.79
5400	5	4x8	smooth gray	none	high rise	12.10	12.60	24.70
5450		8x8				8.80	9.20	18
5500		10x10				13	3.39	16.39
5550		20x10				11.50	3	14.50
5600	6	4x8	smooth gray	none	high rise	12.45	12.95	25.40
5650		8x8				9.10	9.50	18.60
5700		10x10				13.40	3.49	16.89
5750		20x10				11.90	3.09	14.99
5800	6	4x8	smooth gray	2	high rise	13.30	13.40	26.70
5850		8x8				9.95	9.90	19.85
5900		10x10				14.25	3.90	18.15
5950		20x10				12.75	3.50	16.25
6000	7	4x8	smooth gray	none	high rise	12.70	13.25	25.95
6050		8x8				9.35	9.80	19.15
6100		10x10				13.95	3.64	17.59
6150		20x10				12.25	3.19	15.44
6200	7	4x8	smooth gray	2	high rise	13.55	13.70	27.25
6250		8x8				10.25	10.20	20.45
6300		10x10				14.85	4.05	18.90
6350		20x10				13.15	3.60	16.75
6400	8	4x8	smooth gray	none	high rise	12.85	13.45	26.30
6450		8x8				9.60	10	19.60
6500		10x10				14.40	3.75	18.15
6550		20x10				12.85	3.35	16.20
6600	8	4x8	smooth gray	2	high rise	13.75	13.85	27.60
6650		8x8				10.45	10.40	20.85
6700		10x10				15.25	4.16	19.41
6750		20x10				13.75	3.76	17.51

Figure 6.10

B20 Exterior Enclosure

B2030 Exterior Doors

B2030 105				Wood, Steel & Aluminum			

	MATERIAL	TYPE	DOORS	SPECIFICATION	OPENING	COST PER OPNG.		
						MAT.	INST.	TOTAL
6000	Aluminum	combination	storm & screen	hinged	3'-0" x 6'-8"	225	52	277
6050					3'-0" x 7'-0"	248	57	305
6100		overhead	rolling grill	manual oper.	12'-0" x 12'-0"	2,700	1,675	4,375
6150				motor oper.	12'-0" x 12'-0"	3,550	1,875	5,425
6200	Alum. & Fbrgls.	overhead	heavy duty	manual oper.	12'-0" x 12'-0"	1,425	485	1,910
6250				electric oper.	12'-0" x 12'-0"	1,950	665	2,615
6300	Alum. & glass	w/o transom	narrow stile	w/panic Hrdwre.	3'-0" x 7'-0"	1,025	690	1,715
6350				dbl. door, Hrdwre.	6'-0" x 7'-0"	1,875	1,150	3,025
6400			wide stile	hdwre.	3'-0" x 7'-0"	1,350	675	2,025
6450				dbl. door, Hdwre.	6'-0" x 7'-0"	2,650	1,350	4,000
6500			full vision	hdwre.	3'-0" x 7'-0"	1,525	1,075	2,600
6550				dbl. door, Hdwre.	6'-0" x 7'-0"	2,250	1,550	3,800
6600			non-standard	hdwre.	3'-0" x 7'-0"	955	675	1,630
6650				dbl. door, Hdwre.	6'-0" x 7'-0"	1,900	1,350	3,250
6700			bronze fin.	hdwre.	3'-0" x 7'-0"	1,025	675	1,700
6750				dbl. door, Hrdwre.	6'-0" x 7'-0"	2,025	1,350	3,375
6800			black fin.	hdwre.	3'-0" x 7'-0"	1,675	675	2,350
6850				dbl. door, Hdwre.	6'-0" x 7'-0"	3,350	1,350	4,700
6900		w/transom	narrow stile	hdwre.	3'-0" x 10'-0"	1,350	785	2,135
6950				dbl. door, Hdwre.	6'-0" x 10'-0"	1,900	1,375	3,275
7000			wide stile	hdwre.	3'-0" x 10'-0"	1,550	940	2,490
7050				dbl. door, Hdwre.	6'-0" x 10'-0"	2,050	1,625	3,675
7100			full vision	hdwre.	3'-0" x 10'-0"	1,725	1,050	2,775
7150				dbl. door, Hdwre.	6'-0" x 10'-0"	2,200	1,800	4,000
7200			non-standard	hdwre.	3'-0" x 10'-0"	1,000	730	1,730
7250				dbl. door, Hdwre.	6'-0" x 10'-0"	2,025	1,450	3,475
7300			bronze fin.	hdwre.	3'-0" x 10'-0"	1,075	730	1,805
7350				dbl. door, Hdwre.	6'-0" x 10'-0"	2,150	1,450	3,600
7400			black fin.	hdwre.	3'-0" x 10'-0"	1,725	730	2,455
7450				dbl. door, Hdwre.	6'-0" x 10'-0"	3,450	1,450	4,900
7500		revolving	stock design	minimum	6'-10" x 7'-0"	16,300	2,550	18,850
7550				average	6'-0" x 7'-0"	20,100	3,175	23,275
7600				maximum	6'-10" x 7'-0"	26,500	4,250	30,750
7650				min., automatic	6'-10" x 7'-0"	28,200	2,950	31,150
7700				avg., automatic	6'-10" x 7'-0"	32,000	3,575	35,575
7750				max., automatic	6'-10" x 7'-0"	38,400	4,650	43,050
7800		balanced	standard	economy	3'-0" x 7'-0"	5,100	1,050	6,150
7850				premium	3'-0" x 7'-0"	6,350	1,350	7,700
7900		mall front	sliding panels	alum. fin.	16'-0" x 9'-0"	2,475	535	3,010
7950					24'-0" x 9'-0"	3,600	990	4,590
8000				bronze fin.	16'-0" x 9'-0"	2,875	625	3,500
8050					24'-0" x 9'-0"	4,200	1,150	5,350
8100			fixed panels	alum. fin.	48'-0" x 9'-0"	6,675	770	7,445
8150				bronze fin.	48'-0" x 9'-0"	7,775	895	8,670
8200		sliding entrance	5' x 7' door	electric oper.	12'-0" x 7'-6"	6,300	990	7,290
8250		sliding patio		economy	6'-0" x 7'-0"	835	182	1,017
8300			temp. glass	economy	12'-0" x 7'-0"	2,100	243	2,343
8350				premium	6'-0" x 7'-0"	1,250	273	1,523
8400					12'-0" x 7'-0"	3,150	365	3,515

Figure 6.11

B2020 Exterior Windows

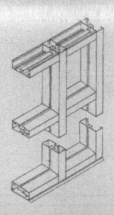

The table below lists costs per S.F of opening for framing with 1-3/4" x 4-1/2" clear anodized tubular aluminum framing. This is the type often used for 1/4" plate glass flush glazing.

For bronze finish, add 18% to material cost. For black finish, add 27% to material cost. For stainless steel, add 75% to material cost. For monumental grade, add 50% to material cost. This tube framing is usually installed by a glazing contractor.

Note: The costs below do not include the glass.

System Components	QUANTITY	UNIT	COST/S.F. OPNG.		
			MAT.	INST.	TOTAL
SYSTEM B2020 118					
ALUM FLUSH TUBE, FOR 1/4"GLASS, 5'X20'OPENING, 3 INTER. HORIZONTALS					
Flush tube frame, alum mill fin, 1-3/4"x4" open headr for 1/4" glass	.450	L.F.	3.69	3.89	7.58
Flush tube frame alum mill fin 1-3/4"x4" open sill for 1/4" glass	.050	L.F.	.36	.42	.78
Flush tube frame alum mill fin,1-3/4"x4"closed back sill, 1/4" glass	.150	L.F.	1.64	1.22	2.86
Aluminum structural shapes, 1" to 10" members, under 1 ton	.040	Lb.	.10	.17	.27
Joints for tube frame, 90° clip type	.100	Ea.	2.05	-	2.05
Caulking/sealants, polysulfide, 1 or 2 part,1/2x1/4"bead 154 ff/gal	.500	L.F.	.16	1.36	1.52
TOTAL			8	7.06	15.06

B2020 118	Tubular Aluminum Framing	COST/S.F. OPNG.		
		MAT.	INST.	TOTAL
1100	Alum flush tube frame,for 1/4"glass,1-3/4"x4",5'x6'opng, no inter horizntls	8.90	8.50	17.40
1150	One intermediate horizontal	12.15	10	22.15
1200	Two intermediate horizontals	15.40	11.45	26.85
1250	5' x 20' opening, three intermediate horizontals	8	7.05	15.05
1400	1-3/4" x 4-1/2", 5' x 6' opening, no intermediate horizontals	10.10	8.50	18.60
1450	One intermediate horizontal	13.50	10	23.50
1500	Two intermediate horizontals	16.90	11.45	28.35
1550	5' x 20' opening, three intermediate horizontals	9	7.05	16.05
1700	For insulating glass, 2"x4-1/2", 5'x6' opening, no intermediate horizontals	11.35	8.95	20.30
1750	One intermediate horizontal	14.90	10.50	25.40
1800	Two intermediate horizontals	18.40	12.05	30.45
1850	5' x 20' opening, three intermediate horizontals	9.95	7.45	17.40
2000	Thermal break frame, 2-1/4"x4-1/2", 5'x6'opng, no intermediate horizontals	12.20	9.05	21.25
2050	One intermediate horizontal	16.40	10.90	27.30
2100	Two intermediate horizontals	20.50	12.75	33.25
2150	5' x 20' opening, three intermediate horizontals	11.15	7.75	18.90

Figure 6.12

B20 Exterior Enclosure

B2020 Exterior Windows

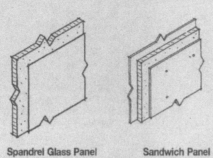

Spandrel Glass Panel   Sandwich Panel

The table below lists costs of curtain wall and spandrel panels per S.F. Costs do not include structural framing used to hang the panels from.

B2020 120	Curtain Wall Panels	COST PER S.F.		
		MAT.	INST.	TOTAL
1000	Glazing panel, insulating, 1/2" thick, 2 lites 1/8" float, clear	6.90	7.30	14.20
1100	Tinted	10.25	7.30	17.55
1200	5/8" thick units, 2 lites 3/16" float, clear	8.30	7.70	16
1400	1" thick units, 2 lites, 1/4" float, clear	11.80	9.25	21.05
1700	Light and heat reflective glass, tinted	19	8.15	27.15
2000	Plate glass, 1/4" thick, clear	5	5.75	10.75
2050	Tempered	5.95	5.75	11.70
2100	Tinted	4.75	5.75	10.50
2200	3/8" thick, clear	7.90	9.25	17.15
2250	Tempered	11.90	9.25	21.15
2300	Tinted	9.50	9.25	18.75
2400	1/2" thick, clear	15.50	12.60	28.10
2450	Tempered	17.80	12.60	30.40
2500	Tinted	16.65	12.60	29.25
2600	3/4" thick, clear	21.50	19.80	41.30
2650	Tempered	25	19.80	44.80
3000	Spandrel glass, panels, 1/4" plate glass insul w/fiberglass, 1" thick	10.15	5.75	15.90
3100	2" thick	11.85	5.75	17.60
3200	Galvanized steel backing, add	3.51		3.51
3300	3/8" plate glass, 1" thick	17	5.75	22.75
3400	2" thick	18.70	5.75	24.45
4000	Polycarbonate, masked, clear or colored, 1/8" thick	5.70	4.08	9.78
4100	3/16" thick	6.90	4.20	11.10
4200	1/4" thick	7.60	4.47	12.07
4300	3/8" thick	14.05	4.62	18.67
5000	Facing panel, textured al, 4' x 8' x 5/16" plywood backing, sgl face	2.40	2.21	4.61
5100	Double face	3.62	2.21	5.83
5200	4' x 10' x 5/16" plywood backing, single face	2.55	2.21	4.76
5300	Double face	3.75	2.21	5.96
5400	4' x 12' x 5/16" plywood backing, single face	3.29	2.21	5.50
5500	Sandwich panel, 22 Ga. galv., both sides 2" insulation, enamel exterior	7.55	3.89	11.44
5600	Polyvinylidene floride exterior finish	7.95	3.89	11.84
5700	26 Ga., galv. both sides, 1" insulation, colored 1 side	3.97	3.69	7.66
5800	Colored 2 sides	5.10	3.69	8.79

Figure 6.13

C3020 Floor Finishes

C3020 410	Tile & Covering	MAT.	INST.	TOTAL
1380	Polyethylene, in rolls, minimum	2.54	1.21	3.75
1400	Maximum	4.94	1.21	6.15
1420	Polyurethane, thermoset, minimum	3.91	3.33	7.24
1440	Maximum	4.64	6.65	11.29
1460	Rubber, sheet goods, minimum	3.50	2.77	6.27
1480	Maximum	5.75	3.70	9.45
1500	Tile, minimum	3.92	.83	4.75
1520	Maximum	6.65	1.21	7.86
1540	Synthetic turf, minimum	2.96	1.59	4.55
1560	Maximum	7.60	1.75	9.35
1580	Vinyl, composition tile, minimum	.78	.67	1.45
1600	Maximum	2.24	.67	2.91
1620	Tile, minimum	1.75	.67	2.42
1640	Maximum	4.93	.67	5.60
1660	Sheet goods, minimum	2.04	1.33	3.37
1680	Maximum	3.93	1.66	5.59
1720	Tile, ceramic natural clay	4.29	3.29	7.58
1730	Marble, synthetic 12"x12"x5/8"	9.65	10	19.65
1740	Porcelain type, minimum	4.61	3.29	7.90
1760	Maximum	6.35	3.88	10.23
1800	Quarry tile, mud set, minimum	3.16	4.30	7.46
1820	Maximum	5.65	5.45	11.10
1840	Thin set, deduct		.86	.86
1860	Terrazzo precast, minimum	3.89	4.44	8.33
1880	Maximum	8.60	4.44	13.04
1900	Non-slip, minimum	16.20	22.50	38.70
1920	Maximum	18.65	31	49.65
1960	Wood, block, end grain factory type, creosoted, 2" thick	3.09	1.24	4.33
1980	2-1/2" thick	3.20	2.92	6.12
2000	3" thick	3.15	2.92	6.07
2020	Natural finish, 2" thick	3.19	2.92	6.11
2040	Fir, vertical grain, 1"x4", no finish, minimum	2.38	1.43	3.81
2060	Maximum	2.53	1.43	3.96
2080	Prefinished white oak, prime grade, 2-1/4" wide	6.45	2.15	8.60
2100	3-1/4" wide	8.25	1.97	10.22
2120	Maple strip, sanded and finished, minimum	3.64	3.12	6.76
2140	Maximum	4.30	3.12	7.42
2160	Oak strip, sanded and finished, minimum	3.69	3.12	6.81
2180	Maximum	4.57	3.12	7.69
2200	Parquetry, sanded and finished, minimum	3.22	3.25	6.47
2220	Maximum	6.35	4.62	10.97
2260	Add for sleepers on concrete, treated, 24" O.C., 1"x2"	.65	1.88	2.53
2280	1"x3"	.81	1.45	2.26
2300	2"x4"	.87	.74	1.61
2320	2"x6"	.99	.56	1.55
2340	Underlayment, plywood, 3/8" thick	.64	.49	1.13
2350	1/2" thick	.81	.50	1.31
2360	5/8" thick	.77	.52	1.29
2370	3/4" thick	.99	.56	1.55
2380	Particle board, 3/8" thick	.42	.49	.91
2390	1/2" thick	.44	.50	.94
2400	5/8" thick	.59	.52	1.11
2410	3/4" thick	.64	.56	1.20
2420	Hardboard, 4' x 4', .215" thick	.45	.49	.94

Figure 6.14

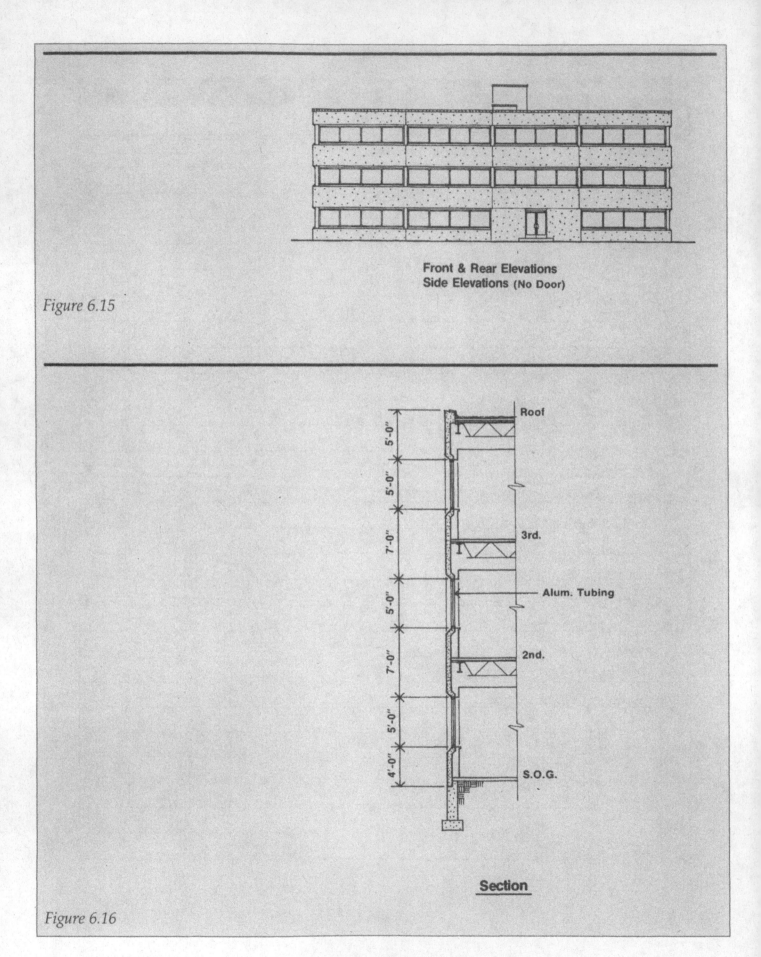

Front & Rear Elevations
Side Elevations (No Door)

Figure 6.15

Roof

5'-0"

5'-0"

7'-0"

3rd.

Alum. Tubing

5'-0"

7'-0"

2nd.

5'-0"

4'-0"

S.O.G.

Section

Figure 6.16

ASSEMBLY NUMBER	DESCRIPTION	QTY	UNIT	TOTAL COST		COST PER S.F.
				UNIT	TOTAL	
DIVISION B	SHELL					
B10	SUPERSTRUCTURE					
B1010-208-1600	Columns, 38K (50K), 25 x 9'	225	V.L.F.	$ 19.90	$ 4,478	
B1010-208-4600	Columns, 191K (200K), 25 x 27.33'	683	V.L.F.	$ 43.45	$ 29,687	
B1010-212-3450	Column Fireproofing, 8" Average, Gyp. Board	859	L.F.	$ 16.45	$ 14,131	
B1010-250-5100	Floor Framing	20,000	S.F.	$ 10.02	$ 200,400	
B1020-112-3300	Roof Framing	10,000	S.F.	$ 4.52	$ 45,200	
	Subtotal, Division B10, Superstructure				$ 293,896	$ 9.80
B20	EXTERIOR CLOSURE					
B2010-109-3310	Insulated CMU at Penthouse, 6" CMU	493	S.F.	$ 6.78	$ 3,343	
B2010-103-5750	Wall Panels, Precast concrete, (170 S.F. Ea.)	9,338	S.F.	$ 14.99	$ 139,977	
B2020-118-2000	Windows, Alum. Flush Tube w/Thermal Break	5,750	S.F.	$ 21.25	$ 122,188	
B2020-120-1000	Glass, Insulating Type	5,750	S.F.	$ 14.20	$ 81,650	
B2030-105-6350	Doors, Aluminum and Glass, 6' x 7'	2	OPNG.	$ 3,025.00	$ 6,050	
C3020-410-1880	Window Sills, Precast Terrazzo	771	S.F.	$ 13.04	$ 10,054	
	Subtotal, Division B20, Exterior Closure				$ 363,262	$ 12.11
B30	ROOFING					

Figure 6.17

Chapter 7

Roofing

Like exterior closure systems, the roof is a critical building element because it provides protection from the weather and insulation from extremes of heat and cold. In comparison to exterior walls, the cost of a roof is relatively low, but roofing systems can have a large cost impact in terms of energy usage and maintenance and repair over time. Any roof defects, such as leaks, can also result in expensive interior damage. The relationship between first, or new construction costs, quality, anticipated energy costs, and maintenance and repair costs over time are factors to be carefully investigated and weighed in the selection process. The value of roof material and installation warranties is significant.

Selecting a Roof Assembly

The type of roofing selected for a project often depends on the knowledge and experience of the designer, as well as the preference of the owner. Many factors govern the choice, including building height, roof slope and weight, weather conditions, wind load, energy code considerations, local fire codes, aesthetics, and architectural or environmental guidelines. Cost usually does not dictate the selection of one roof assembly over another. The roof selected must demonstrate performance in the area where the project will be constructed. A qualified roofing contractor will also be needed to install and, when necessary, make any future repairs.

There are four basic groups of roof covering assemblies:

- Built-up
- Single-ply membranes
- Metal
- Shingles and tiles

The roofing assembly should last 20 or more years, provided it is properly specified, detailed, installed, and cared for according to the manufacturer's guidelines. Assemblies often consist of many items from different manufacturers—items that must be put together to form either a waterproofing membrane or a water shed. Other related items are required for the success of the roofing assembly, including flashing, roof deck insulation, hatches, skylights, gutters, and downspouts.

Built-up Roofing Assemblies

Built-up roofs, one of the oldest types of modern roofing, comprise three different elements: felt, bitumen, and surfacing. The *felts*, which were traditionally organic (rag/paper) but now are almost exclusively inorganic (glass), work much like reinforcing steel in concrete. Felts are necessary as tensile reinforcement to resist the extreme pulling forces on top of buildings. Installed in layers, they also allow more bitumen to be applied. *Bitumen*, either coal-tar pitch or asphalt, is the "glue" that holds the felts together and acts as a waterproofing material in the roofing assembly.

The *surfacings* normally applied to built-up roofs are smooth gravel or slag, mineral granules, or a mineral-coated cap sheet. Gravel, slag, and mineral granules may be embedded in the still-fluid flood coat. Gravel and slag serve as excellent wearing surfaces to protect the membrane from mechanical damage. In some systems, a mineral-coated cap sheet is applied on top of the plies of felt and interply mappings of bitumen. This material is nothing more than a thicker or heavier ply of felt with a mineral granule surface hot-mopped into place.

The most common built-up assemblies contain between three to five felt plies with either asphalt bitumen or coal-tar pitch. Almost all assemblies are available for application to either nailable or non-nailable decks, and may be applied to rigid deck insulation or directly to the structural deck.

See Figure 7.1 for a drawing of a typical built-up roof.

Single-Ply Roofing Assemblies

Since the early 1970s, the use of single-ply roofing membranes in the construction industry has been on the rise. Market surveys show that single-ply assemblies currently account for about 55% of commercial roofing systems. Single-ply roofing is designed to reduce the time and difficulty of installing a waterproofing roof membrane. Single-ply

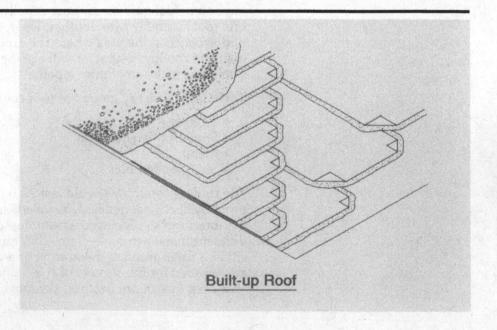

Built-up Roof

Figure 7.1

membranes have proven their durability and cost effectiveness and do not have the problems of safety and environmental hazards associated with the installation of built-up roofs. Re-roofing represents the largest market today for single-ply roofing because of its ease of application.

Single-Ply Roofing Materials

Materials for single-ply assemblies are more expensive than other, more conventional roofing systems. However, labor costs tend to be lower for single ply because the membrane sheets are pre-fabricated, which makes installation less time-consuming. There are three common methods of installation: fully adhered, mechanically fastened, and loosely laid with gravel or stone ballast. Labor costs vary with the method of installation, with loosely laid and ballasted the least costly, and fully adhered the most expensive.

One advantage of single-ply roofing materials is that the assemblies can be installed in the way that best suits a particular project. If the structure cannot take the additional weight of stone ballast (sometimes as much as 12 pounds per square foot), the membrane can be fully adhered to the roof deck or insulation. This method is acceptable, provided that the membrane material is treated to withstand ozone and ultraviolet degradation.

Single-Ply Roofing Types

There are three categories of single-ply roofing assemblies:

- Thermoset membranes
- Thermoplastic membranes
- Composite, or modified bitumen membranes

Figure 7.2 compares the three types and includes information on compatible substrates and attachment and sealing methods. Each of the materials and assemblies has unique requirements, performance characteristics, and warranties. The industry is expanding rapidly, making it a challenge to keep up with the changes.

Single-Ply Roofing Installation

The installation methods for single-ply assemblies generally follow the guidelines below.

- The *loose-laid and ballasted* method is generally the easiest and quickest to install. This system consists of large sheets held in place with stone or paver ballasts, and attached at the roof's perimeter. Special consideration must be given where flashings are attached. The membrane is stretched out flat, fused or glued together at the side and end laps, and then fully ballasted with 3/4" to 1-1/2" stones (10-12 pounds per square foot) to prevent wind blow-off. This extra dead load must be considered during initial structural design stages, and is particularly important when re-roofing over an existing built-up roof that already weighs 10-15 pounds per square foot. A slip sheet or vapor retarder is sometimes required to separate the new roofing membrane from the old. Ballasted systems are usually not recommended for roofs with slopes greater than 2 inches in 12 inches.
- The *mechanically fastened* method uses a series of bar or point attachments that mechanically attach the membrane to the substrate. The membrane manufacturer typically specifies the

method of attachment based on the materials used and the substrate. Partially adhered systems do not require ballast material. A slip sheet may be required for some materials. This method can be used for roofs with steep slopes, and is generally lightweight.

- The *fully adhered* method is by far the most time-consuming and costly installation because it requires contact cement, cold adhesive, or hot bitumen to uniformly adhere the membrane to the substrate. Only manufacturer-approved roof deck insulation board should be used to receive the membrane. A slip sheet may be required to separate the new and old materials.

Metal Roofing Assemblies

Metal roofing materials can be divided into two groups: preformed and formed metal. *Preformed metal* roofs, available in sheets of varying lengths and widths, are most often constructed from aluminum, asphalt, fibrous glass, colored metal, or galvanized steel. They are usually nailed or screwed to the supporting members. Preformed metal roofing is relatively economical, but can be used only where positive drainage is provided because the material, with its mechanical joints, is not watertight.

Formed metal roofing, often selected for aesthetic reasons, is practical on sloped roofs from 1/4" per foot to vertical. It is installed over base materials such as plywood or concrete, and can be joined by either

Single-Ply Roofing Membrane Installation Guide

Generic Materials (Classification)		Compatible Substrates						Attachment Method				Sealing Method				
		Slip-sheet req'd.	Concrete	Exist. asphalt memb.	Insulation board	Plywood	Spray urethane foam	Adhesive	Fully adhered	Loosely laid/ballast	Partially-adhered	Adhesive	Hot air gun	Self sealing	Solvent	Torch heating
Thermo Setting	EPDM (Ethylene, propylene)	●	●	●	●	●	●	●	●	●	●	●		●	●	
	Neoprene (Synthetic rubber)	●	●		●	●		●	●	●		●				
	PIB (Polyisobutylene)	●	●	●	●	●	●			●		●	●		●	
Thermo Plastic	CSPE (Chlorosulfenated polyethylene)	●	●		●	●	●	●	●	●	●	●	●			
	CPE (Chlorinated polyethylene)	●	●		●	●			●	●	●	●	●			
	PVC (Polyvinyl chloride)	●	●		●	●	●		●	●		●			●	
Composites	Glass reinforced EPDM/neoprene	●	●		●	●			●			●				
	Modified bitumen/polyester	●		●	●	●			●			●	●			●
	Modified bitumen/polyethylene & aluminum	●	●		●	●			●			●	●			●
	Modified bitumen/polyethylene sheet	●	●		●	●				●			●			●
	Modified CPE				●	●			●							
	Non-woven glass reinforced PVC							●	●			●				
	Nylon reinforced PVC		●		●	●			●				●		●	
	Nylon reinforced/butyl or neoprene	●							●				●		●	
	Polyester reinforced CPE	●	●	●	●	●	●		●	●	●	●	●		●	
	Polyester reinforced PVC	●	●		●	●	●		●	●		●	●		●	
	Rubber asphalt/plastic sheet	●	●	●	●	●			●					●		

Figure 7.2

batten, flat, or standing seam. The metals most often used are copper, lead, and zinc copper alloy.

Shingles and Tiles

Shingles and tiles have long been used as roofing materials on both residential and commercial buildings. They are considered watershed materials because they direct water away from the building. They require a 3" or more slope per foot to perform properly, as they are not fully waterproof.

Included in the category of shingles are composition asphalt strip shingles, wood shingles and shakes, aluminum and steel shingles, and slate shingles. Tiles include concrete and clay materials, as well as aluminum.

The most significant design considerations are slope and weight. Both shingles and tiles require a minimum slope since they serve only as a watershed. The heavier materials—concrete, clay, and slate—require structural systems capable of carrying their large dead loads.

Calculation of Roof Area

With a sloping roof, the dimensions from ridge to eave, as well as the eave and ridge lengths, must be determined before the actual roof area can be calculated. When the plan dimensions of the roof are known but the sloping dimensions are not, the actual roof area can be estimated if the approximate slope of the roof is known. See Figure 7.3. The multipliers shown may be used to convert slope to linear feet in the horizontal plan.

Figures 7.4 and 7.5 illustrate some examples using the information in Figure 7.3.

Flashing

Flashing and counterflashing are necessary to prevent water from entering into the roofing system via roof discontinuities, such as intersecting roof lines, skylights, chimneys, hatches, and vent pipes.

Factors for Converting Inclined to Horizontal

Roof Slope	Approx. Angle	Factor	Roof Slope	Approx. Angle	Factor
Flat	0°	1.000	12 in 12	45.0	1.414
1 in 12	4.8°	1.003	13 in 12	47.3	1.474
2 in 12	9.5°	1.014	14 in 12	49.4	1.537
3 in 12	14.0°	1.031	15 in 12	51.3	1.601
4 in 12	18.4°	1.054	16 in 12	53.1	1.667
5 in 12	22.6°	1.083	17 in 12	54.8	1.734
6 in 12	26.6°	1.118	18 in 12	56.3	1.803
7 in 12	30.3°	1.158	19 in 12	57.7	1.873
8 in 12	33.7°	1.202	20 in 12	59.0	1.943
9 in 12	36.9°	1.250	21 in 12	60.3	2.015
10 in 12	39.8°	1.302	22 in 12	61.4	2.088
11 in 12	42.5°	1.357	23 in 12	62.4	2.162

Figure 7.3

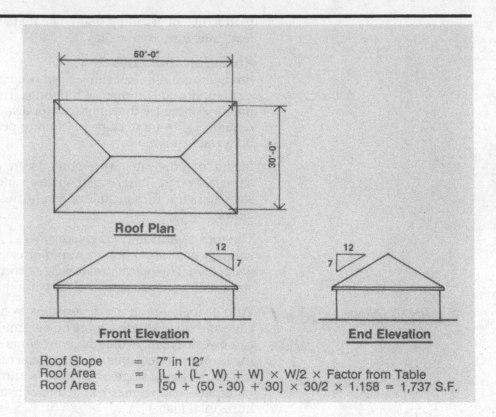

Roof Plan

Front Elevation   **End Elevation**

Roof Slope = 7" in 12"
Roof Area = [L + (L - W) + W] × W/2 × Factor from Table
Roof Area = [50 + (50 - 30) + 30] × 30/2 × 1.158 = 1,737 S.F.

Figure 7.4

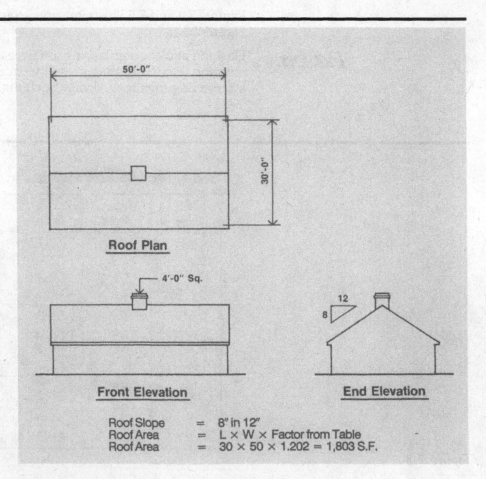

Roof Plan

4'-0" Sq.

Front Elevation   **End Elevation**

Roof Slope = 8" in 12"
Roof Area = L × W × Factor from Table
Roof Area = 30 × 50 × 1.202 = 1,803 S.F.

Figure 7.5

Flashing materials may be preformed or formed on the job to satisfy particular requirements and details. Concealed flashing consists of either sheet metal or membrane material. Exposed flashing materials include:

- Aluminum
- Copper
- Lead coated copper
- Lead
- Polyvinyl chloride
- Butyl rubber
- Neoprene
- Copper-clad stainless steel
- Stainless steel
- Galvanized metal

It is important to be aware of certain design precautions. Check to see that the flashing materials are compatible with the roof selected, and that the details are properly designed for the roofing assembly.

Roof Deck Insulation

Insulation is used to reduce heat transfer through the exterior enclosure of the building. The type and form of insulation varies according to where it is located in the structure. Major insulation types include mineral granules, glass fibers, plastic foams, organic vegetable fibers, solids, and composites of all of these.

Insulation is usually specified by "R" value per inch of thickness (see discussion in Chapter 6). Labor costs vary from one insulation material to another because of fragility, weight, and bulkiness of the materials. Since the insulation used will affect the long-term energy costs of the building, use the most economical material that meets the desired insulation capability and is consistent with the requirements of the roofing system. Wood blocking and nailers that receive fasteners must match the insulation thickness on roof decks. As a result, a thick, inexpensive insulation board may far exceed the cost of a thinner, more expensive one when the additional thickness of the wood blocking and nailer is considered.

Conclusion

In making a cost-conscious selection of roof covering design, it is important to consider the total impact of each variable. Single-ply assemblies cost more for materials than built-up roofing, but are less expensive to install. Ballast for roof membranes adds weight to the roofing system, which requires more structural support—an additional expense to be considered when evaluating the total cost of the system. Initial costs including installation cost are not always the most important consideration in the selection of the roofing system. Life cycle costs may be more of a consideration to the roofing system than to any other component of the building. Higher initial costs can be offset by savings in energy, repair, and re-roofing costs over the life of the building. Roofing considerations frequently overlooked in preliminary estimates include:

- Flashing around mechanical equipment
- Insulation
- Hatches for access
- Skylights
- Gutters

- Downspouts
- Impact on storm drain systems
- Blocking
- Gravel stops

In a single-story building, the roof system, insulation, and roof covering costs have a significant impact on the total cost. They may represent between 2 and 7.5% of the total cost of construction. The time spent considering alternative roofing systems during preliminary design is worthwhile.

Assemblies Estimate: Three-Story Office Building

The office building (Figure 7.6) will have the following included in the roofing system:

- Built-up roof
- Roof deck insulation
- Aluminum flashing

Costs:

Built-up, 3-ply with gravel on non-nailable deck/insulation (Figure 7.7, line #1900)	10,000 S.F.	@ $1.93/S.F.	=	$19,300
Roof Deck Insulation, Composite, 1-1/2" polyisocyanurate, 1" perlite (Figure 7.8, line #4020)	10,000 S.F.	@ $1.36/S.F.	=	13,600
Aluminum Flashing (Figure 7.9, line #0350)	1040 S.F.	@ $2.46/S.F.	=	2,558
Aluminum Gravel Stop (Figure 7.10, line #5200)	92 L.F.	@ $6.92/L.F.	=	$637
Total Cost				$36,095
Cost per Square Foot				$1.20/S.F.

Figure 7.11 summarizes the costs for roofing.

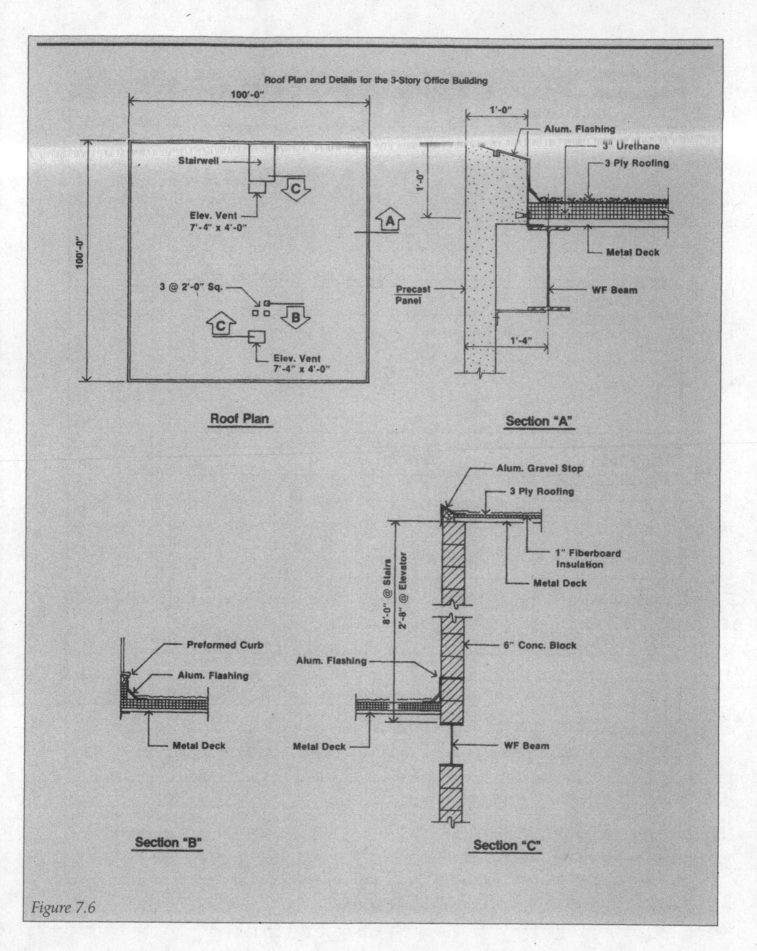

Roof Plan and Details for the 3-Story Office Building

Roof Plan

100'-0"

100'-0"

Stairwell

Elev. Vent
7'-4" x 4'-0"

3 @ 2'-0" Sq.

Elev. Vent
7'-4" x 4'-0"

Section "A"

1'-0"

1'-0"

Alum. Flashing

3" Urethane

3 Ply Roofing

Metal Deck

WF Beam

Precast Panel

1'-4"

Section "B"

Preformed Curb

Alum. Flashing

Metal Deck

Section "C"

Alum. Gravel Stop

3 Ply Roofing

1" Fiberboard Insulation

Metal Deck

8'-0" @ Stairs
2'-8" @ Elevator

Alum. Flashing

6" Conc. Block

Metal Deck

WF Beam

Figure 7.6

B3010 Roof Coverings

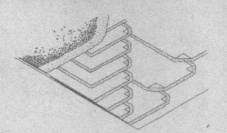

Multiple ply roofing is the most popular covering for minimum pitch roofs. Lines 1200 through 6300 list the costs of the various types, plies and weights per S.F.

System Components	QUANTITY	UNIT	COST PER S.F.		
			MAT.	INST.	TOTAL
SYSTEM B3010 105					
ASPHALT FLOOD COAT, W/GRAVEL, 4 PLY ORGANIC FELT					
Organic #30 base felt	1.000	S.F.	.05	.06	.11
Organic #15 felt, 3 plies	3.000	S.F.	.07	.18	.25
Asphalt mopping of felts	4.000	S.F.	.18	.56	.74
Asphalt flood coat	1.000	S.F.	.11	.45	.56
Gravel aggregate, washed river stone	4.000	Lb.	.05	.10	.15
TOTAL			.46	1.35	1.81

B3010 105	Built-Up	COST PER S.F.		
		MAT.	INST.	TOTAL
1200	Asphalt flood coat w/gravel; not incl. insul, flash., nailers			
1300				
1400	Asphalt base sheets & 3 plies #15 asphalt felt, mopped	.45	1.22	1.67
1500	On nailable deck	.49	1.27	1.76
1600	4 plies #15 asphalt felt, mopped	.63	1.34	1.97
1700	On nailable deck	.57	1.41	1.98
1800	Coated glass base sheet, 2 plies glass (type IV), mopped	.50	1.22	1.72
1900	For 3 plies	.59	1.34	1.93
2000	On nailable deck	.55	1.41	1.96
2300	4 plies glass fiber felt (type IV), mopped	.71	1.34	2.05
2400	On nailable deck	.64	1.41	2.05
2500	Organic base sheet & 3 plies #15 organic felt, mopped	.46	1.35	1.81
2600	On nailable deck	.49	1.41	1.90
2700	4 plies #15 organic felt, mopped	.60	1.22	1.82
2750				
2800	Asphalt flood coat, smooth surface			
2850				
2900	Asphalt base sheet & 3 plies #15 asphalt felt, mopped	.47	1.12	1.59
3000	On nailable deck	.44	1.17	1.61
3100	Coated glass fiber base sheet & 2 plies glass fiber felt, mopped	.44	1.08	1.52
3200	On nailable deck	.42	1.12	1.54
3300	For 3 plies, mopped	.53	1.17	1.70
3400	On nailable deck	.50	1.22	1.72
3700	4 plies glass fiber felt (type IV), mopped	.62	1.17	1.79
3800	On nailable deck	.59	1.22	1.81
3900	Organic base sheet & 3 plies #15 organic felt, mopped	.47	1.12	1.59
4000	On nailable decks	.43	1.17	1.60
4100	4 plies #15 organic felt, mopped	.54	1.22	1.76
4200	Coal tar pitch with gravel surfacing			
4300	4 plies #15 tarred felt, mopped	1.11	1.27	2.38
4400	3 plies glass fiber felt (type IV), mopped	.91	1.41	2.32
4500	Coated glass fiber base sheets 2 plies glass fiber felt, mopped	.92	1.41	2.33

Figure 7.7

B3010 Roof Coverings

B3010 320	Roof Deck Rigid Insulation	COST PER S.F.		
		MAT.	INST.	TOTAL
0100	Fiberboard low density, 1/2" thick, R1.39			
0150	1" thick R2.78	.36	.44	.80
0300	1 1/2" thick R4.17	.54	.44	.98
0350	2" thick R5.56	.73	.44	1.17
0370	Fiberboard high density, 1/2" thick R1.3	.21	.35	.56
0380	1" thick R2.5	.39	.44	.83
0390	1 1/2" thick R3.8	.63	.44	1.07
0410	Fiberglass, 3/4" thick R2.78	.50	.35	.85
0450	15/16" thick R3.70	.65	.35	1
0500	1-1/16" thick R4.17	.81	.35	1.16
0550	1-5/16" thick R5.26	1.12	.35	1.47
0600	2-1/16" thick R8.33	1.20	.44	1.64
0650	2 7/16" thick R10	1.36	.44	1.80
1000	Foamglass, 1 1/2" thick R4.55	1.56	.44	2
1100	3" thick R9.09	3.23	.50	3.73
1200	Tapered for drainage	1	.59	1.59
1260	Perlite, 1/2" thick R1.32	.29	.34	.63
1300	3/4" thick R2.08	.35	.44	.79
1350	1" thick R2.78	.30	.44	.74
1400	1 1/2" thick R4.17	.42	.44	.86
1450	2" thick R5.56	.59	.50	1.09
1510	Polyisocyanurate 2#/CF density, 1" thick R7.14	.34	.25	.59
1550	1 1/2" thick R10.87	.37	.28	.65
1600	2" thick R14.29	.48	.32	.80
1650	2 1/2" thick R16.67	.54	.34	.88
1700	3" thick R21.74	.76	.35	1.11
1750	3 1/2" thick R25	.79	.35	1.14
1800	Tapered for drainage	.41	.25	.66
1810	Expanded polystyrene, 1#/CF density, 3/4" thick R2.89	.20	.24	.44
1820	2" thick R7.69	.37	.28	.65
1825	Extruded Polystyrene			
1830	Extruded polystyrene, 15 PSI compressive strength, 1" thick R5	.24	.24	.48
1835	2" thick R10	.37	.28	.65
1840	3" thick R15	.83	.35	1.18
1900	25 PSI compressive strength, 1" thick R5	.39	.24	.63
1950	2" thick R10	.76	.28	1.04
2000	3" thick R15	1.14	.35	1.49
2050	4" thick R20	1.19	.35	1.54
2150	Tapered for drainage	.43	.24	.67
2550	40 PSI compressive strength, 1" thick R5	.37	.24	.61
2600	2" thick R10	.74	.28	1.02
2650	3" thick R15	1.08	.35	1.43
2700	4" thick R20	1.44	.35	1.79
2750	Tapered for drainage	.54	.25	.79
2810	60 PSI compressive strength, 1" thick R5	.45	.24	.69
2850	2" thick R10	.80	.29	1.09
2900	Tapered for drainage	.65	.25	.90
4000	Composites with 1-1/2" polyisocyanurate			
4010	1" fiberboard	.90	.44	1.34
4020	1" perlite	.94	.42	1.36
4030	7/16" oriented strand board	1.08	.44	1.52

Figure 7.8

B3010 Roof Coverings

B3010 430			Flashing				

	MATERIAL	BACKING	SIDES	SPECIFICATION	QUANTITY	COST PER S.F.		
						MAT.	INST.	TOTAL
0040	Aluminum	none		.019"		.85	2.86	3.71
0050				.032"		1.14	2.86	4
0100				.040"		1.60	2.86	4.46
0150				.050"		2.02	2.86	4.88
0300		fabric	2	.004"		1.02	1.26	2.28
0350				.016"		1.20	1.26	2.46
0400		mastic		.004"		1.02	1.26	2.28
0450				.005"		1.20	1.26	2.46
0500				.016"		1.32	1.26	2.58
0510								
0700	Copper	none		16 oz.	<500 lbs.	2.40	3.60	6
0710					>2000 lbs.	3.20	2.67	5.87
0750				20 oz.	<500 lbs.	4.29	3.77	8.06
0760					>2000 lbs.	3.98	2.86	6.84
0800				24 oz.	<500 lbs.	5.15	3.95	9.10
0810					>2000 lbs.	4.77	3.07	7.84
0850				32 oz.	<500 lbs.	6.85	4.14	10.99
0860					>2000 lbs.	6.40	3.19	9.59
1000		paper backed	1	2 oz.		.95	1.26	2.21
1100				3 oz.		1.23	1.26	2.49
1200			2	2 oz.		.95	1.26	2.21
1300				5 oz.		1.84	1.26	3.10
2000	Copper lead	fabric	1	2 oz.		1.55	1.26	2.81
2100	Coated			5 oz.		1.78	1.26	3.04
2200		mastic	2	2 oz.		1.21	1.26	2.47
2300				5 oz.		1.50	1.26	2.76
2400		paper	1	2 oz.		.95	1.26	2.21
2500				3 oz.		1.23	1.26	2.49
2600			2	2 oz.		.95	1.26	2.21
2700				5 oz.		1.84	1.26	3.10
3000	Lead	none		2.5 lb.	to 12" wide	3.08	2.62	5.70
3100					over 12"	3.58	2.62	6.20
3500	PVC black	none		.010"		.15	1.24	1.39
3600				.020"		.24	1.24	1.48
3700				.030"		.32	1.24	1.56
3800				.056"		.76	1.24	2
4000	Rubber butyl	none		1/32"		.77	1.24	2.01
4100				1/16"		1.16	1.24	2.40
4200	Neoprene			1/16"		1.63	1.24	2.87
4300				1/8"		3.29	1.24	4.53
4500	Stainless steel	none		.015"	<500 lbs.	3.48	3.60	7.08
4600	Copper clad				>2000 lbs.	3.36	2.67	6.03
4700				.018"	<500 lbs.	4.60	4.14	8.74
4800					>2000 lbs.	3.37	2.86	6.23
5000	Plain			32 ga.		2.38	2.67	5.05
5100				28 ga.		2.81	2.67	5.48
5200				26 ga.		3.58	2.67	6.25
5300				24 ga.		4.64	2.67	7.31
5400	Terne coated			28 ga.		4.46	2.67	7.13
5500				26 ga.		5.05	2.67	7.72

Figure 7.9

B3010 Roof Coverings

B3010 630 — Downspouts

	MATERIALS	SECTION	SIZE	FINISH	THICKNESS	COST PER V.L.F. MAT.	INST.	TOTAL
0100	Aluminum	rectangular	2"x3"	embossed mill	.020"	.81	2.18	2.99
0150				enameled	.020"	.99	2.18	3.17
0200				enameled	.024"	1.20	2.30	3.50
0250			3"x4"	enameled	.024"	1.73	2.96	4.69
0300		round corrugated	3"	enameled	.020"	.87	2.18	3.05
0350			4"	enameled	.025"	1.66	2.96	4.62
0500	Copper	rectangular corr.	2"x3"	mill	16 Oz.	3.50	2.18	5.68
0550				lead coated	16 Oz.	5.35	2.18	7.53
0600		smooth		mill	16 Oz.	5.25	2.18	7.43
0650				lead coated	16 Oz.	7.80	2.18	9.98
0700		rectangular corr.	3"x4"	mill	16 Oz.	5.20	2.86	8.06
0750				lead coated	16 Oz.	7.30	2.48	9.78
0800		smooth		mill	16 Oz.	6.90	2.86	9.76
0850				lead coated	16 Oz.	8.45	2.86	11.31
0900		round corrugated	2"	mill	16 Oz.	4.66	2.25	6.91
0950		smooth		lead coated	16 Oz.	6.20	2.18	8.38
1000		corrugated	3"	mill	16 Oz.	4.45	2.26	6.71
1050		smooth		lead coated	16 Oz.	5.35	2.18	7.53
1100		corrugated	4"	mill	16 Oz.	5.35	2.97	8.32
1150		smooth		lead coated	16 Oz.	7.15	2.86	10.01
1200		corrugated	5"	mill	16 Oz.	7.50	3.33	10.83
1250				lead coated	16 Oz.	8.90	3.19	12.09
1300	Steel	rectangular corr.	2"x3"	galvanized	28 Ga.	.59	2.18	2.77
1350				epoxy coated	24 Ga.	1.11	2.18	3.29
1400		smooth		galvanized	28 Ga.	.69	2.18	2.87
1450		rectangular corr.	3"x4"	galvanized	28 Ga.	1.65	2.86	4.51
1500				epoxy coated	24 Ga.	1.84	2.86	4.70
1550		smooth		galvanized	28 Ga.	1.35	2.86	4.21
1600		round corrugated	2"	galvanized	28 Ga.	.80	2.18	2.98
1650			3"	galvanized	28 Ga.	.80	2.18	2.98
1700			4"	galvanized	28 Ga.	1.18	2.86	4.04
1750			5"	galvanized	28 Ga.	1.54	3.19	4.73
1800				galvanized	26 Ga.	1.50	3.19	4.69
1850			6"	galvanized	28 Ga.	2.13	3.95	6.08
1900				galvanized	26 Ga.	1.56	3.95	5.51
2000	Steel pipe	round	4"	black	X.H.	4.93	20.50	25.43
2050			6"	black	X.H.	8.35	23	31.35
2500	Stainless steel	rectangular	2"x3"	mill		17.60	2.18	19.78
2550	Tubing sch.5		3"x4"	mill		22.50	2.86	25.36
2600			4"x5"	mill		46	3.07	49.07
2650		round	3"	mill		17.60	2.18	19.78
2700			4"	mill		22.50	2.86	25.36
2750			5"	mill		46	3.07	49.07

B3010 630 — Gravel Stop

	MATERIALS	SECTION	SIZE	FINISH	THICKNESS	COST PER L.F. MAT.	INST.	TOTAL
5100	Aluminum	extruded	4"	mill	.050"	3.54	2.86	6.40
5200			4"	duranodic	.050"	4.06	2.86	6.92
5300			8"	mill	.050"	4.94	3.32	8.26
5400			8"	duranodic	.050"	5.50	3.32	8.82
5500			12"-2 pc.	mill	.050"	6.80	4.14	10.94
6000			12"-2 pc.	duranodic	.050"	6.90	4.14	11.04

Figure 7.10

ASSEMBLY NUMBER	DESCRIPTION	QTY	UNIT	TOTAL COST		COST PER S.F.
				UNIT	TOTAL	
DIVISION B	SHELL					
B10	SUPERSTRUCTURE					
B1010-208-1600	Columns, 38K (50K), 25 x 9'	225	V.L.F.	$ 19.90	$ 4,478	
B1010-208-4600	Columns, 191K (200K), 25 x 27.33'	683	V.L.F.	$ 43.45	$ 29,687	
B1010-212-3450	Column Fireproofing, 8" Average, Gyp. Board	859	L.F.	$ 16.45	$ 14,131	
B1010-250-5100	Floor Framing	20,000	S.F.	$ 10.02	$ 200,400	
B1020-112-3300	Roof Framing	10,000	S.F.	$ 4.52	$ 45,200	
	Subtotal, Division B10, Superstructure				$ 293,896	$ 9.80
B20	EXTERIOR CLOSURE					
B2010-109-3310	Insulated CMU at Penthouse, 6" CMU	493	S.F.	$ 6.78	$ 3,343	
B2010-103-5750	Wall Panels, Precast concrete, (170 S.F. Ea.)	9,338	S.F.	$ 14.99	$ 139,977	
B2020-118-2000	Windows, Alum. Flush Tube w/Thermal Break	5,750	S.F.	$ 21.25	$ 122,188	
B2020-120-1000	Glass, Insulating Type	5,750	S.F.	$ 14.20	$ 81,650	
B2030-105-6350	Doors, Aluminum & Glass, 6' x 7'	2	OPNG.	$ 3,025.00	$ 6,050	
C3020-410-1880	Window Sills, Precast Terrazzo	771	S.F.	$ 13.04	$ 10,054	
	Subtotal, Division B20, Exterior Closure				$ 363,262	$ 12.11
B30	ROOFING					
B3010-105-1900	Built-up, 3 ply w/gravel, Non-nailable Deck, Insulated	10,000	S.F.	$ 1.93	$ 19,300	
B3010-120-4020	Insulation, Composite w/1-1/2" Polyisocyanurate, 1" Perlite	10,000	S.F.	$ 1.36	$ 13,600	
B3010-430-0350	Aluminum Flashing	1,040	S.F.	$ 2.46	$ 2,558	
B3010-630-5200	Gravel Stop, Aluminum, 4"	92	L.F.	$ 6.92	$ 637	
	Subtotal, Division B30, Roofing				$ 36,095	$ 1.20

Figure 7.11

136

Chapter 8

Interiors

The interior finish includes all the elements that define the building's use and interior aesthetics. UNIFORMAT II Division C, Interiors, includes many different types of partitions, doors, ceilings, floor treatments, and interior finishes. Division C10, Interior Construction, includes partitions, interior doors, and fittings. C20 covers stair construction and finishes, and C30 includes wall, ceiling, and floor finishes. Since the cost of the interior finish in some building types can exceed 25% of total building cost, it is important to develop a careful estimate of this work.

While many of today's commercial buildings seem to have an abundance of various partitions, doors, and floor and wall finish treatments, much of this work is standard. For instance, a typical interior partition consists of 3-5/8" metal studs at 24" on center, acoustical insulation, and 5/8" fire-rated gypsum board on each side. Using an assembly for this wall system can save time in the estimating process without the loss of accuracy.

Partitions

Interior partitions are categorized as either fixed or movable. They serve to separate open spaces and provide privacy within buildings, and can also function as bearing, shaft, or fire walls.

Various types of partitions may be used in any building project, depending on several variables. Some types must bear the load of construction above, requiring structural analysis and selection of materials capable of carrying the load. Partitions may also be required to create a fire separation from one area of the building to another, or create a visual, physical, or sound barrier. The purpose of the partition will determine the selection of materials and construction techniques.

The types of fixed partitions most commonly found in construction today are:

- Concrete and/or masonry
- Drywall on wood or metal framing
- Plaster on masonry or lath (metal or gypsum)

The materials used to construct partitions include masonry brick and concrete block, structural clay facing tile, and gypsum and glass block.

Wood or steel studs with plaster, or gypsum board—and possibly insulation or acoustical batts—can also be used. The combinations of these materials are almost limitless and can impact costs and weights.

The length of interior partitions may be determined from a simple sketch, from the project's detailed drawings, or by approximation from the "Partition/Door Density" table in Figure 8.1. Use this table to calculate the average length of interior partitions based on the particular building square footage for different building types. The table also provides the average number of interior doors, based on the building square footage, necessary to service the partitions. To use this table, divide the total building area by the partition density (S.F./L.F.) to approximate the total length of partitions in the project. Then multiply the partition linear footage by the required partition height to find the total square footage of the partitions in the project. Multiply this figure by the cost per square foot from the selected assembly. (Do not deduct the area of openings under 4 S.F.)

Example One: Apartment Building versus Department Store

The following example compares the number of doors and linear feet of partitions typically required for a three-story apartment building to a one-story department store.

Apartment	3 Stories		30,000 S.F.
Partitions	9 S.F./L.F.	=	3,333 L.F. of partitions
Doors	90 S.F./door	=	333 doors
Department Store	1 Story		50,000 S.F.
Partitions	60 S.F./L.F.	=	833 L.F. of partitions
Doors	600 S.F./door	=	3 doors

These quantities are approximations and should be used only in the early stages of project development. If sketches or drawings are available, they should be consulted, and actual quantities could be used.

Example Two: Partition Weight

This example illustrates the cost and weight differences of various types of partitions. (*Note: The partitions shown are representative types and are not necessarily included in this book's estimate for the three-story office building, located at the end of this chapter.*)

Partition Component	Cost per S.F.	Partition Weight
6" Concrete Masonry Unit (C.M.U.)	$5.92	30 psf to 42 psf
6" C.M.U. with 5/8" gypsum plaster both sides	$10.25	40 psf to 52 psf
Structural clay facing tile glazed both sides (6")	$21.95	47 psf
2" × 4" studs @ 16" O.C. 5/8" fire-rated gypsum board both sides	$3.35	7 psf
2" × 4" studs @ 16" O.C. Gypsum plaster on gypsum lath both sides	$7.53	14 psf

Partition/Door Density

Building Type		Stories	Partition/Density	Doors	Description of Partition
Apartments		1 story	9 SF/LF	90 SF/door	Plaster, wood doors & trim
		2 story	8 SF/LF	80 SF/door	Drywall, wood studs, wood doors & trim
		3 story	9 SF/LF	90 SF/door	Plaster, wood studs, wood doors & trim
		5 story	9 SF/LF	90 SF/door	Plaster, ~~wood studs, wood doors & trim~~
		~~8 story~~	~~8 SF/LF~~	80 SF/door	Drywall, wood studs, wood doors & trim
Bakery		1 story	50 SF/LF	500 SF/door	Conc. block, paint, door & drywall, wood studs
		2 story	50 SF/LF	500 SF/door	Conc. block, paint, door & drywall, wood studs
Bank		1 story	20 SF/LF	200 SF/door	Plaster, wood studs, wood doors & trim
		24 story	15 SF/LF	150 SF/door	Plaster, wood studs, wood doors & trim
Bottling Plant		1 story	50 SF/LF	500 SF/door	Conc. block, drywall, wood studs, wood trim
Bowling Alley		1 story	50 SF/LF	500 SF/door	Conc. block, wood & metal doors, wood trim
Bus Terminal		1 story	15 SF/LF	150 SF/door	Conc. block, ceramic tile, wood trim
Cannery		1 story	100 SF/LF	1000 SF/door	Drywall on metal studs
Car Wash		1 story	18 SF/LF	180 SF/door	Concrete block, painted & hollow metal door
Dairy Plant		1 story	30 SF/LF	300 SF/door	Concrete block, glazed tile, insulated cooler doors
Department Store		1 story	60 SF/LF	600 SF/door	Drywall, wood studs, wood doors & trim
		25 story	60 SF/LF	600 SF/door	30% concrete block, 70% drywall, wood studs
Dormitory		2 story	9 SF/LF	90 SF/door	Plaster, concrete block, wood doors & trim
		35 story	9 SF/LF	90 SF/door	Plaster, concrete block, wood doors & trim
		6-15 story	9 SF/LF	90 SF/door	Plaster, concrete block, wood doors & trim
Funeral Home		1 story	15 SF/LF	150 SF/door	Plaster on concrete block & wood studs, paneling
		2 story	14 SF/LF	140 SF/door	Plaster, wood studs, paneling & wood doors
Garage Sales & Service		1 story	30 SF/LF	300 SF/door	50% conc. block, 50% drywall, wood studs
Hotel		38 story	9 SF/LF	90 SF/door	Plaster, conc. block, wood doors & trim
		9-15 story	9 SF/LF	90 SF/door	Plaster, conc. block, wood doors & trim
Laundromat		1 story	25 SF/LF	250 SF/door	Drywall, wood studs, wood doors & trim
Medical Clinic		1 story	6 SF/LF	60 SF/door	Drywall, wood studs, wood doors & trim
		24 story	6 SF/LF	60 SF/door	Drywall, wood studs, wood doors & trim
Motel		1 story	7 SF/LF	70 SF/door	Drywall, wood studs, wood doors & trim
		23 story	7 SF/LF	70 SF/door	Concrete block, drywall on wood studs, wood paneling
Movie Theater	200-600 seats	1 story	18 SF/LF	180 SF/door	Concrete block, wood, metal, vinyl trim
	601-1400 seats		20 SF/LF	200 SF/door	Concrete block, wood, metal, vinyl trim
	1401-22000 seats		25 SF/LF	250 SF/door	Concrete block, wood, metal, vinyl trim
Nursing Home		1 story	8 SF/LF	80 SF/door	Drywall, wood studs, wood doors & trim
		24 story	8 SF/LF	80 SF/door	Drywall, wood studs, wood doors & trim
Office		1 story	20 SF/LF	200-500 SF/door	30% concrete block, 70% drywall on wood studs
		2 story	20 SF/LF	200-500 SF/door	30% concrete block, 70% drywall on wood studs
		35 story	20 SF/LF	200-500 SF/door	30% concrete block, 70% movable partitions
		610 story	20 SF/LF	200-500 SF/door	30% concrete block, 70% movable partitions
		11-20 story	20 SF/LF	200-500 SF/door	30% concrete block, 70% movable partitions
Parking Ramp (Open)		28 story	60 SF/LF	600 SF/door	Stair and elevator enclosures only
Parking garage		28 story	60 SF/LF	600 SF/door	Stair and elevator enclosures only
Pre-Engineered	Steel	1 story	0		
	Store	1 story	60 SF/LF	600 SF/door	Drywall on wood studs, wood doors & trim
	Office	1 story	15 SF/LF	150 SF/door	Concrete block, movable wood partitions
	Shop	1 story	15 SF/LF	150 SF/door	Movable wood partitions
	Warehouse	1 story	0		
Radio & TV Broadcasting		1 story	25 SF/LF	250 SF/door	Concrete block, metal and wood doors
& TV Transmitter		1 story	40 SF/LF	400 SF/door	Concrete block, metal and wood doors
Self Service Restaurant		1 story	15 SF/LF	150 SF/door	Concrete block, wood and aluminum trim
Cafe & Drive-in Restaurant		1 story	18 SF/LF	180 SF/door	Drywall, wood studs, ceramic & plastic trim
Restaurant with seating		1 story	25 SF/LF	250 SF/door	Concrete block, paneling, wood studs & trim
Supper Club		1 story	25 SF/LF	250 SF/door	Concrete block, paneling, wood studs & trim
Bar or Lounge		1 story	24 SF/LF	240 SF/door	Plaster or gypsum lath, wooded studs
Retail Store or Shop		1 story	60 SF/LF	600 SF/door	Drywall wood studs, wood doors & trim
Service Station	Masonry	1 story	15 SF/LF	150 SF/door	Concrete block, paint, door & drywall, wood studs
	Metal panel	1 story	15 SF/LF	150 SF/door	Concrete block, paint, door & drywall, wood studs
	Frame	1 story	15 SF/LF	150 SF/door	Drywall, wood studs, wood doors & trim
Shopping Center	(strip)	1 story	30 SF/LF	300 SF/door	Drywall, wood studs, wood doors & trim
	(group)	1 story	40 SF/LF	400 SF/door	50% concrete block, 50% drywall, wood studs
		2 story	40 SF/LF	400 SF/door	50% concrete block, 50% drywall, wood studs
Small Food Store		1 story	30 SF/LF	300 SF/door	Concrete block drywall, wood studs, wood trim
Store/Apt. above	Masonry	2 story	10 SF/LF	100 SF/door	Plaster, wood studs, wood doors & trim
	Frame	2 story	10 SF/LF	100 SF/door	Plaster, wood studs, wood doors & trim
	Frame	3 story	10 SF/LF	100 SF/door	Plaster, wood studs, wood doors & trim
Supermarkets		1 story	40 SF/LF	400 SF/door	Concrete block, paint, drywall & porcelain panel
Truck Terminal		1 story	0		
Warehouse		1 story	0		

Figure 8.1

3-5/8", 18 gauge steel studs & track @ 24" O.C.		
5/8" gypsum board both sides	$ 4.17	7 psf
6", 16 gauge steel studs & track @ 16" O.C.		
3 coats gypsum plaster on metal lath	$ 8.88	18 psf

Toilet Partitions

Quality, ease of installation, and maintenance are some of the biggest factors used to select and price toilet partitions. Floor-mounted, painted metal and plastic laminate units are the most economical, whereas ceiling-hung units of the same materials are more costly to install. Wall-hung units cost more to fabricate, thus the material price is higher, but the installation cost is about the same as for floor-mounted units. For buildings requiring handicapped accessibility, do not forget the additional partition accessories, such as handrails, and associated blocking required in the walls.

Stairs

Using the assemblies method of estimating to determine the cost of stairs can save time, which can be applied to other aspects of the project that may have a greater cost impact. The following sequence will help simplify the estimating process for the cost of stairs:

- Decide on the materials and select a description using Figure 8.2.
- Determine whether the stairs will be straight or have a landing.
- Determine the number of risers (from calculations or by using Figure 8.3).
- Select the cost per flight using Figure 8.2. Multiply this figure by the required number of flights.

Remember that most multi-story buildings with stairs will include a flight that leads up to the roof. When selecting stairs, keep similar materials together. For example, use concrete stairs in concrete-framed buildings and steel pan stairs in steel-framed buildings.

Wall and Floor Finishes

Interior finish is difficult to estimate at the conceptual phase of a project. The total cost will be influenced by the quality and type of materials selected. Determine the specific types and/or level of quality of finish materials planned for the building, and estimate accordingly. For instance, the installed cost of a medium-duty, commercial, nylon-level loop carpet is approximately $26.00 per square yard. The cost of medium-traffic, wool carpet is about $75.00 per square yard. Similarly, painting a gypsum board partition may cost $1.00 per square foot. An average quality vinyl wall covering would cost approximately $1.45 per square foot, and fabric may cost $5.50 per square foot. This wide range of costs follows for nearly every type of interior finish material. If the level of quality is unknown, the estimate for wall, floor, and ceiling finishes could be off by 300% or even more.

In assemblies estimating, time can be saved by determining a percentage of the total area of the different wall and floor finishes used. Be sure to include the interior side of exterior walls. It is not necessary to know the exact square foot area of each type of finish. For instance, if a building has 30,000 square feet, and 75% of the area is carpeted, 5% is quarry tile, and 20% is ceramic tile, it is not difficult to determine the square foot area and associated costs of each floor finish. The same technique can be used for wall and ceiling finishes. For simplicity,

C2010 Stair Construction

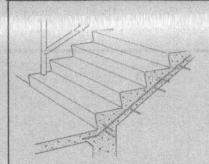

General Design: See reference section for code requirements. Maximum height between landings is 12'; usual stair angle is 20° to 50° with 30° to 35° best. Usual relation of riser to treads is:

    Riser + tread = 17.5.
    2x (Riser) + tread = 25.
    Riser x tread = 70 or 75.

Maximum riser height is 7" for commercial, 8-1/4" for residential.
Usual riser height is 6-1/2" to 7-1/4".

Minimum tread width is 11" for commercial and 9" for residential.

For additional information please see reference section.

Cost Per Flight: Table below lists the cost per flight for 4'-0" wide stairs. Side walls are not included. Railings are included.

System Components	QUANTITY	UNIT	COST PER FLIGHT		
			MAT.	INST.	TOTAL
SYSTEM C2010 100					
STAIRS, C.I.P. CONCRETE WITH LANDING, 12 RISERS					
Concrete in place, free standing stairs not incl. safety treads	48.000	L.F.	321.60	1,296	1,617.60
Concrete in place, free standing stair landing	32.000	S.F.	82.88	355.52	438.40
Stair tread C.I. abrasive 4" wide	48.000	L.F.	369.60	218.88	588.48
Industrial railing, welded, 2 rail 3'-6" high 1-1/2" pipe	18.000	L.F.	285.30	143.28	428.58
Wall railing with returns, steel pipe	17.000	L.F.	114.75	135.32	250.07
TOTAL			1,174.13	2,149	3,323.13

C2010 110	Stairs		COST PER FLIGHT		
			MAT.	INST.	TOTAL
0470	Stairs, C.I.P. concrete, w/o landing, 12 risers, w/o nosing RC2010 -100		720	1,575	2,295
0480	With nosing		1,100	1,775	2,875
0550	W/landing, 12 risers, w/o nosing		805	1,925	2,730
0560	With nosing		1,175	2,150	3,325
0570	16 risers, w/o nosing		1,000	2,425	3,425
0580	With nosing		1,500	2,725	4,225
0590	20 risers, w/o nosing		1,200	2,925	4,125
0600	With nosing		1,825	3,300	5,125
0610	24 risers, w/o nosing		1,400	3,425	4,825
0620	With nosing		2,125	3,850	5,975
0630	Steel, grate type w/nosing & rails, 12 risers, w/o landing		1,800	565	2,365
0640	With landing		2,350	870	3,220
0660	16 risers, with landing		2,950	1,050	4,000
0680	20 risers, with landing		3,550	1,250	4,800
0700	24 risers, with landing		4,150	1,450	5,600
0710	Cement fill metal pan & picket rail, 12 risers, w/o landing		2,350	565	2,915
0720	With landing		3,100	970	4,070
0740	16 risers, with landing		3,875	1,150	5,025
0760	20 risers, with landing		4,650	1,325	5,975
0780	24 risers, with landing		5,450	1,550	7,000
0790	Cast iron tread & pipe rail, 12 risers, w/o landing		2,350	565	2,915
0800	With landing		3,100	970	4,070
0820	16 risers, with landing		3,875	1,150	5,025
0840	20 risers, with landing		4,650	1,325	5,975
0860	24 risers, with landing		5,450	1,550	7,000
0870	Pan tread & flat bar rail, pre-assembled, 12 risers, w/o landing		1,950	620	2,570
0880	With landing		3,525	970	4,495
0900	16 risers, with landing		3,900	1,075	4,975
0920	20 risers, with landing		4,525	1,300	5,825
0940	24 risers, with landing		5,175	1,500	6,675

Figure 8.2

when computing the wall area to be finished, include only the wall area up to the finished ceiling. Do not deduct for openings. A more detailed method will not improve the level of accuracy significantly until exact areas and specific materials are known.

Figure 8.4 is a time-saving method to determine the interior wall finish area when compared to the floor area. Ratio numbers are based on a 10' floor-to-ceiling height.

Typical Range of Risers for Various Story Heights

Story Height	Minimum Risers	Maximum Riser Height	Tread Width	Maximum Risers	Minimum Riser Height	Tread Width	Average Risers	Average Riser Height	Tread Width
7'-6"	12	7.50"	10.00"	14	6.43"	11.07"	13	6.92"	10.58"
8'-0"	13	7.38	10.12	15	6.40	11.10	14	6.86	10.64
8'-6"	14	7.29	10.21	16	6.38	11.12	15	6.80	10.70
9'-0"	15	7.20	10.30	17	6.35	11.15	16	6.75	10.75
9'-6"	16	7.13	10.37	18	6.33	11.17	17	6.71	10.79
10'-0"	16	7.50	10.00	19	6.32	11.18	18	6.67	10.83
10'-6"	17	7.41	10.09	20	6.30	11.20	18	7.00	10.50
11'-0"	18	7.33	10.17	21	6.29	11.21	19	6.95	10.55
11'-6"	19	7.26	10.24	22	6.27	11.23	20	6.90	10.60
12'-0"	20	7.20	10.30	23	6.26	11.24	21	6.86	10.64
12'-6"	20	7.50	10.00	24	6.25	11.25	22	6.82	10.68
13'-0"	21	7.43	10.07	25	6.24	11.26	22	7.09	10.41
13'-6"	22	7.36	10.14	25	6.48	11.02	23	7.04	10.46
14'-0"	23	7.30	10.20	26	6.46	11.04	24	7.00	10.50

Figure 8.3

Area	S.F. of Interior Wall Finish Required as Compared to Floor Area (based on 10' floor to ceiling height)			
	Room Length-To-Width Ratio		Approximate Room Size	
	1:1	2:1	1:1	1:2
100	4.00	4.26	10 x 10	8 x 14
150	3.27	3.48	12 x 12	9 x 17
200	2.82	3.00	15 x 15	10 x 10
250	2.53	2.69	16 x 16	11 x 22
300	2.31	2.45	17 x 17	12 x 25
350	2.14	2.27	19 x 19	13 x 27
400	2.00	2.12	20 x 20	14 x 28
450	1.89	2.00	21 x 21	15 x 30
500	1.79	1.90	22 x 22	16 x 32
550	1.71	1.81	23 x 23	17 x 33
600	1.63	1.73	24 x 24	17 x 35
650	1.57	1.66	26 x 26	28 x 36
700	1.51	1.60	26 x 26	19 x 37
750	1.46	1.55	27 x 27	19 x 39
800	1.41	1.50	28 x 28	20 x 40
850	1.37	1.45	29 x 29	21 x 41
900	1.33	1.41	30 x 30	21 x 42
950	1.30	1.37	31 x 31	22 x 44
1000	1.27	1.34	32 x 32	22 x 45
1500	1.03	1.10	39 x 39	27 x 55
2000	0.89	0.95	45 x 45	32 x 63
Note: For obtaining total S.F. of Interior Partitions, divide ratio multiplier by 2.				

Figure 8.4

Example Three: 500 Square Foot Room

Using Figure 8.4, a square room of 500 S.F. requires approximately 895 S.F. of wall finish treatment.

$$500 \times 1.79 = 895 \text{ S.F.}$$

If the room had a length-to-width ratio of 2:1, the finish requirement would be.

$$500 \times 1.90 = 950 \text{ S.F.}$$

Types of Wall Finish

There are three common types of wall finish:

- Paint
- Wallcovering
- Tile

Painting and finishing work includes preparation of the surface, such as filling nail holes, light sanding, and removing oils or dust. Some surfaces may require additional preparation, including applying block filler or wood conditioner and primer paint. Paints may be applied by brush, roller, or spray. Roller and spray work is usually used on large wall surfaces. Trim is usually painted with a brush.

Wallcovering refers to a wide selection of materials that are applied as a decorative treatment to the surface of interior walls. Wallcoverings are manufactured, printed, or woven in fabric, paper, vinyl, leather, suede, cork, wood veneers, and foils. Wallcoverings are available in different weights, backings, and quality. Preparation of the wall surface prior to installation is necessary to ensure proper bonding. Preparation includes minor sanding, repairing wall defects, and applying sizing or primer for proper adhesion.

Wall tile includes ceramic, stone, or porcelain tiles applied to the wall to create a highly durable, washable, and water-resistant surface. Ceramic tile is available glazed or unglazed. Wall tile is usually glazed. Stone and ceramic tiles can be highly decorative and can be applied to create many distinctive patterns. Wall tiles are usually applied with thin-set mortars over a water-resistant gypsum board or a cementitious backer board. The tile itself usually consists of field tile and trim and/or accent tiles. After the tile is applied to the wall surface, the joints are filled with grout, and a sealer is applied.

Types of Floor Finish

There are many options for floor finishes. The selection depends on the type of traffic expected, the durability required, an acceptable level of maintenance, cost, and aesthetics. Common flooring materials include:

- Wood strip or parquet
- Carpet
- Resilient tile or sheet
- Ceramic tile
- Stone tile
- Wood composition/laminate

Wood Flooring

Wood flooring is available in a variety of styles, grades, species, and patterns. It is manufactured in solid and laminated planks, or solid and laminated parquet. Laminated flooring can be defined as a high-grade wood veneer laminated over a lesser-grade base material. Laminated wood flooring is sold pre-finished. Solid wood flooring is available pre-finished or unfinished, and is classified by grade or quality. There are also several highly durable wood-like products on the market that have a plastic laminate surface that is extremely resistant to abrasion.

Strip wood and plank flooring products can be installed over wood sub-flooring attached to joists. A layer of building paper is installed between the sub-floor and the strip flooring to act as a cushion and to reduce noise. Parquet flooring is manufactured in prefabricated panels that are available with backing materials to protect the flooring from moisture, provide insulation, reduce noise, and add comfort to the walking surface. The backing materials are usually attached to the sub-floor with adhesives.

Carpeting

Carpet is popular for residential and commercial construction because it provides a comfortable, sound-absorbing, and attractive finish surface. It can be economical to install and relatively low maintenance. There are many styles and levels of quality available, depending on fiber, weight, pile thickness, and density. Commercial carpet is available in rolls and as tiles. Carpet can be installed over padding, or can be obtained with a cushioned backing material for direct glue-down to the sub-floor. In estimating the cost of carpet, an important consideration is the roll width and pattern relative to the room size and seaming pattern. There can be considerable waste of material due to these factors.

Resilient Flooring

The category of resilient flooring includes materials such as vinyl tile, vinyl composition tile, cork tile, rubber tile, and sheet materials composed of vinyl or rubber. Resilient flooring is designed for situations where durability, low maintenance, and flooring longevity are of primary consideration. These floor materials are manufactured with or without resilient backing, and are available in a wide variety of widths, colors, patterns, textures, and styles. Rubber and vinyl accessories for resilient flooring include base, thresholds, and transition strips. As with carpet, allowances for waste depend on the layout required. The finish quality of resilient flooring depends on the quality of the sub-floor. Because of the thinness and flexibility of the product, any defect in the sub-floor can "telegraph" through the finished floor, ruining its appearance and durability. Hence, the quality of preparation of the sub-floor is critical, and this preparation should not be overlooked in the estimate.

Ceramic Tile

Ceramic tile products for flooring are similar to those used for walls, with the exception that flooring tiles are most often unglazed. Ceramic floor tile can be mud-set, using a thick layer of concrete mortar, or thin-set with mastic. Most ceramic tile cutting is done using a hand tile cutter. Occasionally harder and more dense types of tile and special-shaped cuts require the use of an electric saw with a water-cooled diamond blade. Projects with a large number of cuts should be noted, as this can affect the productivity of the setter as well as the wear and tear on the saw.

Types of Ceiling Finish

Ceiling systems are designed to absorb and reduce sound transmission within a space, and/or to create a fire separation below the structural framing. The two most common systems are painted gypsum board and acoustical tile. Most acoustical ceiling tile is composed of mineral fibers and is available in a variety of colors, patterns, and textures. For takeoff purposes, acoustical ceilings can be classified by application in three groups: acoustical tiles attached directly to ceiling substrates such as wood or gypsum; suspended acoustical tiles with a concealed spline; and acoustical tiles installed in an exposed suspension system.

Other non-acoustic ceiling tiles can be installed in a suspension system, including decorative thin metal panels and vinyl-coated gypsum panels for areas that require washable surfaces. The cost will vary depending on the room's size, ceiling height, how far the ceiling will hang below the supporting structure, and the type of support structure. For instance, hanging tiles from open web bar joists requires less labor than drilling anchors into precast concrete plank.

Assemblies Estimate: Three-Story Office Building

For the sample three-story office building, the following items are included in Interiors:

- Concrete masonry unit partitions
- Drywall partitions
- Interior doors
- Toilet partitions
- Paint, ceramic tile, and vinyl wall finishes
- Carpet, ceramic, and quarry tile floor finishes
- Suspended acoustical ceiling

Figure 8.5 shows the approximate location of partitions in the proposed three-story office building. The second and third floors are identical, except for the exit doors in the lobbies by the stairways. It is assumed that future tenants will add other partitions to divide the large areas into smaller spaces. Those costs will be borne by the future tenants and are not included in the initial budget estimate.

3-Story Office Building: 100' × 100'

Quantities:

Partitions:	8" Block	534 L.F.
	Steel Stud & Gyp. Board	966 L.F.
Finishes:		
Walls	Paint	85%
	Ceramic Tile	5%
	Vinyl Wall Covering	10%
Floors	Carpet	93%
	Ceramic Tile	4%
	Quarry Tile	3%
Ceiling:	Suspended Acoustic, 9'-0"	

Costs:

Concrete Masonry Unit Partition
(Figure 8.6, line #2000)
5,400 S.F. @ $6.33/S.F.   = $ 34,182

Drywall Partition
(Figure 8.7, line #5400)
14,544 S.F. @ $2.81/S.F.   =   40,869

Exterior Wall—Interior Finish

Furring	(Figure 8.8, line #0649)		
	5,750 S.F. @ $1.26/S.F.	=	7,245
Gypsum Board	(Figure 8.8, line #0700)		
	5,750 S.F. @ $0.68/S.F.	=	3,910
Insulation	(Figure 8.8, line #0920)		
	8,050 S.F. @ $0.72/S.F.	=	5,796
Tape & Finish	(Figure 8.8, line #0960)		
	5,750 S.F. @ $0.40/S.F.	=	2,300
Interior Doors	(Figure 8.9, line #0140)		
	60 ea. @ $513 ea.	=	30,780

Toilet Partitions

Painted Metal	(Figure 8.10, line #0680)		
	15 ea. @ $479 ea.	=	7,185
Handicapped	(Figure 8.10, line #0760)		
	6 ea. @ $278 ea.	=	1,668
Urinal Screens	(Figure 8.10, line #1340)		
	3 ea. @ $276 ea.	=	828

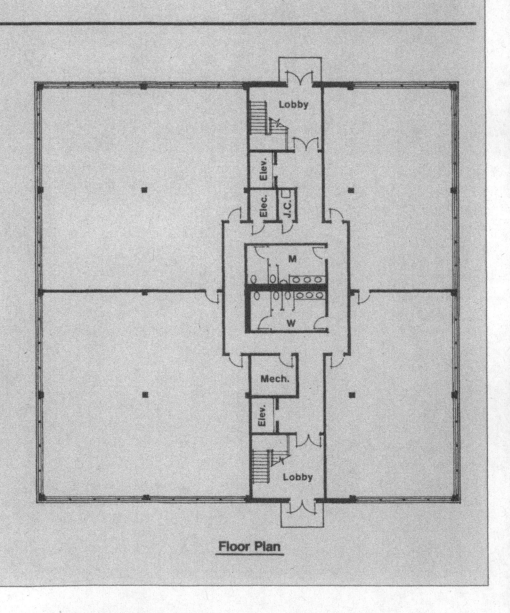

Floor Plan

Figure 8.5

146

Stairs (Figure 8.2, line #0760)

Use steel with concrete filled metal pans, 20 risers, 12'-0" floor-to-floor height (see Figure 8.3). Landings will be used; see the floor plan in Figure 8.5.

	5 flights @ $5,975	=	29,875

Wall Finish

Paint Gypsum Board	(Figure 8.11, line #0140)		
	20,388 S.F. @ $0.57/S.F.	=	11,621
Paint Concrete Block	(Figure 8.11, line #0320)		
	4,725 S.F. @ $1.19/S.F.	=	5,623
V.W.C.	(Figure 8.11, line #1800)		
	2,350 S.F. @ $1.41/S.F.	=	3,314
Ceramic Tile	(Figure 8.11, line #1940)		
	1,175 S.F. @ $5.46/S.F.	=	6,416

Floor Finish

Carpet	(Figure 8.12, line #0160)		
	27,900 S.F. @ $5.47/S.F.	=	152,613
Ceramic Tile	(Figure 8.12, line #1720)		
	1,200 S.F. @ $7.58/S.F.	=	9,096
Quarry Tile	(Figure 8.12, line #1820)		
	900 S.F. @ $11.10/S.F.	=	9,990

Ceiling

Suspended Acous.	(Figure 8.13, line #6000)		
	30,000 S.F. @ $2.69/S.F.	=	80,700
Total Cost			$444,011
Cost per Square Foot			$14.80/S.F.

Figure 8.14 summarizes the costs for interior finish.

C10 Interior Construction

C1010 Partitions

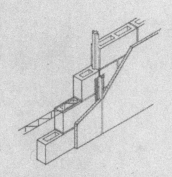

The Concrete Block Partition Systems are defined by weight and type of block, thickness, type of finish and number of sides finished. System components include joint reinforcing on alternate courses and vertical control joints.

System Components	QUANTITY	UNIT	COST PER S.F.		
			MAT.	INST.	TOTAL
SYSTEM C1010 102					
CONC. BLOCK PARTITION, 8" X 16", 4" TK., 2 CT. GYP. PLASTER 2 SIDES					
Conc. block partition, 4" thick	1.000	S.F.	.94	3.87	4.81
Control joint	.050	L.F.	.06	.10	.16
Horizontal joint reinforcing	.800	L.F.	.10	.05	.15
Gypsum plaster, 2 coat, on masonry	2.000	S.F.	.69	3.64	4.33
TOTAL			1.79	7.66	9.45

C1010 102		Concrete Block Partitions - Regular Weight							
	TYPE	THICKNESS (IN.)	TYPE FINISH	SIDES FINISHED		COST PER S.F.			
						MAT.	INST.	TOTAL	
1000	Hollow	4	none	0		1.10	4.02	5.12	
1010			gyp. plaster 2 coat	1		1.44	5.80	7.24	
1020				2		1.79	7.65	9.44	
1100			lime plaster - 2 coat	1		1.31	5.80	7.11	
1150			lime portland - 2 coat	1		1.32	5.80	7.12	
1200			portland - 3 coat	1		1.34	6.15	7.49	
1400			5/8" drywall	1		1.53	5.35	6.88	
1410				2		1.96	6.65	8.61	
1500		6	none	0		1.61	4.31	5.92	
1510			gyp. plaster 2 coat	1		1.95	6.10	8.05	
1520				2		2.30	7.95	10.25	
1600			lime plaster - 2 coat	1		1.82	6.10	7.92	
1650			lime portland - 2 coat	1		1.83	6.10	7.93	
1700			portland - 3 coat	1		1.85	6.40	8.25	
1900			5/8" drywall	1		2.04	5.65	7.69	
1910				2		2.47	6.95	9.42	
2000		8	none	0		1.72	4.61	6.33	
2010			gyp. plaster 2 coat	1		2.06	6.40	8.46	
2020			gyp. plaster 2 coat	2		2.41	8.25	10.66	
2100		8	lime plaster - 2 coat	1		1.93	6.40	8.33	
2150			lime portland - 2 coat	1		1.94	6.40	8.34	
2200			portland - 3 coat	1		1.96	6.70	8.66	
2400			5/8" drywall	1		2.15	5.95	8.10	
2410				2		2.58	7.25	9.83	

Figure 8.6

C1010 Partitions

Wood Stud Framing

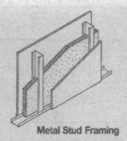

Metal Stud Framing

The Drywall Partitions/Stud Framing Systems are defined by type of drywall and number of layers, type and spacing of stud framing, and treatment on the opposite face. Components include taping and finishing.

Cost differences between regular and fire resistant drywall are negligible, and terminology is interchangeable. In some cases fiberglass insulation is included for additional sound deadening.

System Components			QUANTITY	UNIT	COST PER S.F.		
					MAT.	INST.	TOTAL
SYSTEM C1010 124							
DRYWALL PARTITION,5/8″ F.R.1 SIDE,5/8″ REG.1 SIDE,2″X4″STUDS,16″ O.C.							
Gypsum plasterboard, nailed/screwed to studs, 5/8″F.R. fire resistant			1.000	S.F.	.32	.36	.68
Gypsum plasterboard, nailed/screwed to studs, 5/8″ regular			1.000	S.F.	.32	.36	.68
Taping and finishing joints			2.000	S.F.	.08	.72	.80
Framing, 2 x 4 studs @ 16″ O.C., 10′ high			1.000	S.F.	.46	.73	1.19
		TOTAL			1.18	2.17	3.35

C1010 124		Drywall Partitions/Wood Stud Framing						
	FACE LAYER	BASE LAYER	FRAMING	OPPOSITE FACE	INSULATION	COST PER S.F.		
						MAT.	INST.	TOTAL
1200	5/8″ FR drywall	none	2 x 4, @ 16″ O.C.	same	0	1.18	2.17	3.35
1250				5/8″ reg. drywall	0	1.18	2.17	3.35
1300				nothing	0	.82	1.45	2.27
1400		1/4″ SD gypsum	2 x 4 @ 16″ O.C.	same	1-1/2″ fiberglass	2.14	3.35	5.49
1450				5/8″ FR drywall	1-1/2″ fiberglass	1.88	2.94	4.82
1500				nothing	1-1/2″ fiberglass	1.52	2.22	3.74
1600		resil. channels	2 x 4 @ 16″, O.C.	same	1-1/2″ fiberglass	1.87	4.25	6.12
1650				5/8″ FR drywall	1-1/2″ fiberglass	1.74	3.39	5.13
1700				nothing	1-1/2″ fiberglass	1.38	2.67	4.05
1800		5/8″ FR drywall	2 x 4 @ 24″ O.C.	same	0	1.71	2.75	4.46
1850				5/8″ FR drywall	0	1.39	2.39	3.78
1900				nothing	0	1.03	1.67	2.70
1950		5/8″ FR drywall	2 x 4, 16″ O.C.	same	0	1.82	2.89	4.71
1955				5/8″ FR drywall	0	1.50	2.53	4.03
2000				nothing	0	1.14	1.81	2.95
2010		5/8″ FR drywall	staggered, 6″ plate	same	0	2.30	3.63	5.93
2015				5/8″ FR drywall	0	1.98	3.27	5.25
2020				nothing	0	1.62	2.55	4.17
2200		5/8″ FR drywall	2 rows-2 x 4	same	2″ fiberglass	2.76	3.98	6.74
2250			16″O.C.	5/8″ FR drywall	2″ fiberglass	2.44	3.62	6.06
2300				nothing	2″ fiberglass	2.08	2.90	4.98
2400	5/8″ WR drywall	none	2 x 4, @ 16″ O.C.	same	0	1.48	2.17	3.65
2450				5/8″ FR drywall	0	1.33	2.17	3.50
2500				nothing	0	.97	1.45	2.42
2600		5/8″ FR drywall	2 x 4, @ 24″ O.C.	same	0	2.01	2.75	4.76
2650				5/8″ FR drywall	0	1.54	2.39	3.93
2700				nothing	0	1.18	1.67	2.85

Figure 8.7

C10 Interior Construction

C1010 Partitions

C1010 124 — Drywall Partitions/Wood Stud Framing

	FACE LAYER	BASE LAYER	FRAMING	OPPOSITE FACE	INSULATION	COST PER S.F.		
						MAT.	INST.	TOTAL
2800	5/8 VF drywall	none	2 x 4, @ 16" O.C.	same	0	1.90	2.35	4.25
2850				5/8" FR drywall	0	1.54	2.26	3.80
2900				nothing	0	1.18	1.54	2.72
3000	5/8 VF drywall	5/8 FR drywall	2 x 4, 24" O.C.	same	0	2.43	2.93	5.36
3050				5/8" FR drywall	0	1.75	2.48	4.23
3100				nothing	0	1.39	1.76	3.15
3200	1/2" reg drywall	3/8" reg drywall	2 x 4, @ 16" O.C.	same	0	1.70	2.89	4.59
3250				5/8" FR drywall	0	1.44	2.53	3.97
3300				nothing	0	1.08	1.81	2.89

C1010 124 — Drywall Partitions/Metal Stud Framing

	FACE LAYER	BASE LAYER	FRAMING	OPPOSITE FACE	INSULATION	COST PER S.F.		
						MAT.	INST.	TOTAL
5200	5/8" FR drywall	none	1-5/8" @ 24" O.C.	same	0	.84	1.92	2.76
5250				5/8" reg. drywall	0	.84	1.92	2.76
5300				nothing	0	.48	1.20	1.68
5400			3-5/8" @ 24" O.C.	same	0	.88	1.93	2.81
5450				5/8" reg. drywall	0	.88	1.93	2.81
5500				nothing	0	.52	1.21	1.73
5600		1/4" SD gypsum	1-5/8" @ 24" O.C.	same	0	1.36	2.74	4.10
5650				5/8" FR drywall	0	1.10	2.33	3.43
5700				nothing	0	.74	1.61	2.35
5800			2-1/2" @ 24" O.C.	same	0	1.37	2.75	4.12
5850				5/8" FR drywall	0	1.11	2.34	3.45
5900				nothing	0	.75	1.62	2.37
6000		5/8" FR drywall	2-1/2" @ 16" O.C.	same	0	1.66	3.10	4.76
6050				5/8" FR drywall	0	1.34	2.74	4.08
6100				nothing	0	.98	2.02	3
6200			3-5/8" @ 24" O.C.	same	0	1.52	2.65	4.17
6250				5/8" FR drywall	3-1/2" fiberglass	1.56	2.52	4.08
6300				nothing	0	.84	1.57	2.41
6400	5/8" WR drywall	none	1-5/8" @ 24" O.C.	same	0	1.14	1.92	3.06
6450				5/8" FR drywall	0	.99	1.92	2.91
6500				nothing	0	.63	1.20	1.83
6600			3-5/8" @ 24" O.C.	same	0	1.18	1.93	3.11
6650				5/8" FR drywall	0	1.03	1.93	2.96
6700				nothing	0	.67	1.21	1.88
6800		5/8" FR drywall	2-1/2" @ 16" O.C.	same	0	1.96	3.10	5.06
6850				5/8" FR drywall	0	1.49	2.74	4.23
6900				nothing	0	1.13	2.02	3.15
7000			3-5/8" @ 24" O.C.	same	0	1.82	2.65	4.47
7050				5/8" FR drywall	3-1/2" fiberglass	1.71	2.52	4.23
7100				nothing	0	.99	1.57	2.56
7200	5/8" VF drywall	none	1-5/8" @ 24" O.C.	same	0	1.56	2.10	3.66
7250				5/8" FR drywall	0	1.20	2.01	3.21
7300				nothing	0	.84	1.29	2.13
7400			3-5/8" @ 24" O.C.	same	0	1.60	2.11	3.71
7450				5/8" FR drywall	0	1.24	2.02	3.26
7500				nothing	0	.88	1.30	2.18

Figure 8.7 (cont.)

150

C1010 Partitions

C1010 110	Drywall Components	COST PER S.F.		
		MAT.	INST.	TOTAL
0140	Metal studs, 24" O.C. including track, load bearing, 18 gage, 2-1/2"	.35	.68	1.03
0160	3-5/8"	.42	.70	1.12
0180	4"	.38	.71	1.09
0200	6"	.56	.72	1.28
0220	16 gage, 2-1/2"	.41	.78	1.19
0240	3-5/8"	.49	.80	1.29
0260	4"	.52	.81	1.33
0280	6"	.66	.83	1.49
0300	Non load bearing, 25 gage, 1-5/8"	.12	.48	.60
0340	3-5/8"	.16	.49	.65
0360	4"	.19	.49	.68
0380	6"	.25	.50	.75
0400	20 gage, 2-1/2"	.22	.61	.83
0420	3-5/8"	.24	.62	.86
0440	4"	.30	.62	.92
0460	6"	.35	.63	.98
0540	Wood studs including blocking, shoe and double top plate, 2"x4", 12"O.C.	.58	.91	1.49
0560	16" O.C.	.46	.73	1.19
0580	24" O.C.	.35	.59	.94
0600	2"x6", 12" O.C.	.85	1.04	1.89
0620	16" O.C.	.68	.81	1.49
0640	24" O.C.	.52	.64	1.16
0642	Furring one side only, steel channels, 3/4", 12" O.C.	.18	1.43	1.61
0644	16" O.C.	.16	1.27	1.43
0646	24" O.C.	.11	.96	1.07
0647	1-1/2", 12" O.C.	.27	1.60	1.87
0648	16" O.C.	.24	1.40	1.64
0649	24" O.C.	.16	1.10	1.26
0650	Wood strips, 1" x 3", on wood, 12" O.C.	.21	.66	.87
0651	16" O.C.	.16	.50	.66
0652	On masonry, 12" O.C.	.21	.74	.95
0653	16" O.C.	.16	.56	.72
0654	On concrete, 12" O.C.	.21	1.40	1.61
0655	16" O.C.	.16	1.05	1.21
0665	Gypsum board, one face only, exterior sheathing, 1/2"	.31	.65	.96
0680	Interior, fire resistant, 1/2"	.26	.36	.62
0700	5/8"	.32	.36	.68
0720	Sound deadening board 1/4"	.26	.41	.67
0740	Standard drywall 3/8"	.29	.36	.65
0760	1/2"	.29	.36	.65
0780	5/8"	.32	.36	.68
0800	Tongue & groove coreboard 1"	.50	1.52	2.02
0820	Water resistant, 1/2"	.37	.36	.73
0840	5/8"	.47	.36	.83
0860	Add for the following:, foil backing	.11		.11
0880	Fiberglass insulation, 3-1/2"	.36	.23	.59
0900	6"	.44	.27	.71
0920	Rigid insulation 1"	.36	.36	.72
0940	Resilient furring @ 16" O.C.	.16	1.14	1.30
0960	Taping and finishing	.04	.36	.40
0980	Texture spray	.06	.42	.48
1000	Thin coat plaster	.08	.46	.54
1040	2"x4" staggered studs 2"x6" plates & blocking	.66	.80	1.46

Figure 8.8

C1020 Interior Doors

C1020 114 — Doors/Metal Frames

	MATERIAL	GAUGE	THICKNESS (IN.)	TYPE	SIZE W(FT.)XH(FT.)	COST EACH		
						MAT.	INST.	TOTAL
0120	Hollow metal	20	1-3/4"	flush	2/8 6/8	258	244	502
0140					3/0 7/0	266	247	513
0160				half glass	2/8 6/8	375	267	642
0180					3/0 7/0	380	271	651
0200		18	1-3/4"	flush	2/8 6/8	285	247	532
0220					3/0 7/0	288	247	535
0240				half glass	2/8 6/8	375	271	646
0260					3/0 7/0	380	273	653
0280	Wood		1-3/4	flush-lauan	2/8 6/8	149	151	300
0320	5 ply			birch	2/8 6/8	164	151	315
0340					3/0 7/0	175	162	337
0360				oak	2/8 6/8	173	151	324
0380					3/0 7/0	186	162	348
0400				walnut	2/8 6/8	199	154	353
0420					3/0 7/0	845	600	1,445

C1020 130 — Metal Fire Doors/Metal Frames

	FIRE RATING (HOURS)	GAUGE	THICKNESS (IN.)	TYPE	SIZE W(FT.)XH(FT)	COST EACH		
						MAT.	INST.	TOTAL
0520	1-1/2	20	1-3/4	flush	2/8 6/8	298	244	542
0540					3/0 7/0	305	247	552
0560				composite	2/8 6/8	360	247	607
0580					3/0 7/0	370	249	619
0600		18	1-3/4	flush	3/0 6/8	320	249	569
0620					3/0 7/0	325	249	574
0720	3	18		composite	2/8 6/8	385	252	637
0740					3/0 7/0	390	252	642

C1020 130 — Wood Fire Doors/Metal Frames

	FIRE RATING (HOURS)	CORE MATERIAL	THICKNESS (IN.)	FACE MATERIAL	SIZE W(FT)XH(FT)	COST EACH		
						MAT.	INST.	TOTAL
0840	1	mineral	1-3/4"	birch	2/8 6/8	340	154	494
0860		3 ply stile			3/0 7/0	360	167	527
0880				oak	2/8 6/8	340	154	494
0900					3/0 7/0	360	167	527
0920				walnut	2/8 6/8	415	154	569
0940					3/0 7/0	435	167	602
0960				m.d. overlay	2/8 6/8	315	252	567
0980					3/0 7/0	340	260	600
1000	1-1/2	mineral	1-3/4"	birch	2/8 6/8	335	154	489
1020		3 ply stile			3/0 7/0	355	167	522
1040				oak	2/8 6/8	325	154	479
1060					3/0 7/0	345	167	512
1080				walnut	2/8 6/8	405	154	559
1100					3/0 7/0	430	167	597
1120				m.d. overlay	2/8 6/8	350	252	602
1140					3/0 7/0	365	260	625

Figure 8.9

C1030 Fittings

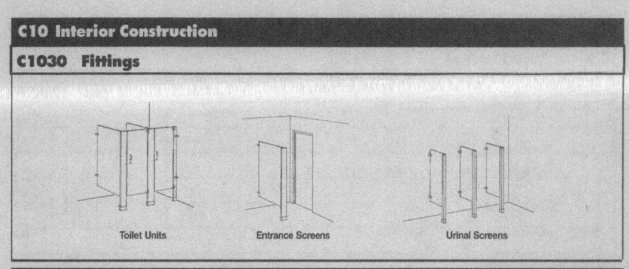

Toilet Units **Entrance Screens** **Urinal Screens**

C1030 110	Toilet Partitions	COST PER UNIT		
		MAT.	INST.	TOTAL
0380	Toilet partitions, cubicles, ceiling hung, marble	1,450	360	1,810
0400	Painted metal	400	182	582
0420	Plastic laminate	545	182	727
0460	Stainless steel	1,100	182	1,282
0480	Handicap addition	320		320
0520	Floor and ceiling anchored, marble	1,575	289	1,864
0540	Painted metal	405	146	551
0560	Plastic laminate	515	146	661
0600	Stainless steel	1,275	146	1,421
0620	Handicap addition	278		278
0660	Floor mounted marble	935	241	1,176
0680	Painted metal	375	104	479
0700	Plastic laminate	510	104	614
0740	Stainless steel	1,300	104	1,404
0760	Handicap addition	278		278
0780	Juvenile deduction	38.50		38.50
0820	Floor mounted with handrail marble	900	241	1,141
0840	Painted metal	405	122	527
0860	Plastic laminate	695	122	817
0900	Stainless steel	1,250	122	1,372
0920	Handicap addition	278		278
0960	Wall hung, painted metal	510	104	614
1020	Stainless steel	1,225	104	1,329
1040	Handicap addition	278		278
1080	Entrance screens, floor mounted, 54" high, marble	585	80	665
1100	Painted metal	182	48.50	230.50
1140	Stainless steel	670	48.50	718.50
1300	Urinal screens, floor mounted, 24" wide, laminated plastic	232	91	323
1320	Marble	520	108	628
1340	Painted metal	185	91	276
1380	Stainless steel	530	91	621
1428	Wall mounted wedge type, painted metal	221	73	294
1460	Stainless steel	455	73	528

Figure 8.10

C3010 Wall Finishes

C3010 105	Paint & Covering	COST PER S.F.		
		MAT.	INST.	TOTAL
0060	Painting, interior on plaster and drywall, brushwork, primer & 1 coat	.10	.49	.59
0080	Primer & 2 coats	.15	.65	.80
0100	Primer & 3 coats	.20	.80	1
0120	Walls & ceilings, roller work, primer & 1 coat	.10	.33	.43
0140	Primer & 2 coats	.15	.42	.57
0160	Woodwork incl. puttying, brushwork, primer & 1 coat	.10	.71	.81
0180	Primer & 2 coats	.15	.94	1.09
0200	Primer & 3 coats	.20	1.27	1.47
0260	Cabinets and casework, enamel, primer & 1 coat	.10	.80	.90
0280	Primer & 2 coats	.15	.98	1.13
0300	Masonry or concrete, latex, brushwork, primer & 1 coat	.18	.66	.84
0320	Primer & 2 coats	.24	.95	1.19
0340	Addition for block filler	.11	.82	.93
0380	Fireproof paints, intumescent, 1/8" thick 3/4 hour	1.77	.65	2.42
0400	3/16" thick 1 hour	3.94	.98	4.92
0420	7/16" thick 2 hour	5.10	2.27	7.37
0440	1-1/16" thick 3 hour	8.30	4.54	12.84
0480	Miscellaneous metal brushwork, exposed metal, primer & 1 coat	.08	.80	.88
0500	Gratings, primer & 1 coat	.19	.99	1.18
0600	Pipes over 12" diameter	.48	3.18	3.66
0700	Structural steel, brushwork, light framing 300-500 S.F./Ton	.06	1.27	1.33
0720	Heavy framing 50-100 S.F./Ton	.06	.64	.70
0740	Spraywork, light framing 300-500 S.F./Ton	.06	.28	.34
0760	Heavy framing 50-100 S.F./Ton	.06	.31	.37
0800	Varnish, interior wood trim, no sanding sealer & 1 coat	.07	.80	.87
0820	Hardwood floor, no sanding 2 coats	.13	.17	.30
0840	Wall coatings, acrylic glazed coatings, minimum	.28	.61	.89
0860	Maximum	.57	1.04	1.61
0880	Epoxy coatings, minimum	.35	.61	.96
0900	Maximum	1.09	1.87	2.96
0940	Exposed epoxy aggregate, troweled on, 1/16" to 1/4" aggregate, minimum	.54	1.35	1.89
0960	Maximum	1.17	2.45	3.62
0980	1/2" to 5/8" aggregate, minimum	1.08	2.45	3.53
1000	Maximum	1.84	3.98	5.82
1020	1" aggregate, minimum	1.87	3.53	5.40
1040	Maximum	2.86	5.80	8.66
1060	Sprayed on, minimum	.51	1.08	1.59
1080	Maximum	.92	2.19	3.11
1100	High build epoxy 50 mil, minimum	.59	.82	1.41
1120	Maximum	1.02	3.35	4.37
1140	Laminated epoxy with fiberglass minimum	.65	1.08	1.73
1160	Maximum	1.16	2.19	3.35
1180	Sprayed perlite or vermiculite 1/16" thick, minimum	.23	.11	.34
1200	Maximum	.66	.50	1.16
1260	Wall coatings, vinyl plastic, minimum	.30	.43	.73
1280	Maximum	.73	1.32	2.05
1300	Urethane on smooth surface, 2 coats, minimum	.22	.28	.50
1320	Maximum	.50	.48	.98
1340	3 coats, minimum	.30	.38	.68
1360	Maximum	.66	.68	1.34
1380	Ceramic-like glazed coating, cementitious, minimum	.43	.72	1.15
1400	Maximum	.72	.92	1.64
1420	Resin base, minimum	.30	.50	.80
1440	Maximum	.48	.96	1.44
1460	Wall coverings, aluminum foil	.88	1.16	2.04
1480	Copper sheets, .025" thick, phenolic backing	6.10	1.33	7.43
1500	Vinyl backing	4.70	1.33	6.03
1520	Cork tiles, 12"x12", light or dark, 3/16" thick	2.68	1.33	4.01

Figure 8.11

C3010 Wall Finishes

C3010 105	Paint & Covering	MAT.	INST.	TOTAL
1540	5/16" thick	3.12	1.36	4.48
1560	Basketweave, 1/4" thick	4.80	1.33	6.13
1580	Natural, non-directional, 1/2" thick	6.75	1.33	8.08
1600	12"x36", granular, 3/16" thick	1.03	.83	1.86
1620	1" thick	1.34	.86	2.20
1640	12"x12", polyurethane coated, 3/16" thick	3.23	1.33	4.56
1660	5/16" thick	4.60	1.36	5.96
1661	Paneling, prefinished plywood, birch	1.23	1.74	2.97
1662	Mahogany, African	2.31	1.82	4.13
1663	Philippine (lauan)	.99	1.46	2.45
1664	Oak or cherry	2.97	1.82	4.79
1665	Rosewood	4.21	2.28	6.49
1666	Teak	2.97	1.82	4.79
1667	Chestnut	4.39	1.95	6.34
1668	Pecan	1.89	1.82	3.71
1669	Walnut	4.80	1.82	6.62
1670	Wood board, knotty pine, finished	1.67	3.04	4.71
1671	Rough sawn cedar	2.05	3.04	5.09
1672	Redwood	4.42	3.04	7.46
1673	Aromatic cedar	3.47	3.26	6.73
1680	Cork wallpaper, paper backed, natural	1.84	.67	2.51
1700	Color	2.28	.67	2.95
1720	Gypsum based, fabric backed, minimum	.72	.40	1.12
1740	Average	1.06	.44	1.50
1760	Maximum	1.18	.50	1.68
1780	Vinyl wall covering, fabric back, light weight	.59	.50	1.09
1800	Medium weight	.74	.67	1.41
1820	Heavy weight	1.19	.74	1.93
1840	Wall paper, double roll, solid pattern, avg. workmanship	.31	.50	.81
1860	Basic pattern, avg. workmanship	.67	.60	1.27
1880	Basic pattern, quality workmanship	1.61	.74	2.35
1900	Grass cloths with lining paper, minimum	.67	.80	1.47
1920	Maximum	2.15	.91	3.06
1940	Ceramic tile, thin set, 4-1/4" x 4-1/4"	2.30	3.16	5.46

C3010 105	Paint Trim	COST PER L.F.		
		MAT.	INST.	TOTAL
2040	Painting, wood trim, to 6" wide, enamel, primer & 1 coat	.10	.40	.50
2060	Primer & 2 coats	.15	.50	.65
2080	Misc. metal brushwork, ladders	.39	3.98	4.37
2100	Pipes, to 4" dia.	.06	.84	.90
2120	6" to 8" dia.	.13	1.67	1.80
2140	10" to 12" dia.	.38	2.50	2.88
2160	Railings, 2" pipe	.14	1.99	2.13
2180	Handrail, single	.10	.80	.90

Figure 8.11 (cont.)

C30 Interior Finishes

C3020 Floor Finishes

C3020 410	Tile & Covering	COST PER S.F.		
		MAT.	INST.	TOTAL
0060	Carpet tile, nylon, fusion bonded, 18" x 18" or 24" x 24", 24 oz.	3.44	.46	3.90
0080	35 oz.	4	.46	4.46
0100	42 oz.	4.77	.46	5.23
0140	Carpet, tufted, nylon, roll goods, 12' wide, 26 oz.	3.11	.53	3.64
0160	36 oz.	4.94	.53	5.47
0180	Woven, wool, 36 oz.	7.85	.53	8.38
0200	42 oz.	7.95	.53	8.48
0220	Padding, add to above, minimum	.42	.25	.67
0240	Maximum	.73	.25	.98
0260	Composition flooring, acrylic, 1/4" thick	1.36	3.60	4.96
0280	3/8" thick	1.79	4.16	5.95
0300	Epoxy, minimum	2.41	2.77	5.18
0320	Maximum	2.93	3.82	6.75
0340	Epoxy terrazzo, minimum	4.88	3.81	8.69
0360	Maximum	8	5.10	13.10
0380	Mastic, hot laid, 1-1/2" thick, minimum	3.53	2.71	6.24
0400	Maximum	4.53	3.60	8.13
0420	Neoprene 1/4" thick, minimum	3.48	3.43	6.91
0440	Maximum	4.79	4.35	9.14
0460	Polyacrylate with ground granite 1/4", minimum	2.72	2.54	5.26
0480	Maximum	5.25	3.89	9.14
0500	Polyester with colored quart 2 chips 1/16", minimum	2.64	1.76	4.40
0520	Maximum	4.43	2.77	7.20
0540	Polyurethane with vinyl chips, minimum	6.80	1.76	8.56
0560	Maximum	9.75	2.18	11.93
0600	Concrete topping, granolithic concrete, 1/2" thick	.18	1.61	1.79
0620	1" thick	.36	1.65	2.01
0640	2" thick	.72	1.90	2.62
0660	Heavy duty 3/4" thick, minimum	2.31	2.50	4.81
0680	Maximum	2.31	2.97	5.28
0700	For colors, add to above, minimum	.33	.58	.91
0720	Maximum	.66	.63	1.29
0740	Exposed aggregate finish, minimum	3.96	.53	4.49
0760	Maximum	12.55	.71	13.26
0780	Abrasives, .25 P.S.F. add to above, minimum	.17	.39	.56
0800	Maximum	.20	.39	.59
0820	Dust on coloring, add, minimum	.64	.26	.90
0840	Maximum	2.23	.53	2.76
0860	1/2" integral, minimum	.21	1.61	1.82
0880	Maximum	.21	1.61	1.82
0900	Dustproofing, add, minimum	.08	.17	.25
0920	Maximum	.14	.26	.40
0930	Paint	.24	.95	1.19
0940	Hardeners, metallic add, minimum	.17	.39	.56
0960	Maximum	.38	.58	.96
0980	Non-metallic, minimum	.19	.39	.58
1000	Maximum	.52	.58	1.10
1020	Integral topping, 3/16" thick	.06	.95	1.01
1040	1/2" thick	.17	1	1.17
1060	3/4" thick	.26	1.12	1.38
1080	1" thick	.34	1.27	1.61
1100	Terrazzo, minimum	2.68	5.85	8.53
1120	Maximum	5	12.75	17.75
1280	Resilient, asphalt tile, 1/8" thick on concrete, minimum	1.06	.83	1.89
1300	Maximum	1.17	.83	2
1320	On wood, add for felt underlay	.20		.20
1340	Cork tile, minimum	3.80	1.06	4.86
1360	Maximum	9.05	1.06	10.11

Figure 8.12

C3020 410	Tile & Covering	MAT.	INST.	TOTAL
1380	Polyethylene, in rolls, minimum	2.54	1.21	3.75
1400	Maximum	4.94	1.21	6.15
1420	Polyurethane, thermoset, minimum	3.91	3.33	7.24
1440	Maximum	4.64	6.65	11.29
1460	Rubber, sheet goods, minimum	3.50	2.77	6.27
1480	Maximum	5.75	3.70	9.45
1500	Tile, minimum	3.92	.83	4.75
1520	Maximum	6.65	1.21	7.86
1540	Synthetic turf, minimum	2.96	1.59	4.55
1560	Maximum	7.60	1.75	9.35
1580	Vinyl, composition tile, minimum	.78	.67	1.45
1600	Maximum	2.24	.67	2.91
1620	Tile, minimum	1.75	.67	2.42
1640	Maximum	4.93	.67	5.60
1660	Sheet goods, minimum	2.04	1.33	3.37
1680	Maximum	3.93	1.66	5.59
1720	Tile, ceramic natural clay	4.29	3.29	7.58
1730	Marble, synthetic 12"x12"x5/8"	9.65	10	19.65
1740	Porcelain type, minimum	4.61	3.29	7.90
1760	Maximum	6.35	3.88	10.23
1800	Quarry tile, mud set, minimum	3.16	4.30	7.46
1820	Maximum	5.65	5.45	11.10
1840	Thin set, deduct		.86	.86
1860	Terrazzo precast, minimum	3.89	4.44	8.33
1880	Maximum	8.60	4.44	13.04
1900	Non-slip, minimum	16.20	22.50	38.70
1920	Maximum	18.65	31	49.65
1960	Wood, block, end grain factory type, creosoted, 2" thick	3.09	1.24	4.33
1980	2-1/2" thick	3.20	2.92	6.12
2000	3" thick	3.15	2.92	6.07
2020	Natural finish, 2" thick	3.19	2.92	6.11
2040	Fir, vertical grain, 1"x4", no finish, minimum	2.38	1.43	3.81
2060	Maximum	2.53	1.43	3.96
2080	Prefinished white oak, prime grade, 2-1/4" wide	6.45	2.15	8.60
2100	3-1/4" wide	8.25	1.97	10.22
2120	Maple strip, sanded and finished, minimum	3.64	3.12	6.76
2140	Maximum	4.30	3.12	7.42
2160	Oak strip, sanded and finished, minimum	3.69	3.12	6.81
2180	Maximum	4.57	3.12	7.69
2200	Parquetry, sanded and finished, minimum	3.22	3.25	6.47
2220	Maximum	6.35	4.62	10.97
2260	Add for sleepers on concrete, treated, 24" O.C., 1"x2"	.65	1.88	2.53
2280	1"x3"	.81	1.45	2.26
2300	2"x4"	.87	.74	1.61
2320	2"x6"	.99	.56	1.55
2340	Underlayment, plywood, 3/8" thick	.64	.49	1.13
2350	1/2" thick	.81	.50	1.31
2360	5/8" thick	.77	.52	1.29
2370	3/4" thick	.99	.56	1.55
2380	Particle board, 3/8" thick	.42	.49	.91
2390	1/2" thick	.44	.50	.94
2400	5/8" thick	.59	.52	1.11
2410	3/4" thick	.64	.56	1.20
2420	Hardboard, 4' x 4', .215" thick	.45	.49	.94

Figure 8.12 (cont.)

C3030 Ceiling Finishes

C3030 105 — Drywall Ceilings

	TYPE	FINISH	FURRING	SUPPORT		COST PER S.F.		
						MAT.	INST.	TOTAL
4800	1/2" F.R. drywall	painted and textured	1"x3" wood, 16" O.C.	wood		.63	2.29	2.92
4900				masonry		.63	2.37	3
5000				concrete		.63	2.82	3.45
5100	5/8" F.R. drywall	painted and textured	1"x3" wood, 16" O.C.	wood		.68	2.29	2.97
5200				masonry		.68	2.37	3.05
5300				concrete		.68	2.82	3.50
5400	1/2" F.R. drywall	painted and textured	7/8"resil. channels	24" O.C.		.57	2.23	2.80
5500			1"x2" wood	stud clips		.72	2.15	2.87
5600			1-5/8"metal studs	24" O.C.		.59	1.99	2.58
5700	5/8" F.R. drywall	painted and textured	1-5/8"metal studs	24" O.C.		.64	1.99	2.63
5702								

C3030 105 — Acoustical Ceilings

	TYPE	TILE	GRID	SUPPORT		COST PER S.F.		
						MAT.	INST.	TOTAL
5800	5/8" fiberglass board	24" x 48"	tee	suspended		.94	1.05	1.99
5900		24" x 24"	tee	suspended		1.02	1.15	2.17
6000	3/4" fiberglass board	24" x 48"	tee	suspended		1.61	1.08	2.69
6100		24" x 24"	tee	suspended		1.69	1.18	2.87
6500	5/8" mineral fiber	12" x 12"	1"x3" wood, 12" O.C.	wood		1.05	1.40	2.45
6600				masonry		1.05	1.50	2.55
6700				concrete		1.05	2.10	3.15
6800	3/4" mineral fiber	12" x 12"	1"x3" wood, 12" O.C.	wood		1.54	1.40	2.94
6900				masonry		1.54	1.40	2.94
7000				concrete		1.54	1.40	2.94
7100	3/4"mineral fiber on	12" x 12"	25 ga. channels	runners		1.96	1.91	3.87
7102	5/8" F.R. drywall							
7200	5/8" plastic coated	12" x 12"		adhesive backed		1.20	.36	1.56
7201	Mineral fiber							
7202								
7300	3/4" plastic coated	12" x 12"		adhesive backed		1.69	.36	2.05
7301	Mineral fiber							
7302								
7400	3/4" mineral fiber	12" x 12"	conceal 2" bar &	suspended		1.78	1.85	3.63
7401			channels					
7402								

Figure 8.13

158

ASSEMBLY NUMBER	DESCRIPTION	QTY	UNIT	TOTAL COST UNIT	TOTAL COST TOTAL	COST PER S.F.
DIVISION C	**INTERIOR CONSTRUCTION**					
	Partitions					
C1010-102-2000	8" CMU at Stairs & Elevator	5,400	S.F.	$ 6.33	$ 34,182	
C1010-124-5400	Drywall, 5/8" Gyp. Bd., 3-5/8" Steel Studs	14,544	S.F.	$ 2.81	$ 40,869	
	Exterior Wall					
C1010-128-0649	Furring, 1-1/2" @ 24" O.C.	5,750	S.F.	$ 1.26	$ 7,245	
C1010-128-0920	Insulation, Rigid	8,050	S.F.	$ 0.72	$ 5,796	
C1010-128-0700	Gypsum Board, 5/8"	5,750	S.F.	$ 0.68	$ 3,910	
C1010-128-0960	Tape and Finish	5,750	S.F.	$ 0.40	$ 2,300	
C1020-114-0140	Interior Doors	60	Ea.	$ 513.00	$ 30,780	
	Toilet Partitions					
C1030-110-0680	Painted Metal	15	Ea.	$ 479.00	$ 7,185	
C1030-110-0760	Handicapped Accessories	6	Ea.	$ 278.00	$ 1,668	
C1030-110-1340	Urinal Screens	3	Ea.	$ 276.00	$ 828	
	Stairs					
C2010-110-0760	Stairs, Steel Pan	5	FLIGHT	$ 5,975.00	$ 29,875	
	Wall Finish					
C3010-105-0140	Painted Gypsum Board	20,388	S.F.	$ 0.57	$ 11,621	
C3010-105-0320	Painted CMU	4,725	S.F.	$ 1.19	$ 5,623	
C3010-105-1940	Ceramic Tile	1,175	S.F.	$ 5.46	$ 6,416	
C3010-105-1800	Vinyl Wall Covering	2,350	S.F.	$ 1.41	$ 3,314	
	Floor Finish					
C3020-410-0160	Carpet	27,900	S.F.	$ 5.47	$ 152,613	
C3020-410-1720	Ceramic Tile	1,200	S.F.	$ 7.58	$ 9,096	
C3020-410-1820	Quarry Tile	900	S.F.	$ 11.10	$ 9,990	
	Ceiling Finish					
C3030-105-6000	Ceiling, Suspended Acoustical	30,000	S.F.	$ 2.69	$ 80,700	
	Subtotal, Division C, Interior Finish				$ 444,011	$ 14.80

Figure 8.14

Chapter 9

Conveying Systems

Conveying systems, Division D10 of UNIFORMAT II, are used to transport people or materials vertically, on an incline, or horizontally through the building. These systems are typically provided and installed by the company that manufactures the system, and often require certified installers. They are highly specialized and can have a dramatic effect on the overall cost of a building. The following is a list of common conveying systems. Each has individual design requirements and special applications.

- Passenger elevators
- Freight and service elevators
- Escalators
- Moving stairs and walkways
- Dumbwaiters
- Conveyors
- Hoists & cranes
- Lifts (wheelchair, book, etc.)
- Material handling systems
- Pneumatic tube systems

Specific factors must be considered before selecting a particular assembly, as outlined in Figure 9.1.

Selecting and Estimating Elevators

This chapter focuses on elevators, because they are the most common conveying system in construction today. However, many of the points that are made apply to other systems found in this division. Before the building design is finalized, manufacturers and consultants should be contacted for details.

Elevator Types

The two general types of elevators are *hydraulic* and *electric*. Each has distinct uses, advantages, and disadvantages. Hydraulic elevators are directly driven from below by a pump and cylinder mechanism. Electric elevators use motors and steel cables with counterweights, which work on the traction principle.

Hydraulic Elevators

Hydraulic elevators are used most often in low-rise buildings—70' maximum—with travel speeds between 25 and 150 feet per minute. Because they are rather slow and travel short distances, hydraulic elevators are most often used in low-rise commercial, industrial, and residential buildings; motels and some hotels; and for moving freight.

The most significant factor to consider is the actual assembly operation. The hydraulic piston is attached to a bolster plank, which supports the elevator cab's platform. The piston operates in a cylinder that extends as far into the ground as the cab rises. For example, typically a hydraulic elevator that travels 55' up must have a piston and casing 55' deep. The casing is installed by a drilling rig. Hydraulic elevators are not often installed in existing buildings because the drilling rig and equipment require a lot of room for maneuvering. Hydraulic elevators usually cost less than their electric counterparts for a low-rise building. Figure 9.2 is a sketch of a typical hydraulic elevator.

Conveying Systems Selection Considerations

	Dumbwaiters	Elevators	Escalators	Material Handling Systems	Moving Stairs & Walks	Pneumatic Tube Systems	Vertical Conveyers
Hydraulic or Electric		x					
Geared or Gearless		x					
Capacity	x	x	x		x		
Size	x	x	x		x	x	
# Floors	x	x	x		x		
Story Height			x		x		
Stops	x	x					
Speed	x	x	x		x		
Finish		x	x		x		
# Required		x	x		x		
Machinery Location		x	x		x		
Door Type		x					
Signals		x					
Special Require.		x	x		x	x	
Operating System		x					
Incline Angle			x		x		
Automated				x			
Non-automated				x			
Automatic						x	x
Manual						x	x
# Stations						x	
Length						x	

Figure 9.1

Electric Elevators

Electric elevators are found most often in commercial and institutional buildings that are mid- to high-rise (over 5 floors). Above 50 stories, they can attain speeds up to 1800 fpm. Though not considered freight haulers, electric elevators can handle capacities up to 10,000 pounds.

The lifting power in electric elevators is provided by traction to steel hoisting cables. One end of the cables is attached to the elevator cab and the other to counterweights, which can weigh up to 40% of the load capacity. The counterweights are needed to help reduce the power requirements of the motors and provide constant traction.

There are two types of electric motor and drum assemblies: geared and gearless. Gearless assemblies are preferred for high-speed installations, while geared assemblies are suitable for low-speed applications.

Figure 9.3 is an illustration of a typical electric elevator.

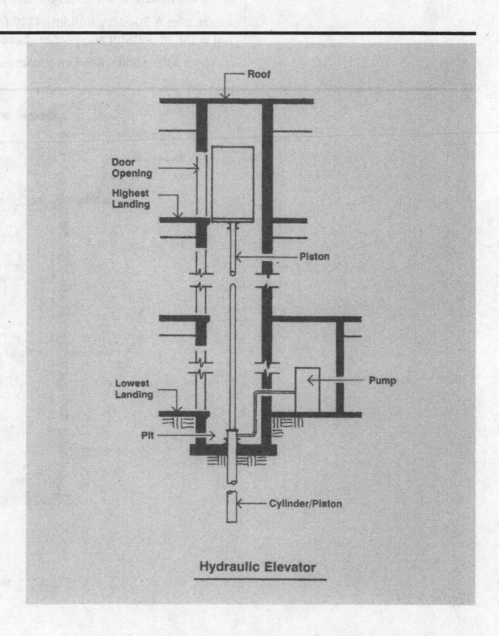

Hydraulic Elevator

Figure 9.2

Safety

Safety is a byword in elevator manufacturing, installation, and operation. Elevator systems are subject to many safety regulations and frequent safety inspections. As demanding as these safety requirements may seem, they are necessary, particularly when considering the heights and speeds that these machines must travel and their required reliability.

Speed and Capacity

The speed and capacity of elevators vary depending on the intended use, travel height, overall quality of the building, and the characteristics of the activities of the occupants. Normally, a small office building requires a minimum size elevator rated at 2,500 pounds capacity with a 6'-8" × 4'-6" car.

Elevator speeds should be sufficient to provide timely, efficient service. The maximum usable speed is limited by the building height, which includes an adequate allowance for acceleration and deceleration distances. A safe rule-of-thumb is that the rated elevator speed in fpm may be calculated as 1.6 times the rise in feet, plus 350.

For example: A 20-story building (310 feet) requires several elevators. What is the maximum acceptable speed?

$$(1.6 \times 310) + 350 = 846 \text{ fpm (maximum)}$$

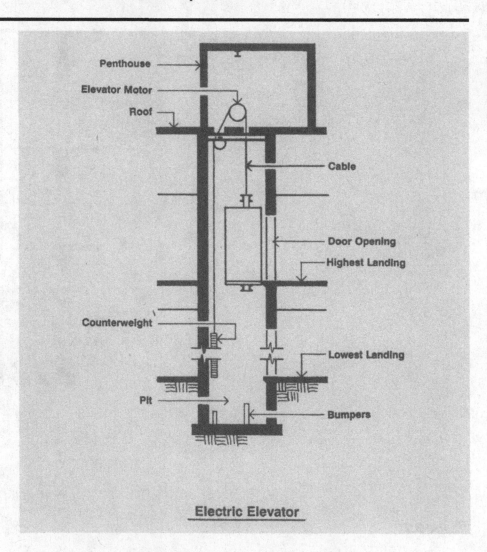

Electric Elevator

Figure 9.3

164

From this calculation, the elevators can be selected from manufacturers' data.

Figure 9.4 indicates the recommended elevator speed for office and general purpose buildings. Local building and elevator codes may mandate different requirements, so it is important to verify what is necessary.

Figure 9.5 indicates the height requirements based on speed and building type with recommended capacities. This information is broken down further in Figure 9.6.

Traffic   Other considerations include size, number, and location of elevators for effective traffic patterns. All of these factors depend on the building use, population, volume of traffic, building height, number of floors, car capacity, and elevator speed.

When computing elevator traffic, use the peak or critical traffic periods, which can vary greatly for different building types and occupancies. For

Number of Floors	Recommended Speeds in Feet Per Minute			
	Small	Average	Prestige	Service
2-5	200-250	300-350	350-400	200
5-10	300-350	350-500	500	300
10-15	500	500-700	700	350-500
15-25	700	800	800	500
25-35	—	1000	1000	500
35-45	—	1000-2000	1200	700-800
45-60	—	1200-1400	1400-1600	800-1000
over 60	—	—	1800	1000

Figure 9.4

Elevator Speed vs. Height Requirements

Building Type and Elevator Capacities	Travel Speeds in Feet per Minute								
	100 fpm	200 fpm	250 fpm	350 fpm	400 fpm	500 fpm	700 fpm	800 fpm	1000 fpm
Apartments 1200 lb. to 2500 lb.	to 70	to 100	to 125	to 150	to 175	to 250	to 350		
Department Stores 1200 lb. to 2500 lb.		100		125	175	250	350		
Hospitals 3500 lb. to 4000 lb.	70	100	125	150	175	250	350		
Office Buildings 2000 lb. to 4000 lb.		100	125	150		175	250	to 350	over 350

Note: Vertical transportation capacity may be determined by code occupancy requirements of the number of square feet per person divided into the total square feet of building type. If we are contemplating an office building, we find that the Occupancy Code Requirement is 100 S.F. per person. For a 20,000 S.F. building, we would have a legal capacity of two hundred people. Elevator handling capacity is subject to the five minute evacuation recommendation, but it may vary from 11 to 18%. Speed required is a function of the travel height, number of stops and capacity of the elevator.

Figure 9.5

example, traffic in office buildings peaks three times: during the morning when people arrive for work, during lunch time, and in the late afternoon when people go home. Hospital traffic usually reaches its peak during evening visiting hours.

Traffic is measured by the number of people handled during a five-minute peak interval. When this figure is divided by the five-minute handling capacity of an elevator, the minimum number of elevators required can be determined.

Handling capacity is computed based on a car's size (number of people carried) and its round-trip time. Analyze the car size first.

Sizing Elevators

The first step in sizing an elevator is to determine the building population. For a commercial building with a diversified occupancy, allow 120 to 150 net square feet per person. For a similar building with a single purpose or common occupancy, the number may be 100 to 120 net square feet per person. The net square feet figure should include only rentable or occupied space and exclude toilet rooms, lobbies, pipe chases, columns, and so forth. In apartments, hotels, and motels, the populations can be estimated by allowing two people per sleeping room. If the floor layout is unknown, use 200 square feet per person. The five-minute capacity requirement for elevators is a percentage of the building population and is shown in Figure 9.7.

Under normal circumstances, the average person requires two square feet of space to feel comfortable in an elevator. (This amount of space may vary for different countries and social climates.) During rush hour, particularly in densely populated areas, the amount of required space may be reduced to 1.3 square feet per person.

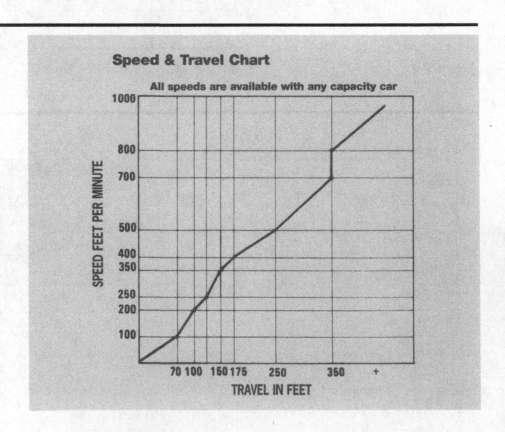

Figure 9.6

166

Use Figure 9.8 to predict elevator passenger capacity during peak periods, provided that the car load capacity (in pounds) is already known. The time required to make a round trip is indicated in Figure 9.9, and includes all the factors involved with a continuous, full speed, round-trip journey.

The waiting time between elevators should be 20 to 30 seconds or less. A waiting period that exceeds 35 seconds is unsatisfactory as far as most passengers are concerned. In residential buildings, passengers tend to be a bit more patient and may wait for up to 60 seconds.

Elevator zoning is a popular alternative for high-rise buildings. Elevators are initially sized in two or three groups of elevators, each

Elevator, 5-Minute Capacity		
Building Type	No. Square Feet/Person	5-Minute Capacity
Commercial, Diversified Occupancy	120 to 150	10% to 12%*
Commercial, Single Purpose Occupancy	100 to 120	15% to 20%*
Apartment	200**	5% to 10%*
Hotel	200**	10% to 15%*
* Percentage of building occupancy.		
** Or 2 persons per sleeping room.		

Figure 9.7

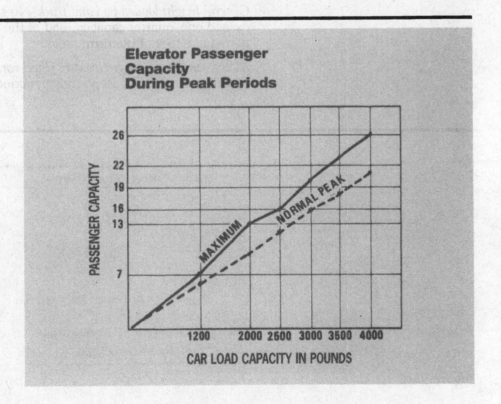

Figure 9.8

167

serving only specified floors. The next group of cars may serve the next group of floors, with no hall doors provided for the lower group of floors (except the lobby, of course). By omitting hall doors at floors not assigned to the group, the cars run express or non-stop from the lobby to their lowest floor, and become local thereafter.

Consider the following example: A proposed 30-story office tower is in the budget-setting stages. The architect has sized the elevators based on the peak traffic periods and has determined that 10 passenger elevators are needed. The 30th floor of the building will contain the company cafeteria.

Elevators: 1-3   Serve floors 1-10 and 30
4-6   Serve floors 1, 11-21, and 30
7-9   Serve floors 1 and 22-30
10   Serves floor 30 only

This example shows that the time saved by having a series of express elevators can result in increased productivity.

Freight Elevators

Freight elevators are specifically designed to vertically transport equipment, materials, and goods rather than passengers. In low-rise buildings, hydraulic elevators are more appropriate, but electrical are perhaps economically better for lifts exceeding 50 feet.

Freight elevators have a more rugged, less finished appearance. They also have a lower speed and greater capacity. Depending on the building height, standard speeds vary from 50 to 200 fpm, but may be greater for smaller capacities. Freight elevators are available in capacities from 2,500 to 25,000 pounds, or more if needed.

The American National Standard Safety Code for Elevators classifies freight elevators in the following categories:

General freight loaded by band truck: No single piece of freight may exceed one-quarter capacity load of the elevator, based on 50 pounds per square foot of platform area.

Motor vehicle or garage elevator: Used for automobiles and trucks only, based on capacity of 30 pounds per square foot of platform area.

Action During Round-Trip	Time (in seconds)
1. Passenger operates call button, doors open	3
2. Passenger enters car, operates floor button	2
3. Doors close	3
4. Car travels to selected floor	8
5. Doors reopen	2
6. Passenger leaves car	2
Total Time to Serve One Passenger	20
7. Doors close	3
8. Car returns to original floor	8
Total Round-Trip Time	31

Figure 9.9

Industrial truck loading: Must be capable of supporting impact load and weight of trucks. Capacity may not be less than 50 pounds per square foot of platform.

Service Elevators

Service elevators usually handle oversized loads, trucks, and four wheel carts. This type of system is usually found in commercial, industrial, apartment, hotel/motel, and institutional buildings. Normally, service elevators are equipped with abuse-resistant materials and have horizontal doors that slide up and down.

Service elevators must comply with the code requirements for carrying passengers. The size is normally determined by the largest or bulkiest anticipated load.

Federal Requirements

The Americans with Disabilities Act (ADA), which provides civil rights protection from discrimination to people with disabilities, contains specific requirements to ensure equal opportunity to access employment, public accommodations, and transportation. The ADA includes detailed requirements for elevators covering construction, dimensions, signage, controls, timing of signals, and access. These must be met by every elevator system and installation.

Cost of Elevators

Conveying systems represent a significant cost in a new building. The more that is known about specific requirements, the closer the estimated cost of the system will be to the final installed cost. Building codes must be complied with first. The next step is to decide between electric and hydraulic systems. When the installation is in a new, low-rise building, a hydraulic elevator is the most cost-effective choice. There are a number of cost-saving trade-offs to consider when estimating the capacity, number of units, speed, and quality of elevators. The time spent evaluating the variables that affect the final system selection is well worth it. Work done in this area during the preliminary estimate can result in a significant cost reduction.

Installed Costs

In almost all situations, elevators are provided and installed by the manufacturer and licensed installers. Typically, the costs quoted from the installer do not include such items as:

- The elevator pit
- Any necessary sump pumps
- Shaft walls
- Temporary doors or shaft fronts
- Steel framing other than rails that are considered part of the elevator equipment

Drilling into common earth is typically included in the installer's price, as is the piston casing for hydraulic elevators. Be sure to verify exactly what is and is not included in the installation price.

During the preliminary stages of a project, specific costs may be difficult to determine because sufficient design detail is not available. Wherever possible, current prices should be obtained from the manufacturer or certified installer. Verify who is responsible for inspections, tests, and permits. Some manufacturers include in their price a monthly maintenance program for the first year of operation with provisions to extend the program each year afterward for a specified price. Most, if not all, municipalities require some sort of annual inspection and certification of elevator systems.

Assemblies Estimate: Three-Story Office Building

From the floor plan in Figure 8.5, it is apparent that two elevators are required in the sample office building, one at each entry lobby. Since the building is only three stories tall, hydraulic elevators with a 3000-pound capacity are most practical.

Figure 9.10 provides several choices of hydraulic elevators. However, prices are listed only for two- and five-story systems. It is possible to accurately interpolate between the two systems to arrive at a reasonable price for a three-story system.

Hydraulic, 3000 lb.
2 Floors, 100 FPM   (Figure 9.10, line #2200)   = $45,125 ea.
Hydraulic, 3000 lb.
5 Floors, 100 FPM   (Figure 9.10, line #2300)   ~~71,300 ea.~~
Difference   = $26,175

Take the difference between the two systems and divide by three. Add the product to the two-floor system cost. The result will be a reasonably accurate system price for the three-floor elevator.

$$\frac{\$26,175}{3} = \$8,725 + \$45,125 = \$53,850 \text{ Ea.}$$

Since two elevators are needed, the total cost for conveying assemblies is:

$53,850 ea. × 2 = $107,000
Cost per Square Foot = $3.59/S.F.

Note that if there is subsurface rock at the building site, drilling and excavation costs will be substantial. "Rock Shafts," as they are known, are very expensive and must be dealt with in the preliminary stages of the estimate. The presence and depth of rocks are easily determined from analysis of nearby soil borings. Soil boring information may be obtained from local architects, engineers, or building or city inspectors. The costs for drilling or excavation of any subsurface rock must be provided for in the elevator estimate.

Figure 9.11 summarizes the costs for conveying systems in the three-story office building.

D10 Conveying

D1010 Elevators and Lifts

The hydraulic elevator obtains its motion from the movement of liquid under pressure in the piston connected to the car bottom. These pistons can provide travel to a maximum rise of 70' and are sized for the intended load. As the rise reaches the upper limits the cost tends to exceed that of a geared electric unit.

System Components	QUANTITY	UNIT	COST EACH		
			MAT.	INST.	TOTAL
SYSTEM D1010 110					
PASS. ELEV. HYDRAULIC 2500 LB. 5 FLOORS, 100 FPM					
Passenger elevator, hydraulic, 1500 lb capacity, 2 stop, 100 FPM	1.000	Ea.	29,700	8,600	38,300
Over 10' travel height, passenger elevator, hydraulic, add	40.000	V.L.F.	9,080	4,760	13,840
Passenger elevator, hydraulic, 2500lb capacity over standard, add	1.000	Ea.	1,175		1,175
Over 2 stops, passenger elevator, hydraulic, add	3.000	Stop	1,380	9,600	10,980
Hall lantern	5.000	Ea.	1,650	540	2,190
Maintenance agreement for pass. elev. 9 months	1.000	Ea.	3,025		3,025
Position indicator at lobby	1.000	Ea.	64	27	91
TOTAL			46,074	23,527	69,601

D1010 110	Hydraulic		COST EACH		
			MAT.	INST.	TOTAL
1300	Pass. elev., 1500 lb., 2 Floors, 100 FPM	RD1010 -010	33,400	8,825	42,225
1400	5 Floors, 100 FPM		44,900	23,500	68,400
1600	2000 lb., 2 Floors, 100 FPM		33,900	8,825	42,725
1700	5 floors, 100 FPM		45,400	23,500	68,900
1900	2500 lb., 2 Floors, 100 FPM		34,600	8,825	43,425
2000	5 floors, 100 FPM		46,100	23,500	69,600
2200	3000 lb., 2 Floors, 100 FPM		36,300	8,825	45,125
2300	5 floors, 100 FPM		47,800	23,500	71,300
2500	3500 lb., 2 Floors, 100 FPM		38,400	8,825	47,225
2600	5 floors, 100 FPM		49,900	23,500	73,400
2800	4000 lb., 2 Floors, 100 FPM		39,200	8,825	48,025
2900	5 floors, 100 FPM		50,500	23,500	74,000
3100	4500 lb., 2 Floors, 100 FPM		40,400	8,825	49,225
3200	5 floors, 100 FPM		52,000	23,500	75,500
4000	Hospital elevators, 3500 lb., 2 Floors, 100 FPM		49,700	8,825	58,525
4100	5 floors, 100 FPM		74,000	23,500	97,500
4300	4000 lb., 2 Floors, 100 FPM		49,700	8,825	58,525
4400	5 floors, 100 FPM		74,000	23,500	97,500
4600	4500 lb., 2 Floors, 100 FPM		55,500	8,825	64,325
4800	5 floors, 100 FPM		79,500	23,500	103,000
4900	5000 lb., 2 Floors, 100 FPM		58,000	8,825	66,825
5000	5 floors, 100 FPM		82,500	23,500	106,000
6700	Freight elevators (Class "B"), 3000 lb., 2 Floors, 50 FPM		51,500	11,200	62,700
6800	5 floors, 100 FPM		75,000	29,500	104,500
7000	4000 lb., 2 Floors, 50 FPM		54,000	11,200	65,200
7100	5 floors, 100 FPM		77,500	29,500	107,000
7500	10,000 lb., 2 Floors, 50 FPM		67,500	11,200	78,700
7600	5 floors, 100 FPM		91,000	29,500	120,500
8100	20,000 lb., 2 Floors, 50 FPM		82,500	11,200	93,700
8200	5 Floors, 100 FPM		106,000	29,500	135,500

Figure 9.10

ASSEMBLY NUMBER	DESCRIPTION	QTY	UNIT	TOTAL COST		COST PER S.F.
				UNIT	TOTAL	
DIVISION D	**SERVICES**					
D10	**CONVEYING**					
D1010-110-2200 D1010-110-2300	Elevator, Hydraulic, 3000# Capacity, 3 Floors	2	EACH	$ 53,850.00	$ 107,700	$ 3.59
	Interpolate Unit Cost:					
	2 Floors = $45,125					
	5 Floors = $71,300					
	Difference = $26,175					
	(Divided by 3 Floors) /3					
	Cost per Floor $8,725					
	(+ 2 Floor Cost) + $45,125					
	Cost for 3 Floors $53,850					
	Subtotal, Division D10, Conveying				$ 107,700	$ 3.59
D20	**PLUMBING**					
						3.59

Figure 9.11

Chapter 10

Plumbing

All buildings that will be occupied require some type of plumbing, from simple toilet facilities in warehouses, to sophisticated plumbing systems in hospitals or restaurants, to elaborate bathrooms in upscale residences. In UNIFORMAT II, Section D20 covers building plumbing. The amount of plumbing in a building is determined primarily by building code requirements related to occupancy and use. Building codes can be used as a starting point to determine the required plumbing if no plans are available.

Estimating Plumbing Systems

In an assemblies estimate, the estimator usually costs out the fixtures and rough plumbing, and uses factors to cover the cost of water service, control, waste, and vents. To properly develop plumbing assemblies in the assemblies square foot estimate format, it is necessary to first define certain parameters of the building, including:

- Building shape
- Building volume
- Number of floors
- Floor area
- Expected occupancy (number of people and male/female ratio)
- A general idea of the desired systems and controls

The following steps should be taken when developing the estimate for plumbing systems:

1. Determine the type and number of fixtures.
2. Select the appropriate price for each fixture system.
3. Develop the necessary piping costs.
4. Size and price the water heating components.
5. Size and price the costs for the storm and roof drainage system.

Once you know the building parameters, you can roughly determine the necessary fixtures based on building type and occupancy. See Figure 10.1, approximated minimum fixture requirements for various building types. Actual minimum fixture requirements are shown in plumbing codes, which vary by code jurisdiction. If the building

173

population were unknown, the information in Figure 10.1 would be difficult to use. If the building size is known, probable occupancy can be determined using Figure 2.6.

The following two examples illustrate the method used to estimate the cost of plumbing in a three-story office building and the cost of roof drainage.

Example One: Three-Story Office Building

(Note: The items in this example are representative and are not necessarily included in this book's estimate for the Three-Story Office Building, located at the end of this chapter.) Determine the amount of occupied space in a three-story office building with 10,000 square feet per floor. The Floor Area Ratios table in Figure 2.1 provides the data to make an approximation. From this table, office buildings have a 75% net to gross ratio. The remaining 25% represents the area occupied by hallways, lobbies, restrooms, janitor closets, and other non-rentable spaces. The three-story office building has:

10,000 S.F./floor (gross) × 75% = 7,500 S.F./floor (net)

Figure 2.6 can now be used to determine the building occupancy at 100 S.F./person. To determine the number of persons per floor, divide the net square feet by the square feet per person:

$$\frac{7,500 \text{ S.F./floor}}{100 \text{ S.F./person}} = 75 \text{ persons/floor}$$

Now, using Figure 10.1, determine the minimum plumbing fixtures required on each floor for an office building with 75 persons per floor. *(See also Figure 10.2.)*

When planning the restrooms, make certain that all ADA requirements are met. Figure 10.3 shows guidelines for toilet room layouts, including facilities for people with disabilities. At least one of each type of fixture should be handicap accessible in each toilet room.

Hot Water Requirements

Domestic water heating systems should be sized according to the consumption rates for maximum and average daily demand for the specific building type and occupancy. Water heating systems should be sized so that the anticipated maximum hourly demand can be met, in addition to an allowance for heat loss from the piping and storage tank.

Water heating systems can range from a small, under-the-counter unit heater, to the more common upright tank found in homes and offices, to a boiler in the building's mechanical room, and to a central heating plant that serves several buildings in a complex. Each system has its own set of requirements and limitations. This discussion focuses on unit- or tank-type water heating systems.

For a quick sizing method, use Figure 10.4, which lists hot water consumption rates for different types of buildings. Select the building type and size factor, then find the corresponding Maximum Hourly Demand and Average Day Demand.

For example, in a three-story office building with a population of 75 people/floor, from Figure 10.4, office buildings have a Maximum Hourly Demand of 0.4 gallons per person.

Minimum Plumbing Fixture Requirements

Type of Building or Occupancy (2)	Water Closets (14) (Fixtures per Person)		Urinals (5,10) (Fixtures per Person)		Lavatories (Fixtures per Person)		Bathtubs or Showers (Fixtures per Person)	Drinking Fountains (Fixtures per Person) (1,13)
	Male	Female	Male	Female	Male	Female		
Assembly Places- Theatres, Auditoriums, Convention Halls, etc.-for permanent employee use	1: 1 - 15 2: 16 - 35 3: 36 - 55 Over 55, add 1 fixture for each additional 40 persons	1: 1 - 15 2: 16 - 35 3: 36 - 55	0: 1 - 9 1: 10 - 50 Add one fixture for each additional 50 males		1 per 40	1 per 40		
Assembly Places- Theatres, Auditoriums, Convention Halls, etc. - for public use	1: 1 - 100 2: 101 - 200 3: 201 - 400 Over 400, add 1 fixture for each additional 500 males and 1 for each additional 125 females	3: 1 - 50 4: 51 - 100 8: 101 - 200 11: 201 - 400	1: 1 - 100 2: 101 - 200 3: 201 - 400 4: 401 - 600 Over 600, add 1 fixture for each additional 300 males		1: 1 - 200 2: 201 - 400 3: 401 - 750 Over 750, add 1 fixture for each additional 500 persons	1: 1 - 200 2: 201 - 400 3: 401 - 750		1: 1 - 150 2: 151 - 400 3: 401 - 750 Over 750, add one fixture for each additional 500 persons
Dormitories (9) School or Labor	1 per 10 Add 1 fixture for each additional 25 males (over 10) and 1 for each additional 20 females (over 8)	1 per 8	1 per 25 Over 150, add 1 fixture for each additional 50 males		1 per 12 Over 12 add 1 fixture for each additional 20 males and 1 for each 15 additional females	1 per 12	1 per 8 For females add 1 bathtub per 30. Over 150, add 1 per 20	1 per 150 (12)
Dormitories- for Staff Use	1: 1 - 15 2: 16 - 35 3: 36 - 55 Over 55, add 1 fixture for each additional 40 persons	1: 1 - 15 3: 16 - 35 4: 36 - 55	1 per 50		1 per 40	1 per 40	1 per 8	
Dwellings: Single Dwelling Multiple Dwelling or Apartment House	1 per dwelling 1 per dwelling or apartment unit				1 per dwelling 1 per dwelling or apartment unit		1 per dwelling 1 per dwelling or apartment unit	
Hospital Waiting rooms	1 per room				1 per room			1 per 150 (12)
Hospitals- for employee use	1: 1 - 15 2: 16 - 35 3: 36 - 55 Over 55, add 1 fixture for each additional 40 persons	1: 1 - 15 3: 16 - 35 4: 36 - 55	0: 1 - 9 1: 10 - 50 Add 1 fixture for each additional 50 males		1 per 40	1 per 40		
Hospitals: Individual Room Ward Room	1 per room 1 per 8 patients				1 per room 1 per 10 patients		1 per room 1 per 20 patients	1 per 150 (12)
Industrial (6) Warehouses Workshops, Foundries and similar establishments- for employee use	1: 1 -10 2: 11 - 25 3: 26 - 50 4: 51 - 75 5: 76 - 100 Over 100, add 1 fixture for each additional 30 persons	1: 1 -10 2: 11 - 25 3: 26 - 50 4: 51 - 75 5: 76 - 100			Up to 100, per 10 persons Over 100, 1 per 15 persons (7, 8)		1 shower for each 15 persons exposed to excessive heat or to skin contamination with poisonous, infectious or irritating material	1 per 150 (12)
Institutional - Other than Hospitals or Penal Institutions (on each occupied floor)	1 per 25	1 per 20	0: 1 - 9 1: 10 - 50 Add 1 fixture for each additional 50 males		1 per 10	1 per 10	1 per 8	1 per 150 (12)
Institutional - Other than Hospitals or Penal Institutions (on each occupied floor)- for employee use	1: 1 - 15 2: 16 - 35 3: 36 - 55 Over 55, add 1 fixture for each additional 40 persons	1: 1 - 15 3: 16 - 35 4: 36 - 55	0: 1 - 9 1: 10 - 50 Add 1 fixture for each additional 50 males		1 per 40	1 per 40	1 per 8	1 per 150 (12)
Office or Public Buildings	1: 1 - 100 2: 101 - 200 3: 201 - 400 Over 400, add 1 fixture for each additional 500 males and 1 for each additional 150 females	3: 1 - 50 4: 51 - 100 8: 101 - 200 11: 201 - 400	1: 1 - 100 2: 101 - 200 3: 201 - 400 4: 401 - 600 Over 600, add 1 fixture for each additional 300 males		1: 1 - 200 2: 201 - 400 3: 401 - 750 Over 750, add 1 fixture for each additional 500 persons	1: 1 - 200 2: 201 - 400 3: 401 - 750		1 per 150 (12)
Office or Public Buildings - for employee use	1: 1 - 15 2: 16 - 35 3: 36 - 55 Over 55, add 1 fixture for each additional 40 persons	1: 1 - 15 3: 16 - 35 4: 36 - 55	0: 1 - 9 1: 10 - 50 Add 1 fixture for each additional 50 males		1 per 40	1 per 40		

Figure 10.1

Type of Building or Occupancy	Water Closets (14) (Fixtures per Person)		Urinals (5, 10) (Fixtures per Person)		Lavatories (Fixtures per Person)		Bathtubs or Showers (Fixtures per Person)	Drinking Fountains (Fixtures per Person) (3, 13)
	Male	Female	Male	Female	Male	Female		
Penal Institutions - for employee use	1: 1 - 15 2: 16 - 35 3: 36 - 55 Over 55, add 1 fixture for each additional 40 persons	1: 1 - 15 3: 16 - 35 4: 36 - 55	0: 1 - 9 1: 10 - 50 Add 1 fixture for each additional 50 males		1 per 40	1 per 40		1 per 150 (12)
Penal Institutions - for prison use Cell	1 per cell				1 per cell			1 per cellblock floor
Exercise room	1 per exercise room		1 per exercise room		1 per exercise room			1 per exercise room
Restaurants, Pubs and Lounges (11)	1: 1 - 50 2: 51 - 150 3: 151 - 300 Over 300, add 1 fixture for each additional 200 persons	1: 1 - 50 2: 51 - 150 4: 151 - 300	1: 1 - 150 Over 150, add 1 fixture for each additional 150 males		1: 1 - 150 2: 151 - 200 3: 201 - 400 Over 400, add 1 fixture for each additional 400 persons	1: 1 - 150 2: 151 - 200 3: 201 - 400		
Schools - for staff use All Schools	1: 1 - 15 2: 16 - 35 3: 36 - 55 Over 55, add 1 fixture for each additional 40 persons	1: 1 - 15 3: 16 - 35 4: 36 - 55	1 per 50		1 per 40	1 per 40		
Schools - for student use: Nursery	1: 1 - 20 2: 21 - 50 Over 50, add 1 fixture for each additional 50 persons	1: 1 - 20 2: 21 - 50			1: 1 - 25 2: 26 - 50 Over 50, add 1 fixture for each additional 50 persons	1: 1 - 25 2: 26 - 50		1 per 150 (12)
Elementary	1 per 30	1 per 25	1 per 75		1 per 35	1 per 35		1 per 150 (12)
Secondary	1 per 40	1 per 30	1 per 35		1 per 40	1 per 40		1 per 150 (12)
Others (Colleges, Universities, Adult Centers, etc.	1 per 40	1 per 30	1 per 35		1 per 40	1 per 40		1 per 150 (12)
Worship Places Educational and Activities Unit	1 per 150	1 per 75	1 per 150		1 per 2 water closets			1 per 150 (12)
Worship Places Principal Assembly Place	1 per 150	1 per 75	1 per 150		1 per 2 water closets			1 per 150 (12)

Notes:

1. The figures shown are based upon one (1) fixture being the minimum required for the number of persons indicated or any fraction thereof.
2. Building categories not shown on this table shall be considered separately by the Administrative Authority.
3. Drinking fountains shall not be installed in toilet rooms.
4. Laundry trays. One (1) laundry tray or one (1) automatic washer standpipe for each dwelling unit or one (1) laundry trays or one (1) automatic washer standpipes, or combination thereof, for each twelve (12) apartments. Kitchen sinks, one (1) for each dwelling or apartment unit.
5. For each urinal added in excess of the minimum required, one water closet may be deducted. The number of water closets shall not be reduced to less than two-thirds (2/3) of the minimum requirement.
6. As required by ANSI Z4.1-1968, Sanitation in Places of Employment.
7. Where there is exposure to skin contamination with poisonous, infectious, or irritating materials, provide one (1) lavatory for each five (5) persons.
8. Twenty-four (24) lineal inches of wash sink or eighteen (18) inches of a circular basin, when provided with water outlets for such space shall be considered equivalent to one (1) lavatory.
9. Laundry trays, one (1) for each fifty (50) persons. Service sinks, one (1) for each hundred (100) persons.
10. General. In applying this schedule of facilities, consideration shall be given to the accessibility of the fixtures. Conformity purely on a numerical basis may not result in an installation suited to the need of the individual establishment. For example, schools should be provided with toilet facilities on each floor having classrooms.
    a. Surrounding materials, wall and floor space to a point two (2) feet in front of urinal lip and four (4) feet above the floor, and at least two (2) feet to each side of the urinal shall be lined with non-absorbent materials.
    b. Trough urinals shall be prohibited.
11. A restaurant is defined as a business which sells food to be consumed on the premises.
    a. The number of occupants for a drive-in restaurant shall be considered as equal to the number of parking stalls.
    b. Employee toilet facilities shall not be included in the above restaurant requirements. Hand washing facilities shall be available in the kitchen for employees.
12. Where food is consumed indoors, water stations may be substituted for drinking fountains. Offices, or public buildings for use by more than six (6) persons shall have one (1) drinking fountain for the first one hundred fifty (150) persons and one additional fountain for each three hundred (300) persons thereafter.
13. There shall be a minimum of one (1) drinking fountain per occupied floor in schools, theaters, auditoriums, dormitories, offices of public building.
14. The total number of water closets for females shall be at least equal to the total number of water closets and urinals required for males.

Reprinted from the Uniform Plumbing Code 2000, maintained by the International Association of Plumbing & Mechanical Officials.

Figure 10.1 (cont.)

75 people/floor × 0.4 gallons/person = 30 gallons/floor

Water heaters may be sized using one per floor or as a total for all floors.

Example Two: Roof Drainage

Adequate drainage of storm water away from the roof is necessary to prevent building damage from leaks and structural failure due to the buildup of weight on the roof. Assume a three-story office building is located in a city where the maximum rainfall is four inches per hour.

Storm water can be removed by the use of external drainage by pitching the roof to exterior scuppers and downspouts. For flat roofs, water is also drained internally through roof drains within the building.

Step 1: Sketch the Roof

See Figures 10.5 and 10.6.

Step 2: Compute the Roof Area to be Drained

Roof: 100' × 100'	=	10,000 S.F.
Parapet: (100' × 4) × 1'	=	400 S.F.
Stair and elevator shaft walls	=	580 S.F.
Total area to be drained	=	10,980 S.F. or 11,000 S.F.

For simplicity, the roof is divided into four equal zones of 2,750 S.F./each.

Step 3: Select the Leader and Horizontal Drain Size

Assume drains and leaders will be the same size for all areas.

Leader (see Figure 10.7): 4"
Horizontal Drain (see Figure 10.8): 4" @ 1/2" slope/foot

Two horizontal drains will join at a common vertical leader. The contributing area now becomes 5,500 S.F.

Leader (Figure 10.7): 5"
Horizontal Drain (Figure 10.8): 5" @ 1/2" slope/foot

The two horizontal drains will join at a common building horizontal drain, which will carry the roof drainage to the storm drainage system. The contributing area now becomes 11,000 S.F.

Horizontal drain (Figure 10.8): Select 6" @ 1/2" slope/foot since it is close to 11,000 S.F.

	Fixtures Per Floor			
Fixture	Minimum Required	Women	Men	Handicapped
Water Closets	4	3	1	2
Urinals	1	–	1	–
Lavatories	2	1	1	2
Drinking Fountain	1	–	–	–
Service Sink	1	–	–	–

Figure 10.2

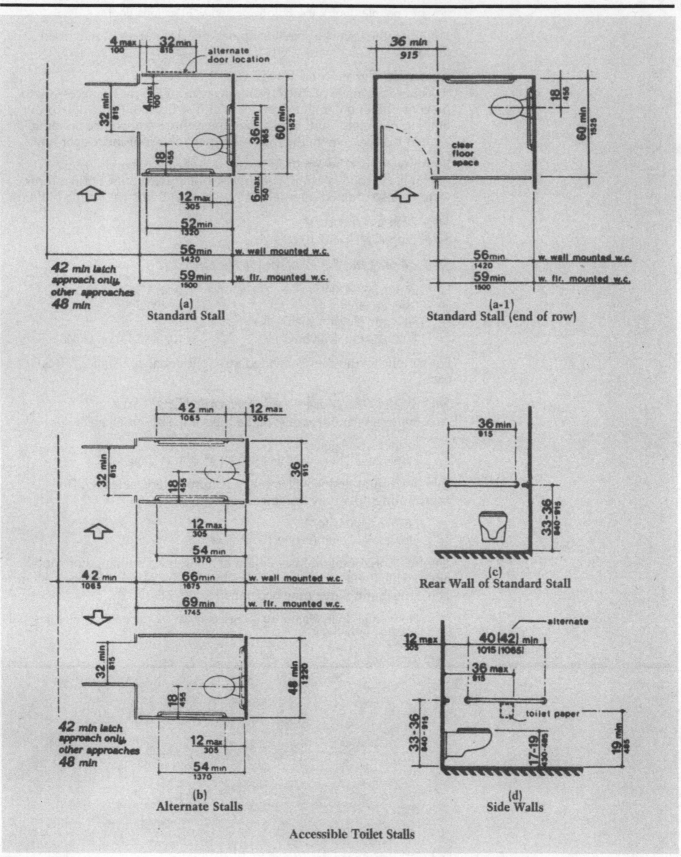

Accessible Toilet Stalls

(a) Standard Stall

(a-1) Standard Stall (end of row)

(b) Alternate Stalls

(c) Rear Wall of Standard Stall

(d) Side Walls

Figure 10.3

178

Hot Water Consumption Rates

Type of Building	Size Factor	Maximum Hourly Demand	Average Day Demand
Apartment Dwellings	No. of Apartments: Up to 20 21 to 50 51 to 75 76 to 100 101 to 200 201 up	12.0 Gal. per apt. 10.0 Gal. per apt. 8.5 Gal. per apt. 7.0 Gal. per apt. 6.0 Gal. per apt. 5.0 Gal. per apt.	42.0 Gal. per apt. 40.0 Gal. per apt. 38.0 Gal. per apt. 37.0 Gal. per apt. 36.0 Gal. per apt. 35.0 Gal. per apt.
Dormitories	Men Women	3.8 Gal. per man 5.0 Gal. per woman	13.1 Gal. per man 12.3 Gal. per woman
Hospitals	Per bed	23.0 Gal. per patient	90.0 Gal. per patient
Hotels	Single room with bath Double room with bath	17.0 Gal. per unit 27.0 Gal. per unit	50.0 Gal. per unit 80.0 Gal. per unit
Motels	No. of units: Up to 20 21 to 100 101 Up	6.0 Gal. per unit 5.0 Gal. per unit 4.0 Gal. per unit	20.0 Gal. per unit 14.0 Gal. per unit 10.0 Gal. per unit
Nursing Homes		4.5 Gal. per bed	18.4 Gal. per bed
Office buildings		0.4 Gal. per person	1.0 Gal. per person
Restaurants	Full meal type Drive-in snack type	1.5 Gal./max. meals/hr. 0.7 Gal./max. meals/hr.	2.4 Gal. per meal 0.7 Gal. per meal
Schools	Elementary Secondary & High	0.6 Gal. per student 1.0 Gal. per student	0.6 Gal. per student 1.8 Gal. per student

For evaluation purposes, recovery rate and storage capacity are inversely proportional. Water heaters should be sized so that the maximum hourly demand anticipated can be met in addition to allowance for the heat loss from the pipes and storage tank.

Figure 10.4

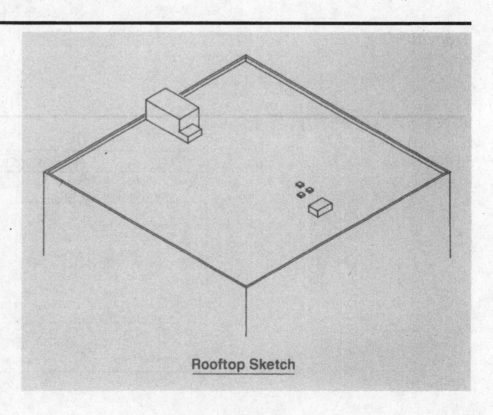

Rooftop Sketch

Figure 10.5

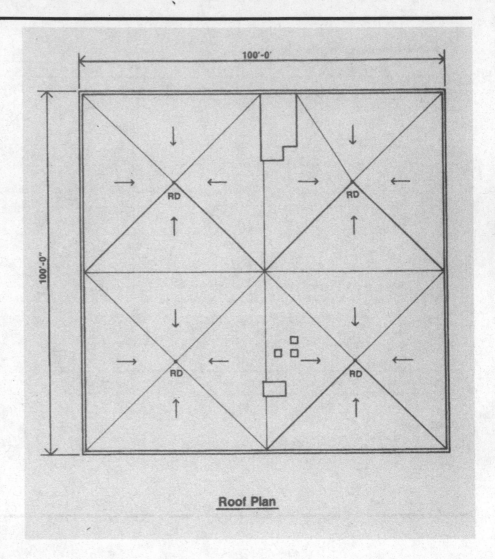

Roof Plan

Figure 10.6

Size of Vertical Leaders	
Size of Leader or Conductor, In.*	Maximum Projected Roof Area, S.F.
2	720
2-1/2	1,300
3	2,200
4	4,600
5	8,600
6	13,500
8	29,000
*The equivalent diameter of square or rectangular leader may be taken as the diameter of the circle that may be inscribed within the cross-sectional area of the leader.	

Figure 10.7

180

Determining the Carrying Capacity of Leaders

When computing the carrying capacity of leaders, there are three factors to consider: dimensions, cross-sectional area, and inside perimeter (area of water contact).

A rectangular leader, because of its four walls and corners, offers greater friction loss, thereby diminishing its carrying capacity. To compensate for this loss, a rectangular leader needs to be about 10% larger than a round leader to carry the same load.

The sizes for rectangular leaders (shown in Figure 10.9 as equivalents of the round leaders) include the 10% adjustment. In some cases, the

Size of Horizontal Storm Drains			
Diameter of Drain, In.	Maximum Projected Roof Area for Drains at Various Slopes, S.F.		
	1/8" Slope	1/4" Slope	1/2" Slope
3	822	1,160	1,644
4	1,880	2,650	3,760
5	3,340	4,720	6,680
6	5,350	7,550	10,700
8	11,500	16,300	23,000
10	20,700	29,200	41,400
12	33,300	47,000	66,600
15	59,500	84,000	119,000

Figure 10.8

Equivalent Size of Round & Rectangular Leaders					
Round Leader			Rectangular Leader		
Diameter, Inches	Cross Sectional Area, Sq. In.	Water Contact Area, In.	Dimension, Inches	Cross Sectional Area, Sq. In.	Water Contact Area, In.
2	3.14	6.28	2 x 2 / 1-1/2 x 1-1/2	4 / 3.75	88
3	7.07	9.42	2 x 4 / 2-1/2 x 3	8 / 7.50	12 / 11
4	12.57	12.57	3 x 4-1/4 / 3-1/2 x 4	12.75 / 14	14.5 / 14
5	19.06	15.07	4 x 5 / 4-1/2 x 4-1/2	20 / 20.25	18 / 18
6	28.27	18.85	5 x 6 / 5-1/2 x 5-1/2	30 / 30.25	22 / 22

Circumference as Straight Line

Rectangle as Straight Line

Figure 10.9

rectangular sizes are more than 10% larger. Where the computation of the equivalent size has resulted in an unavailable size, the next larger stock size is given. Always use "Water Contact Area, Inches" as the basis of equivalency.

Assemblies Estimate: Three-Story Office Building

The Plumbing Estimate Worksheet in Figure 10.10 is useful for preparing the estimate for the plumbing systems. It has sufficient space for development of the plumbing assemblies costs, with notations of cost source, quantities, extensions, and other information. Use this sheet prior to transferring the total figures to the estimate form to help streamline the estimating process.

Plumbing Fixtures

The minimum plumbing fixtures required on each floor of a three-story office building were determined in Figure 10.2. Once the fixtures are selected, price each one by finding the appropriate system cost from the tables in Figures 10.11 through 10.15. Note that these systems from *Means Assemblies Cost Data* include approximately 10' of the supply and drain piping, as well as associated items such as stops, traps, and tail pieces. This allowance assumes that the fixtures are connected to branches from a main.

Piping

If drawings of the proposed building are available, and the estimator has enough knowledge and experience with plumbing, then it is possible to put together an estimate for whatever piping, controls, valves, and other items are necessary for the various fixtures to work properly. The mechanical contractor and plumber should have the required knowledge and experience.

The "Plumbing Approximations" table in Figure 10.16 can be used to estimate the costs for water control, pipe and fittings, and project quality/complexity. The factors are a percentage of the total costs of all the plumbing fixtures in the project. They serve only as allowances and are not sufficient for special or process piping applications. Since the percentages are only guides, decide whether they are appropriate, too low, or too high for the particular project. Use the lower percentages for compact buildings, or for structures where the plumbing is in one area, or the fixtures are vertically stacked. The larger percentages should be used when the plumbing is spread throughout the building. The client's requirements and preliminary drawings will determine which of the quality/complexity multipliers to use. The percentages can be used separately for both material and installation or applied just once against the total cost.

Plumbing Costs:

Lavatory, Vanity Top			
(Figure 10.11, line #1600)	18 ea. @ $ 679 ea.	=	$12,222
Service Sink			
(Figure 10.12, line #4380)	3 ea. @ $1,430 ea.	=	4,290
Urinal			
(Figure 10.13, line #2000)	3 ea. @ $925 ea.	=	2,775
Water Cooler			
(Figure 10.14, line #1880)	3 ea. @ $1,320 ea.	=	3,960
Water Closet			
(Figure 10.15, line #1840)	15 ea. @ $ 1,185 ea.	=	17,775
Subtotal		=	$41,022

Plumbing Estimate Worksheet

Fixture	Table Number	Quantity	Unit Cost	Total Costs	Calculations
Bathtubs					
Drinking fountain					
Kitchen sink					
Laundry sink					
Lavatory Vanity top					
Service sink					
Shower					
Urinal					
Water cooler					
Water closet					
W/C group					
Wash fountain group					
Bathrooms					
Water heater					

Roof drains
    (Additional Length)
    "					

SUBTOTAL					
Water control					
Pipe & fittings					
Quality complexity					
Other					
TOTAL					

Figure 10.10

D2010 Plumbing Fixtures

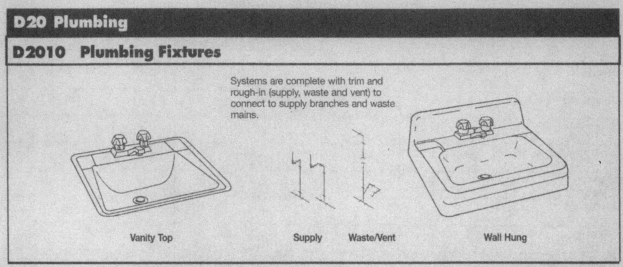

Systems are complete with trim and rough-in (supply, waste and vent) to connect to supply branches and waste mains.

Vanity Top   Supply   Waste/Vent   Wall Hung

System Components	QUANTITY	UNIT	COST EACH		
			MAT.	INST.	TOTAL
SYSTEM D2010 310					
LAVATORY W/TRIM, VANITY TOP, P.E. ON C.I., 20" X 18"					
Lavatory w/trim, PE on CI, white, vanity top, 20" x 18" oval	1.000	Ea.	172	117	289
Pipe, steel, galvanized, schedule 40, threaded, 1-1/4" diam	4.000	L.F.	9.48	33.60	43.08
Copper tubing type DWV, solder joint, hanger 10'OC 1-1/4" diam	4.000	L.F.	9.44	27.80	37.24
Wrought copper DWV, Tee, sanitary, 1-1/4" diam	1.000	Ea.	4.26	46	50.26
P trap w/cleanout, 20 ga, 1-1/4" diam	1.000	Ea.	13.10	23	36.10
Copper tubing type L, solder joint, hanger 10' OC 1/2" diam	10.000	L.F.	13.80	51.50	65.30
Wrought copper 90° elbow for solder joints 1/2" diam	2.000	Ea.	.48	42	42.48
Wrought copper Tee for solder joints, 1/2" diam	2.000	Ea.	.80	64	64.80
Stop, chrome, angle supply, 1/2" diam	2.000	Ea.	10.40	37.80	48.20
TOTAL			233.76	442.70	676.46

D2010 310	Lavatory Systems	COST EACH		
		MAT.	INST.	TOTAL
1560	Lavatory w/trim, vanity top, PE on CI, 20" x 18", Vanity top by others.	234	445	679
1600	19" x 16" oval	234	445	679
1640	18" round	221	445	666
1680	Cultured marble, 19" x 17"	173	445	618
1720	25" x 19"	203	445	648
1760	Stainless, self-rimming, 25" x 22"	240	445	685
1800	17" x 22"	235	445	680
1840	Steel enameled, 20" x 17"	155	455	610
1880	19" round	153	455	608
1920	Vitreous china, 20" x 16"	246	465	711
1960	19" x 16"	246	465	711
2000	22" x 13"	256	465	721
2040	Wall hung, PE on CI, 18" x 15"	535	490	1,025
2080	19" x 17"	435	490	925
2120	20" x 18"	410	490	900
2160	Vitreous china, 18" x 15"	355	500	855
2200	19" x 17"	340	500	840
2240	24" x 20"	465	500	965

(Note at 1640/1680: RD2010 -400)

Figure 10.11

D2010 Plumbing Fixtures

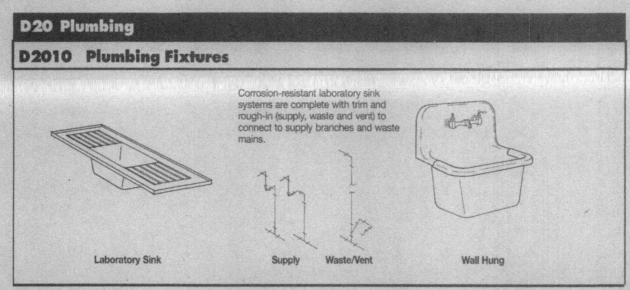

Corrosion-resistant laboratory sink systems are complete with trim and rough-in (supply, waste and vent) to connect to supply branches and waste mains.

Laboratory Sink   Supply   Waste/Vent   Wall Hung

System Components	QUANTITY	UNIT	COST EACH		
			MAT.	INST.	TOTAL
SYSTEM D2010 430					
LABORATORY SINK W/TRIM, POLYETHYLENE, SINGLE BOWL					
DOUBLE DRAINBOARD, 54" X 24" OD					
Sink w/trim, polyethylene, 1 bowl, 2 drainboards 54" x 24" OD	1.000	Ea.	855	250	1,105
Pipe, polypropylene, schedule 40, acid resistant 1-1/2" diam	10.000	L.F.	10.50	110	120.50
Tee, sanitary, polypropylene, acid resistant, 1-1/2" diam	1.000	Ea.	16.90	41.50	58.40
P trap, polypropylene, acid resistant, 1-1/2" diam	1.000	Ea.	29.50	24.50	54
Copper tubing type L, solder joint, hanger 10' OC 1/2" diam	10.000	L.F.	13.80	51.50	65.30
Wrought copper 90° elbow for solder joints 1/2" diam	2.000	Ea.	.48	42	42.48
Wrought copper Tee for solder joints, 1/2" diam	2.000	Ea.	.80	64	64.80
Stop, angle supply, chrome, 1/2" diam	2.000	Ea.	10.40	37.80	48.20
TOTAL			937.38	621.30	1,558.68

D2010 430	Laboratory & Service Sink Systems		COST EACH		
			MAT.	INST.	TOTAL
1580	Laboratory sink w/trim, polyethylene, single bowl,				
1600	Double drainboard, 54" x 24" O.D.		935	620	1,555
1640	Single drainboard, 47" x 24"O.D.	RD2010 -400	975	620	1,595
1680	70" x 24" O.D.		995	620	1,615
1760	Flanged, 14-1/2" x 14-1/2" O.D.		370	560	930
1800	18-1/2" x 18-1/2" O.D.		325	560	885
1840	23-1/2" x 20-1/2" O.D.		380	560	940
1920	Polypropylene, cup sink, oval, 7" x 4" O.D.		168	495	663
1960	10" x 4-1/2" O.D.		179	495	674
2000					
4260	Service sink w/trim, PE on CI, corner floor, 28" x 28", w/rim guard		665	625	1,290
4300	Wall hung w/rim guard, 22" x 18"		730	725	1,455
4340	24" x 20"		775	725	1,500
4380	Vitreous china, wall hung 22" x 20"		705	725	1,430

Figure 10.12

D2010 Plumbing Fixtures

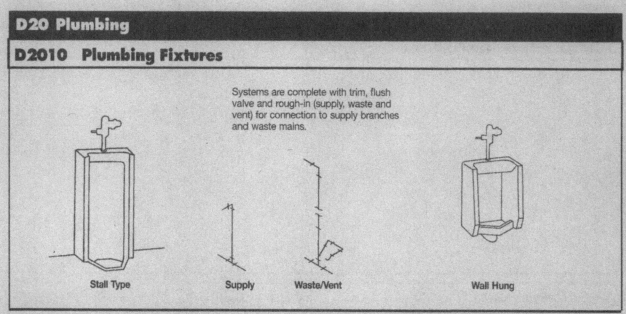

Systems are complete with trim, flush valve and rough-in (supply, waste and vent) for connection to supply branches and waste mains.

Stall Type   Supply   Waste/Vent   Wall Hung

System Components	QUANTITY	UNIT	COST EACH		
			MAT.	INST.	TOTAL
SYSTEM D2010 210					
URINAL, VITREOUS CHINA, WALL HUNG					
Urinal, wall hung, vitreous china, incl. hanger	1.000	Ea.	305	250	555
Pipe, steel, galvanized, schedule 40, threaded, 1-1/2" diam	5.000	L.F.	13.15	46.75	59.90
Copper tubing type DWV, solder joint, hangers 10'OC, 2" diam	3.000	L.F.	10.53	28.35	38.88
Combination Y & 1/8 bend for CI soil pipe, no hub, 3" diam	1.000	Ea.	9		9
Pipe, CI, no hub, cplg 10' OC, hanger 5' OC, 3" diam	4.000	L.F.	29	46.80	75.80
Pipe coupling standard, CI soil, no hub, 3" diam	3.000	Ea.	9.98	39.40	49.38
Copper tubing type L, solder joint, hanger 10' OC 3/4" diam	5.000	L.F.	8.70	27.25	35.95
Wrought copper 90° elbow for solder joints 3/4" diam	1.000	Ea.	.53	22	22.53
Wrought copper Tee for solder joints, 3/4" diam	1.000	Ea.	.98	34.50	35.48
TOTAL			391.08	536.55	927.63

D2010 210	Urinal Systems	COST EACH		
		MAT.	INST.	TOTAL
2000	Urinal, vitreous china, wall hung	390	535	925
2040	Stall type	605	590	1,195

Figure 10.13

D2010   Plumbing Fixtures

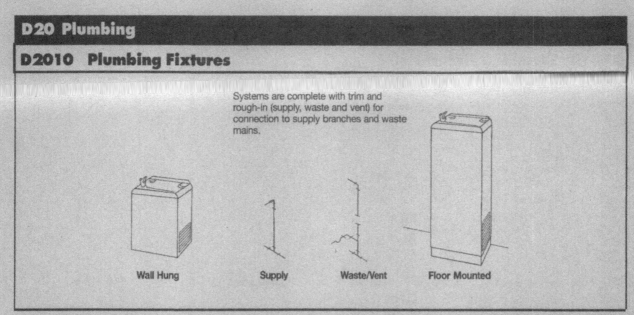

Systems are complete with trim and rough-in (supply, waste and vent) for connection to supply branches and waste mains.

Wall Hung   Supply   Waste/Vent   Floor Mounted

System Components	QUANTITY	UNIT	COST EACH		
			MAT.	INST.	TOTAL
SYSTEM D2010 820					
WATER COOLER, ELECTRIC, SELF CONTAINED, WALL HUNG, 8.2 GPH					
Water cooler, wall mounted, 8.2 GPH	1.000	Ea.	590	187	777
Copper tubing type DWV, solder joint, hanger 10'OC 1-1/4" diam	4.000	L.F.	9.44	27.80	37.24
Wrought copper DWV, Tee, sanitary 1-1/4" diam	1.000	Ea.	4.26	46	50.26
P trap, copper drainage, 1-1/4" diam	1.000	Ea.	13.10	23	36.10
Copper tubing type L, solder joint, hanger 10' OC 3/8" diam	5.000	L.F.	6.20	24.75	30.95
Wrought copper 90° elbow for solder joints 3/8" diam	1.000	Ea.	.66	18.90	19.56
Wrought copper Tee for solder joints, 3/8" diam	1.000	Ea.	1.13	29.50	30.63
Stop and waste, straightway, bronze, solder, 3/8" diam	1.000	Ea.	3.95	17.35	21.30
TOTAL			628.74	374.30	1,003.04

D2010 820	Water Cooler Systems		COST EACH		
			MAT.	INST.	TOTAL
1840	Water cooler, electric, wall hung, 8.2 GPH		630	375	1,005
1880	Dual height, 14.3 GPH		935	385	1,320
1920	Wheelchair type, 7.5 G.P.H.	RD2010 -400	1,575	375	1,950
1960	Semi recessed, 8.1 G.P.H.		900	375	1,275
2000	Full recessed, 8 G.P.H.		1,375	400	1,775
2040	Floor mounted, 14.3 G.P.H.		665	325	990
2080	Dual height, 14.3 G.P.H.		985	395	1,380
2120	Refrigerated compartment type, 1.5 G.P.H.		1,225	325	1,550

Figure 10.14

D2010   Plumbing Fixtures

Systems are complete with trim seat and rough-in (supply, waste and vent) for connection to supply branches and waste mains.

One Piece Wall Hung	Supply	Waste/Vent	Floor Mount

System Components			COST EACH		
	QUANTITY	UNIT	MAT.	INST.	TOTAL
SYSTEM D2010 110					
WATER CLOSET, VITREOUS CHINA, ELONGATED					
TANK TYPE, WALL HUNG, ONE PIECE					
Wtr closet tank type vit china wall hung 1 pc w/seat supply & stop	1.000	Ea.	470	141	611
Pipe steel galvanized, schedule 40, threaded, 2" diam	4.000	L.F.	13.36	46.80	60.16
Pipe, CI soil, no hub, cplg 10' OC, hanger 5' OC, 4" diam	2.000	L.F.	17.90	25.80	43.70
Pipe, coupling, standard coupling, CI soil, no hub, 4" diam	2.000	Ea.	11.90	45	56.90
Copper tubing type L, solder joint, hanger 10'OC, 1/2" diam	6.000	L.F.	8.28	30.90	39.18
Wrought copper 90° elbow for solder joints 1/2" diam	2.000	Ea.	.48	42	42.48
Wrought copper Tee for solder joints 1/2" diam	1.000	Ea.	.40	32	32.40
Support/carrier, for water closet, siphon jet, horiz, single, 4" waste	1.000	Ea.	226	69.50	295.50
TOTAL			748.32	433	1,181.32

D2010 110	Water Closet Systems		COST EACH		
			MAT.	INST.	TOTAL
1800	Water closet, vitreous china, elongated				
1840	Tank type, wall hung, one piece		750	435	1,185
1880	Close coupled two piece	RD2010 -400	805	435	1,240
1920	Floor mount, one piece		675	470	1,145
1960	One piece low profile		735	470	1,205
2000	Two piece close coupled		291	470	761
2040	Bowl only with flush valve				
2080	Wall hung		670	490	1,160
2120	Floor mount		480	475	955

Figure 10.15

Water Heaters

The three-story office building will need 30 gallons/floor for domestic hot water. Figure 10.17 (line #1820) shows an electric, 50-gallon tank with 37 gallons per hour heating capacity.

Cost:

(Figure 10.17, line #1820) ea. @ $3,075 ea =		$9,225
Cost per Square Foot =		$0.31/S.F.

Storm Drainage/Roof Drains

The key to pricing the roof drainage system is to remember the drainage pipe. Most estimators remember to include the roof drains, but fail to include the vertical and horizontal drain pipes.

Costs:

4" Drain with 10' of pipe (Cast Iron)			
(Figure 10.18, line #4200) 4 ea. @ $945 ea.		=	$3,780
5" Drain Pipe (Cast Iron)			
(Figure 10.18, line #4320) 60 L.F. @ $27.00/L.F.		=	1,620
6" Drain Pipe (Cast Iron)			
(Figure 10.18, line #4400) 200 L.F. @ $26.00/L.F.		=	5,200
Total Cost			$10,600
Cost per Square Foot		=	$0.35/S.F.
Water Control			
(Figure 10.16)	10% of Subtotal	=	$6,085
Pipe and Fittings			
(Figure 10.16)	30% of Subtotal	=	18,254
Quality/Complexity			
(Figure 10.16)	10% of Subtotal	=	6,085
Total Cost			$91,271
Cost per Square Foot		=	$3.04/S.F.

Figures 10.19 and 10.20 summarize the square foot costs for plumbing for the project.

Table D2010-031   Plumbing Approximations for Quick Estimating

Water Control
Water Meter; Backflow Preventer,

Shock Absorbers; Vacuum
Breakers; Mixer .... 10 to 15% of Fixtures
Pipe And Fittings .... 30 to 60% of Fixtures

Note: Lower percentage for compact buildings or larger buildings with plumbing in one area.
Larger percentage for large buildings with plumbing spread out.
In extreme cases pipe may be more than 100% of fixtures.
Percentages **do not** include special purpose or process piping.

Plumbing Labor
1 & 2 Story Residential .... Rough-in Labor = 80% of Materials
Apartment Buildings .... Rough-in Labor = 90 to 100% of Materials
Labor for handling and placing fixtures is approximately 25 to 30% of fixtures
Quality/Complexity Multiplier (for all installations)
Economy installation, add .... 0 to 5%
Good quality, medium complexity, add .... 5 to 15%
Above average quality and complexity, add .... 15 to 25%

Figure 10.16

D2020 Domestic Water Distribution

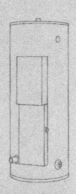

Systems below include piping and fittings within 10' of heater. Electric water heaters do not require venting.

System Components			COST EACH		
	QUANTITY	UNIT	MAT.	INST.	TOTAL
SYSTEM D2020 240					
ELECTRIC WATER HEATER, COMMERCIAL, 100° F RISE					
50 GALLON TANK, 9 KW, 37 GPH					
Water heater, commercial, electric, 50 Gal, 9 KW, 37 GPH	1.000	Ea.	2,175	231	2,406
Copper tubing, type L, solder joint, hanger 10' OC, 3/4" diam	34.000	L.F.	59.16	185.30	244.46
Wrought copper 90° elbow for solder joints 3/4" diam	5.000	Ea.	2.65	110	112.65
Wrought copper Tee for solder joints, 3/4" diam	2.000	Ea.	1.96	69	70.96
Wrought copper union for soldered joints, 3/4" diam	2.000	Ea.	5.78	46	51.78
Valve, gate, bronze, 125 lb, NRS, soldered 3/4" diam	2.000	Ea.	37.10	42	79.10
Relief valve, bronze, press & temp, self-close, 3/4" IPS	1.000	Ea.	71.50	14.85	86.35
Wrought copper adapter, copper tubing to male, 3/4" IPS	1.000	Ea.	.84	24.50	25.34
TOTAL			2,353.99	722.65	3,076.64

D2020 240	Electric Water Heaters - Commercial Systems	COST EACH		
		MAT.	INST.	TOTAL
1800	Electric water heater, commercial, 100° F rise			
1820	50 gallon tank, 9 KW 37 GPH [RD2010-400]	2,350	725	3,075
1860	80 gal, 12 KW 49 GPH	2,950	890	3,840
1900	36 KW 147 GPH	4,025	965	4,990
1940	120 gal, 36 KW 147 GPH	4,325	1,050	5,375
1980	150 gal, 120 KW 490 GPH	13,500	1,125	14,625
2020	200 gal, 120 KW 490 GPH	14,300	1,150	15,450
2060	250 gal, 150 KW 615 GPH	15,800	1,350	17,150
2100	300 gal, 180 KW 738 GPH	17,300	1,400	18,700
2140	350 gal, 30 KW 123 GPH	12,800	1,525	14,325
2180	180 KW 738 GPH	17,800	1,525	19,325
2220	500 gal, 30 KW 123 GPH	16,700	1,800	18,500
2260	240 KW 984 GPH	24,300	1,800	26,100
2300	700 gal, 30 KW 123 GPH	20,100	2,050	22,150
2340	300 KW 1230 GPH	29,900	2,050	31,950
2380	1000 gal, 60 KW 245 GPH	24,200	2,850	27,050
2420	480 KW 1970 GPH	24,800	2,875	27,675
2460	1500 gal, 60 KW 245 GPH	35,500	3,525	39,025
2500	480 KW 1970 GPH	48,800	3,525	52,325

Figure 10.17

D2040   Rain Water Drainage

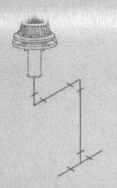

Design Assumptions: vertical conductor size is based on a maximum rate of rainfall of 4" per hour. To convert roof area to other rates multiply "Max. S.F. Roof Area" shown by four and divide the result by desired local rate. The answer is the local roof area that may be handled by the indicated pipe diameter.

Basic cost is for roof drain, 10' of vertical leader and 10' of horizontal, plus connection to the main.

Pipe Dia.	Max. S.F. Roof Area	Gallons per Min.
2"	544	23
3"	1610	67
4"	3460	144
5"	6280	261
6"	10,200	424
8"	22,000	913

System Components			COST EACH		
	QUANTITY	UNIT	MAT.	INST.	TOTAL
SYSTEM D2040 210					
ROOF DRAIN, DWV PVC PIPE, 2" DIAM., 10' HIGH					
Drain, roof, main, PVC, dome type 2" pipe size	1.000	Ea.	32	53.50	85.50
Clamp, roof drain, underdeck	1.000	Ea.	17.60	31	48.60
Pipe, Tee, PVC DWV, schedule 40, 2" pipe size	1.000	Ea.	1.86	37.50	39.36
Pipe, PVC, DWV, schedule 40, 2" diam.	20.000	L.F.	55.20	254	309.20
Pipe, elbow, PVC schedule 40, 2" diam.	2.000	Ea.	3.44	41	44.44
TOTAL			110.10	417	527.10

D2040 210	Roof Drain Systems	COST PER SYSTEM		
		MAT.	INST.	TOTAL
1880	Roof drain, DWV PVC, 2" diam., piping, 10' high	110	415	525
1920	For each additional foot add	2.76	12.70	15.46
1960	3" diam., 10' high	163	485	648
2000	For each additional foot add	4.78	14.15	18.93
2040	4" diam., 10' high	186	545	731
2080	For each additional foot add	5.20	15.60	20.80
2120	5" diam., 10' high	395	630	1,025
2160	For each additional foot add	8.50	17.40	25.90
2200	6" diam., 10' high	585	700	1,285
2240	For each additional foot add	8.95	19.20	28.15
2280	8" diam., 10' high	1,375	1,200	2,575
2320	For each additional foot add	14.40	24.50	38.90
3940	C.I., soil, single hub, service wt., 2" diam. piping, 10' high	247	455	702
3980	For each additional foot add	5.20	11.90	17.10
4120	3" diam., 10' high	345	495	840
4160	For each additional foot add	6.95	12.50	19.45
4200	4" diam., 10' high	405	540	945
4240	For each additional foot add	8.75	13.60	22.35
4280	5" diam., 10' high	450	595	1,045
4320	For each additional foot add	11.65	15.35	27
4360	6" diam., 10' high	690	635	1,325
4400	For each additional foot add	10.05	15.95	26
4440	8" diam., 10' high	1,500	1,300	2,800
4480	For each additional foot add	22	27	49
6040	Steel galv. sch 40 threaded, 2" diam. piping, 10' high	246	440	686
6080	For each additional foot add	3.34	11.70	15.04
6120	3" diam., 10' high	460	635	1,095
6160	For each additional foot add	6.15	17.40	23.55

Figure 10.18

Plumbing Estimate Worksheet

Fixture	Table Number	Quantity	Unit Cost	Total Costs	Calculations
Bathtubs					
Drinking fountain					
Kitchen sink					
Laundry sink					
Lavatory Vanity top	D2020-310-1600	18 EA	$ 679.00	$ 12,222.00	
Service sink	D2010-420-4380	3 EA	$ 1,430.00	$ 4,290.00	
Shower					
Urinal	D2010-210-2000	3 EA	$ 925.00	$ 2,775.00	
Water cooler	D2010-820-1880	3 EA	$ 1,320.00	$ 3,960.00	
Water closet	D2010-110-1840	15 EA	$ 1,185.00	$ 17,775.00	
W/C group					
Wash fountain group					
Bathrooms					
Water heater	D2020-240-1820	3 EA	$ 3,075.00	$ 9,225.00	
Roof drains	D2040-210-4200	4 EA	$ 945.00	$ 3,780.00	
(Additional Length)	D2040-210-4320	60 EA	$ 27.00	$ 1,620.00	
"	D2040-210-4400	200 LF	$ 26.00	$ 5,200.00	
SUBTOTAL				$ 60,847.00	
Water control	RD2010-031	10%		$ 6,084.70	% of Subtotal
Pipe & fittings	RD2010-031	30%		$ 18,254.10	% of Subtotal
Quality complexity	RD2010-031	10%		$ 6,084.70	% of Subtotal
Other					
TOTAL				$ 91,270.50	

Figure 10.19

ASSEMBLY NUMBER	DESCRIPTION	QTY	UNIT	TOTAL COST		COST PER S.F.
				UNIT	TOTAL	
DIVISION D	SERVICES					
D10	CONVEYING					
D1010-110-2200	Elevator, Hydraulic, 3000# Capacity, 3 Foors	2	EACH	$ 53,850.00	$ 107,700	$ 3.59
D1010-110-2300						
	Interpolate Unit Cost:					
	2 Floors = $45,125					
	5 Floors = $71,300					
	Difference = $26,175					
	(Divided by 3 Floors) /3					
	Cost per Foor $8,725					
	(+ 2 Floor Cost) + $45,125					
	Cost for 3 Floors $53,850					
	Subtotal, Division D10, Conveying				$ 107,700	$ 3.59
D20	PLUMBING					
D2020-310-1600	Lavatory, Vanity Top	18	Ea.	$ 679.00	$ 12,222	
D2010-420-4380	Service Sink	3	Ea.	$ 1,430.00	$ 4,290	
D2010-210-2000	Urinal	3	Ea.	$ 925.00	$ 2,775	
D2010-820-1880	Water Cooler	3	Ea.	$ 1,320.00	$ 3,960	
D2010-110-1840	Water Closet	15	Ea.	$ 1,185.00	$ 17,775	
D2020-240-1820	Water Heater, 50 Gal./37GPH	3	Ea.	$ 3,075.00	$ 9,225	
D2040-210-4200	Roof Drains, 4" CI	4	Ea.	$ 945.00	$ 3,780	
D2040-210-4320	5" Pipe, CI	60	Ea.	$ 27.00	$ 1,620	
D2040-210-4400	6" Pipe, CI	200	Ea.	$ 26.00	$ 5,200	
	Plumbing Subtotal				$ 60,847	
	Control (10%), Fittings (30%), Quality/Comp. (10%)				$ 30,424	
	Subtotal, Division D20, Plumbing				$ 91,271	$ 3.04

Figure 10.20

Chapter 11

Heating, Ventilation, and Air Conditioning

Mechanical systems in most buildings represent as much as 20% of the total building cost. In special building types such as hospitals, laboratories, and some manufacturing plants, mechanical costs can be even higher. As a result, designers and estimators must carefully evaluate the options and consider both initial installation costs and future operating costs. In some cases, the type of HVAC system is dictated by the owner's preference or specific requirements. Often an engineer can recommend the most suitable system based on initial design assumptions and calculations.

It is important to have the most accurate information available at the time of the estimate. For example, in the conceptual stages of a project, it may be known that the building is commercial, for general lease space. The HVAC system selected is a rooftop, multi-zone type. A little further into the design process, the client decides that the building will be used by a specific industry as a regional headquarters. This new purpose will require the HVAC system to be much more responsive to changing occupancies and uses on a daily basis. In addition, several areas in the building will generate an inordinate amount of waste heat, while others will require specific and critical environmental control. The original costs must be quickly reassessed to determine if there is enough money in the HVAC budget for the new building use.

The basic function of an HVAC system is to heat or cool an enclosed space to a desired temperature, and maintain it within a reasonable range. The system must be able to offset the heat transmission losses from the temperature difference between the interior and exterior of the enclosing walls and the roof/ceiling assembly, as well as the losses from cold air infiltration through crevices around doors and windows. The amount of heat or cooling needed depends on the building's size and use, construction quality, potential maximum temperature difference, air leakage, and the building's orientation and exposure. Air circulation is also an important consideration, as it prevents stratification, a condition that can result in heat loss through uneven temperatures at various levels, not to mention discomfort for occupants.

Heating

While the most accurate estimates of heating requirements are based on detailed information, it is possible to approximate the equipment size/capacity for costing purposes by using the following procedure:

1. Calculate the volume of the building.
2. Select the appropriate heat loss factor from Figure 11.1. Note that these factors apply only to the inside design temperatures listed in the first column and to a 0°F outside design temperature.
3. If the building has undesirable north and west exposures, multiply the heat loss factor by 1.1. Poor exposures can be determined if there is substantial shade on the north or west elevations and exposure to intense winter winds.
4. If the lowest expected outside design temperature is other than 0°F, multiply the factor from Figure 11.1 by the appropriate factor from Figure 11.2.
5. Multiply the cubic volume by the factor selected from Figure 11.1. This will result in the estimated BTU per hour heat loss that must be made up to maintain the inside design temperature.

(Note: The recommended low design temperature can be found in the ASHRAE [American Society of Heating, Refrigerating and Air-Conditioning Engineers, Inc.] book of Fundamentals under the heading "Climatic Design Information" and in some building codes. If these are not available, select a seasonally low temperature for estimating purposes.)

Factor for Determining Heat Loss for Various Types of Buildings

Building Type	Conditions	Qualifications	Loss Factor*
Factories & Industrial Plants General Office Areas at 70°F	One Story	Skylight in Roof	6.2
		No Skylight in Roof	5.7
	Multiple Story	Two Story	4.6
		Three Story	4.3
		Four Story	4.1
		Five Story	3.9
		Six Story	3.6
	All Walls Exposed	Flat Roof	6.9
		Heated Space Above	5.2
	One Long Warm Common Wall	Flat Roof	6.3
		Heated Space Above	4.7
	Warm Common Walls on Both Long Sides	Flat Roof	5.8
		Heated Space Above	4.1
Warehouses at 60°F	All Walls Exposed	Skylights in Roof	5.5
		No Skylight in Roof	5.1
		Heated Space Above	4.0
	One Long Warm Common Wall	Skylight in Roof	5.0
		No Skylight in Roof	4.9
		Heated Space Above	3.4
	Warm Common Walls on Both Long Sides	Skylight in Roof	4.7
		No Skylight in Roof	4.4
		Heated Space Above	3.0

*Note: This table tends to be conservative particularly for new buildings designed for minimum energy consumption.

Figure 11.1

Air Conditioning

The purpose of air conditioning is to control the environment of an enclosed space for human comfort or for equipment operation. System objectives must be defined and evaluated by several factors, including:

- Temperature control
- Humidity control
- Cleanliness
- Odor, smoke, and fumes
- Ventilation

Figure 11.3 provides a general rule of thumb for the air conditioning requirements of various building types. The numbers include the BTUs per hour per square foot of floor area and square feet per ton of air conditioning.

Air Conditioning Assemblies

Heating and air conditioning assemblies are frequently linked together in one system. One exception is perimeter fin tube radiation heating with a separate forced-air cooling system.

Air conditioning system prices in *Means Assemblies Cost Data* are listed by the cost per square foot for several building types, as shown in

Outside Design Temperature Correction Factor (for Degrees Fahrenheit)

Outside Design Temperature	50	40	30	20	10	0	−10	−20	−30
Correction Factor	0.29	0.43	0.57	0.72	0.86	1.00	1.14	1.28	1.43

Figure 11.2

Air Conditioning Requirements

BTU's per hour per S.F. of floor area and S.F. per ton of air conditioning.

Type of Building	BTU per S.F.	S.F. per Ton	Type of Building	BTU per S.F.	S.F. per Ton	Type of Building	BTU per S.F.	S.F. per Ton
Apartments, Individual	26	450	Dormitory, Rooms	40	300	Libraries	50	240
Corridors	22	550	Corridors	30	400	Low Rise Office, Exterior	38	320
Auditoriums & Theaters	40	300/18*	Dress Shops	43	280	Interior	33	360
Banks	50	240	Drug Stores	80	150	Medical Centers	28	425
Barber Shops	48	250	Factories	40	300	Motels	28	425
Bars & Taverns	133	90	High Rise Office—Ext. Rms.	46	263	Office (small suite)	43	280
Beauty Parlors	66	180	Interior Rooms	37	325	Post Office, Individual Office	42	285
Bowling Alleys	68	175	Hospitals, Core	43	280	Central Area	46	260
Churches	36	330/20*	Perimeter	46	260	Residences	20	600
Cocktail Lounges	68	175	Hotel, Guest Rooms	44	275	Restaurants	60	200
Computer Rooms	141	85	Corridors	30	400	Schools & Colleges	46	260
Dental Offices	52	230	Public Spaces	55	220	Shoe Stores	55	220
Dept. Stores, Basement	34	350	Industrial Plants, Offices	38	320	Shop'g. Ctrs., Supermarkets	34	350
Main Floor	40	300	General Offices	34	350	Retail Stores	48	250
Upper Floor	30	400	Plant Areas	40	300	Specialty	60	200

*Persons per ton
12,000 BTU = 1 ton of air conditioning

Figure 11.3

Figure 11.4. The buildings include apartments, banks and libraries, bars and taverns, bowling alleys, drug stores, offices, restaurants, and schools and colleges. For each system, there are several building sizes to choose from. If a particular building type or use is not listed, substitute another that is as similar as possible. The costs arrived at from interpolation and extrapolation are still reasonable enough for the accuracy level of an assemblies estimate. Large systems can be estimated as multiples of the systems shown at approximately the same price per square foot.

The most common air conditioning systems are listed below. (Heating may be included directly or through supplemental re-heat.)

- Chilled water with air-cooled condenser
- Chilled water with cooling tower
- Rooftop single-zone units
- Rooftop multi-zone units
- Self-contained water-cooled units
- Self-contained air-cooled units
- Split systems with air-cooled condensing

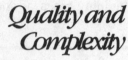

Quality and Complexity

The quality of the systems and prices in *Means Assemblies Cost Data* reflect recognized national building codes and represent sound construction. Average complexity is assumed. When using the assemblies format for estimating, some allowances may be made for situations that differ from the published or assumed quality and complexity. When using cost guides, be sure to read any descriptive or qualifying information that may indicate the quality/complexity allowed for in the prices quoted.

Figure 11.5 provides percentage factors that can be added to the HVAC estimate for varying quality and complexity.

Comparison of Systems

The following example compares the cost of a chilled water system with separate boiler for heat with single-zone and multi-zone heating/cooling rooftop systems.

Assume the building shown in Figure 11.6.

Lowest expected outside design temperature: 0°F
Heat Loss Factor from Figure 11.1: 4.3
Mechanical Penthouse: 70′ × 30′ = 2,100 S.F.
Office: 90′ × 210′ × 3 floors = 56,700 S.F.
Garage: 90′ × 210′ = 18,900 S.F.
Floor-to-floor height at offices: 14′
Volume of heated area 14′ × 56,700 S.F. = 793,800 C.F.

Chilled Water System with Boiler

The chilled water cooling system with a separate boiler for heat is used where building space for distribution is limited. Water is a more economical system space-wise for the transfer of heat or cooling. A small diameter pipe can be used instead of a larger dimension air duct. Air distribution in this system is achieved with fan coil units.

793,800 C.F. × 4.3 (heat loss factor) = 3,413,340 BTUH = 3.4 MBH

Using Figure 11.7, select the gas boiler nearest to the heat demand. The heat loss table is extremely conservative for new, highly insulated

D3030 Cooling Generating Systems

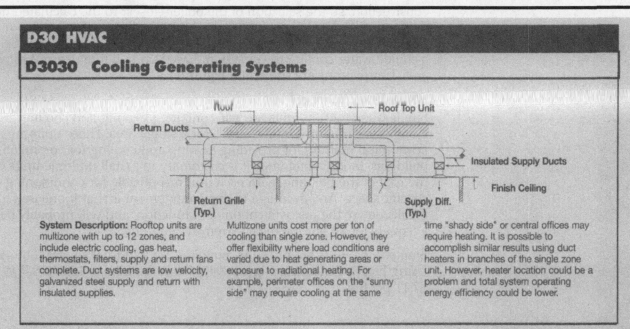

Return Ducts — Roof — Roof Top Unit — Insulated Supply Ducts — Finish Ceiling — Return Grille (Typ.) — Supply Diff. (Typ.)

System Description: Rooftop units are multizone with up to 12 zones, and include electric cooling, gas heat, thermostats, filters, supply and return fans complete. Duct systems are low velocity, galvanized steel supply and return with insulated supplies.

Multizone units cost more per ton of cooling than single zone. However, they offer flexibility where load conditions are varied due to heat generating areas or exposure to radiational heating. For example, perimeter offices on the "sunny side" may require cooling at the same time "shady side" or central offices may require heating. It is possible to accomplish similar results using duct heaters in branches of the single zone unit. However, heater location could be a problem and total system operating energy efficiency could be lower.

System Components			COST EACH		
	QUANTITY	UNIT	MAT.	INST.	TOTAL
SYSTEM D3030 206 ROOFTOP, MULTIZONE, AIR CONDITIONER APARTMENT CORRIDORS, 3,000 S.F., 5.50 TON					
Rooftop multizone unit, standard controls, curb	1.000	Ea.	19,316	1,155	20,471
Ductwork package for rooftop multizone units	1.000	System	2,282.50	8,250	10,532.50
TOTAL			21,598.50	9,405	31,003.50
COST PER S.F.			7.20	3.14	10.34
Note A: Small single zone unit recommended					

D3030 206	Rooftop Multizone Unit Systems	COST PER S.F.		
		MAT.	INST.	TOTAL
1240	Rooftop, multizone, air conditioner			
1260	Apartment corridors, 1,500 S.F., 2.75 ton. See Note A.			
1280	3,000 S.F., 5.50 ton	7.20	3.14	10.34
1440	25,000 S.F., 45.80 ton	4.40	3.03	7.43
1520	Banks or libraries, 1,500 S.F., 6.25 ton	16.35	7.15	23.50
1640	15,000 S.F., 62.50 ton	8.60	6.85	15.45
1720	25,000 S.F., 104.00 ton	8.15	6.80	14.95
1800	Bars and taverns, 1,500 S.F., 16.62 ton	34.50	10.25	44.75
1840	3,000 S.F., 33.24 ton	25.50	9.90	35.40
1880	10,000 S.F., 110.83 ton	19.35	9.85	29.20
2080	Bowling alleys, 1,500 S.F., 8.50 ton	22.50	9.70	32.20
2160	10,000 S.F., 56.70 ton	13.60	9.40	23
2240	20,000 S.F., 113.00 ton	11.05	9.25	20.30
2640	Drug stores, 1,500 S.F., 10.00 ton	26	11.40	37.40
2680	3,000 S.F., 20.00 ton	19.45	11	30.45
2760	15,000 S.F., 100.00 ton	13	10.95	23.95
3760	Offices, 1,500 S.F., 4.75 ton, See Note A			
3880	15,000 S.F., 47.50 ton	7.60	5.25	12.85
3960	25,000 S.F., 79.16 ton	6.55	5.20	11.75
4000	Restaurants, 1,500 S.F., 7.50 ton	19.65	8.55	28.20
4080	10,000 S.F., 50.00 ton	12	8.25	20.25
4160	20,000 S.F., 100.00 ton	9.75	8.20	17.95
4240	Schools and colleges, 1,500 S.F., 5.75 ton	15.05	6.55	21.60
4360	15,000 S.F., 57.50 ton	7.90	6.30	14.20

Note on row 1280: RD3030 -010

Figure 11.4

construction, so the selection of the boiler closest to the calculated heat demand will provide an adequate and reasonable cost projection. Calculations and component selection are shown on Figure 11.8. Select a chilled water cooling tower system based on building use and total area. Calculations are shown on Figure 11.9.

Single-Zone Rooftop Unit

Single-zone rooftop units are self-contained heating and cooling systems with thermostatic controls for a single area. These units are an economical method for providing heating and cooling for commercial buildings. In large buildings it is customary to install multiple units on the roof of the building, with each unit responsible for conditioning a specific space. An advantage to this arrangement is that if one unit breaks down, the rest will continue to function, and will probably be adequate for all but the most extreme conditions.

Single-zone units are priced by building use and total square footage. Using Figure 11.10, select line #3960, Office buildings, 10,000 S.F. @ $7.11 per S.F.

Heating Approximations for Quick Estimating

Oil Piping & Boiler Room Piping:
Small System .... 20 to 30% of Boiler
Complex System
with Pumps, Headers, Etc. .... 80 to 110% of Boiler
Breeching With Insulation:
Small .... 10 to 15% of Boiler
Large .... 15 to 25% of Boiler
Coils: .... 15 to 30% of Containing Unit
Balancing (Independent) .... 1/2% of H.V.A.C. Estimating

Quality/Complexity Adjustment: For all heating installations add these adjustments to the estimate to more closely allow for the equipment and conditions of the particular job under consideration.
Economy installation, add .... 0 to 5% of System
Good quality, medium complexity, add .... 5 to 15% of System
Above average quality and complexity, add .... 15 to 25% of System

Figure 11.5

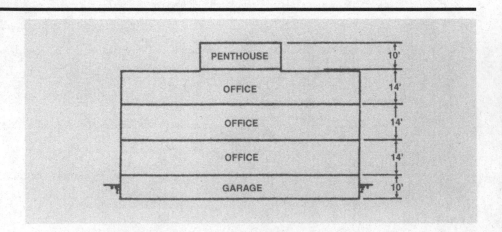

Figure 11.6

D3020 Heat Generating Systems

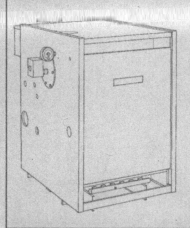

Boiler Selection: The maximum allowable working pressures are limited by ASME "Code for Heating Boilers" to 15 PSI for steam and 160 PSI for hot water heating boilers, with a maximum temperature limitation of 250°F. Hot water boilers are generally rated for a working pressure of 30 PSI. High pressure boilers are governed by the ASME "Code for Power Boilers" which is used almost universally for boilers operating over 15 PSIG. High pressure boilers used for a combination of heating/process loads are usually designed for 150 PSIG.

Boiler ratings are usually indicated as either Gross or Net Output. The Gross Load is equal to the Net Load plus a piping and pickup allowance. When this allowance cannot be determined, divide the gross output rating by 1.25 for a value equal to or greater than the next heat loss requirement of the building.

Table below lists installed cost per boiler and includes insulating jacket, standard controls, burner and safety controls. Costs do not include piping or boiler base pad. Outputs are Gross.

D3020 106	Boilers, Hot Water & Steam		COST EACH		
			MAT.	INST.	TOTAL
0600	Boiler, electric, steel, hot water, 12 K.W., 41 M.B.H.		3,725	900	4,625
0620	30 K.W., 103 M.B.H.		4,100	975	5,075
0640	60 K.W., 205 M.B.H.	RD3010 -010	5,200	1,050	6,250
0660	120 K.W., 410 M.B.H.		6,625	1,300	7,925
0680	210 K.W., 716 M.B.H.	RD3020 -020	11,000	1,950	12,950
0700	510 K.W., 1,739 M.B.H.		20,800	3,600	24,400
0720	720 K.W., 2,452 M.B.H.		26,200	4,075	30,275
0740	1,200 K.W., 4,095 M.B.H.		37,000	4,675	41,675
0760	2,100 K.W., 7,167 M.B.H.		58,000	5,875	63,875
0780	3,600 K.W., 12,283 M.B.H.		79,500	9,925	89,425
0820	Steam, 6 K.W., 20.5 M.B.H.		8,700	975	9,675
0840	24 K.W., 81.8 M.B.H.		9,250	1,050	10,300
0860	60 K.W., 205 M.B.H.		10,700	1,175	11,875
0880	150 K.W., 512 M.B.H.		13,900	1,800	15,700
0900	510 K.W., 1,740 M.B.H.		22,600	4,400	27,000
0920	1,080 K.W., 3,685 M.B.H.		34,700	6,350	41,050
0940	2,340 K.W., 7,984 M.B.H.		65,500	9,925	75,425
0980	Gas, cast iron, hot water, 80 M.B.H.		1,125	1,100	2,225
1000	100 M.B.H.		1,300	1,200	2,500
1020	163 M.B.H.		1,750	1,600	3,350
1040	280 M.B.H.		2,475	1,775	4,250
1060	544 M.B.H.		4,175	3,150	7,325
1080	1,088 M.B.H.		7,025	4,025	11,050
1100	2,000 M.B.H.		12,100	6,275	18,375
1120	2,856 M.B.H.		16,700	8,025	24,725
1140	4,720 M.B.H.		48,300	11,100	59,400
1160	6,970 M.B.H.		61,500	18,000	79,500
1180	For steam systems under 2,856 M.B.H., add 8%				
1240	Steel, hot water, 72 M.B.H.		1,925	590	2,515
1260	101 M.B.H.		2,200	655	2,855
1280	132 M.B.H.		2,500	695	3,195
1300	150 M.B.H.		2,900	785	3,685
1320	240 M.B.H.		4,450	905	5,355
1340	400 M.B.H.		6,350	1,475	7,825
1360	640 M.B.H.		8,625	1,975	10,600
1380	800 M.B.H.		10,100	2,350	12,450
1400	960 M.B.H.		12,500	2,625	15,125
1420	1,440 M.B.H.		17,400	3,375	20,775

Figure 11.7

D3030 Cooling Generating Systems

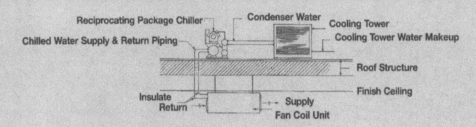

General: Water cooled chillers are available in the same sizes as air cooled units. They are also available in larger capacities.

Design Assumptions: The chilled water systems with water cooled condenser

include reciprocating hermetic compressors, water cooling tower, pumps, piping and expansion tanks and are based on a two pipe system. Chilled water piping is insulated. No ducts are included and fan-coil units are cooling only. Area

distribution is through use of multiple fan coil units. Fewer but larger fan coil units with duct distribution would be approximately the same S.F. cost. Water treatment and balancing are not included.

System Components	QUANTITY	UNIT	COST EACH		
			MAT.	INST.	TOTAL
SYSTEM D3030 115					
PACKAGED CHILLER, WATER COOLED, WITH FAN COIL UNIT •					
APARTMENT CORRIDORS, 4,000 S.F., 7.33 TON					
Fan coil air conditioner unit, cabinet mounted & filters, chilled water	2.000	Ea.	4,458.48	464.17	4,922.65
Water chiller, reciprocating, water cooled, 1 compressor semihermetic	1.000	Ea.	9,892.80	2,267.10	12,159.90
Cooling tower, draw thru single flow, belt drive	1.000	Ea.	630.38	96.02	726.40
Cooling tower pumps & piping	1.000	System	315.19	227.23	542.42
Chilled water unit coil connections	2.000	Ea.	1,030	1,900	2,930
Chilled water distribution piping	520.000	L.F.	5,070	15,080	20,150
TOTAL			21,396.85	20,034.52	41,431.37
COST PER S.F.			5.35	5.01	10.36

*Cooling requirements would lead to choosing a water cooled unit

D3030 115	Chilled Water, Cooling Tower Systems		COST PER S.F.		
			MAT.	INST.	TOTAL
1300	Packaged chiller, water cooled, with fan coil unit				
1320	Apartment corridors, 4,000 S.F., 7.33 ton		5.35	5	10.35
1600	Banks and libraries, 4,000 S.F., 16.66 ton	RD3030 -010	8.05	5.50	13.55
1800	60,000 S.F., 250.00 ton		5.55	4.57	10.12
1880	Bars and taverns, 4,000 S.F., 44.33 ton		12.65	6.95	19.60
2000	20,000 S.F., 221.66 ton		11.90	6.20	18.10
2160	Bowling alleys, 4,000 S.F., 22.66 ton		9.45	6.05	15.50
2320	40,000 S.F., 226.66 ton		6.65	4.45	11.10
2440	Department stores, 4,000 S.F., 11.66 ton		6.05	5.45	11.50
2640	60,000 S.F., 175.00 ton		4.49	4.28	8.77
2720	Drug stores, 4,000 S.F., 26.66 ton		9.90	6.25	16.15
2880	40,000 S.F., 266.67 ton		7.75	4.90	12.65
3000	Factories, 4,000 S.F., 13.33 ton		6.80	5.25	12.05
3200	60,000 S.F., 200.00 ton		4.85	4.38	9.23
3280	Food supermarkets, 4,000 S.F., 11.33 ton		5.95	5.40	11.35
3480	60,000 S.F., 170.00 ton		4.42	4.27	8.69
3560	Medical centers, 4,000 S.F., 9.33 ton		5.05	4.95	10
3760	60,000 S.F., 140.00 ton		4.06	4.21	8.27
3840	Offices, 4,000 S.F., 12.66 ton		6.55	5.20	11.75
4040	60,000 S.F., 190.00 ton		4.68	4.34	9.02
4120	Restaurants, 4,000 S.F., 20.00 ton		8.45	5.60	14.05
4320	60,000 S.F., 300.00 ton		6.25	4.73	10.98

Figure 11.8

Heating & Air Conditioning Estimate Worksheet
Chiller/Boiler

Equipment	Type	Table Number	Quantity	Unit Cost	Total Costs	Calculations
Heat source	Gas	D3020-106-1120	1 Ea.	$ 24,725.00	$ 24,725	Volume = 18,900 S.F. x 14' x 3 floors - 793,800 C.F. x 4.3 = 3,413,340 BTUH
Pipe	With A / C					
Duct	With A / C					
Terminals	With A / C					
Other						
Cold source		D3030-115-4040	56,700 S.F.	$ 9.02	$ 511,434.00	18,900 S.F. x 3 = 56,700 S.F. By taking the higher value, gives allowance for connect htg. pipe
Pipe	Included					
Duct	Included					
Terminals	Included					
Other						
SUBTOTAL					$ 536,159.00	
Quality complexity		RD3010-102	10 %		$ 53,615.90	
TOTAL					$ 589,774.90	

Figure 11.9

D3030 Cooling Generating Systems

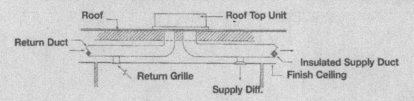

System Description: Rooftop single zone units are electric cooling and gas heat. Duct systems are low velocity, galvanized steel supply and return. Price variations between sizes are due to several factors. Jumps in the cost of the rooftop unit occur when the manufacturer shifts from the largest capacity unit on a small frame to the smallest capacity on the next larger frame, or changes from one compressor to two. As the unit capacity increases for larger areas the duct distribution grows in proportion. For most applications there is a tradeoff point where it is less expensive and more efficient to utilize smaller units with short simple distribution systems. Larger units also require larger initial supply and return ducts which can create a space problem. Supplemental heat may be desired in colder locations. The table below is based on one unit supplying the area listed. The 10,000 S.F. unit for bars and taverns is not listed because a nominal 110 ton unit would be required and this is above the normal single zone rooftop capacity.

System Components	QUANTITY	UNIT	COST EACH		
			MAT.	INST.	TOTAL
SYSTEM D3030 202					
ROOFTOP, SINGLE ZONE, AIR CONDITIONER					
APARTMENT CORRIDORS, 500 S.F., .92 TON					
Rooftop air-conditioner, 1 zone, electric cool, standard controls, curb	1.000	Ea.	2,127.50	494.50	2,622
Ductwork package for rooftop single zone units	1.000	System	265.88	713	978.88
TOTAL			2,393.38	1,207.50	3,600.88
COST PER S.F.			4.79	2.41	7.20

D3030 202	Rooftop Single Zone Unit Systems	COST PER S.F.		
		MAT.	INST.	TOTAL
1260	Rooftop, single zone, air conditioner			
1280	Apartment corridors, 500 S.F., .92 ton	4.80	2.40	7.20
1480	10,000 S.F., 18.33 ton	2.60	1.64	4.24
1560	Banks or libraries, 500 S.F., 2.08 ton	10.80	5.45	16.25
1760	10,000 S.F., 41.67 ton	5.55	3.71	9.26
1840	Bars and taverns, 500 S.F. 5.54 ton	13	7.30	20.30
2000	5,000 S.F., 55.42 ton	12.80	5.55	18.35
2080	Bowling alleys, 500 S.F., 2.83 ton	8.65	6.15	14.80
2280	10,000 S.F., 56.67 ton	7.35	5.05	12.40
2360	Department stores, 500 S.F., 1.46 ton	7.60	3.83	11.43
2560	10,000 S.F., 29.17 ton	3.95	2.59	6.54
2640	Drug stores, 500 S.F., 3.33 ton	10.15	7.25	17.40
2840	10,000 S.F., 66.67 ton	8.65	5.95	14.60
2920	Factories, 500 S.F., 1.67 ton	8.65	4.37	13.02
3120	10,000 S.F., 33.33 ton	4.52	2.96	7.48
3200	Food supermarkets, 500 S.F., 1.42 ton	7.35	3.72	11.07
3400	10,000 S.F., 28.33 ton	3.84	2.52	6.36
3480	Medical centers, 500 S.F., 1.17 ton	6.05	3.06	9.11
3680	10,000 S.F., 23.33 ton	3.31	2.09	5.40
3760	Offices, 500 S.F., 1.58 ton	8.25	4.15	12.40
3960	10,000 S.F., 31.67 ton	4.30	2.81	7.11
4000	Restaurants, 500 S.F., 2.50 ton	13	6.55	19.55
4200	10,000 S.F., 50.00 ton	6.50	4.45	10.95
4240	Schools and colleges, 500 S.F., 1.92 ton	9.95	5.05	15
4440	10,000 S.F., 38.33 ton	5.10	3.41	8.51

R03030
-010

Figure 11.10

Total cost = 56,700 S.F. × $7.11/S.F.	=	$403,137
Quality/Complexity allowance @ 10%	=	40,314
Total System Cost	=	$443,451

See Figure 11.11 for calculations.

Multi-Zone Rooftop Unit

The multi-zone rooftop unit is a sophisticated source for heating and cooling commercial space. In this system, one rooftop unit is controlled by multiple thermostats. Both supply and return air are ducted. Because of the complexity of the control system, the rooftop multi-zone unit is more costly than the use of multiple single-zone units for the same size building.

Multi-zone units are priced by building use and total square footage. Using Figure 11.4, select line #3960, Office buildings, 25,000 S.F. @ $11.75 per S.F.

Total cost = 56,700 S.F. × $11.75/S.F.	=	$666.225
Quality/Complexity allowance @ 10%	=	66,623
Total System Cost	=	$732,848

See Figure 11.12 for calculations.

Assemblies Estimate: Three-Story Office Building

The HVAC equipment in the sample office building consists of a rooftop multi-zone system, with a 7% allowance added for quality and complexity. Figure 11.13 is a heating and air conditioning estimate worksheet. Figure 11.14 summarizes the costs for this division.

Heating & Air Conditioning Estimate Worksheet
Single Zone Rooftop Unit

Equipment	Type	Table Number	Quantity	Unit Cost	Total Costs	Calculations
Heat source	single zone, gas With A / C					
Pipe	With A / C					
Duct	With A / C					
Terminals	With A / C					
Other	With A / C					
Cold source		D3030-202-3960	56,700 S.F.	$ 7.11	$ 403,137.00	
Pipe		-				
Duct						
Terminals						
Other						
SUBTOTAL					$ 403,137.00	
Quality complexity		RD3010-102	10 %		$ 40,313.70	
TOTAL					$ 443,451	

Figure 11.11

Heating & Air Conditioning Estimate Worksheet
Multi-zone Rooftop Unit

Equipment	Type	Table Number	Quantity	Unit Cost	Total Costs	Calculations
Heat source	multizone, gas With A / C					
Pipe	With A / C					
Duct	With A / C					
Terminals	With A / C					
Other	With A / C					
Cold source		D3030-206-3960	56,700 S.F.	$ 11.75	$ 666,225.00	
Pipe						
Duct						
Terminals						
Other						
SUBTOTAL					$ 666,225.00	
Quality complexity		RD3010-102	10 %		$ 66,622.50	
TOTAL					$ 732,848	

Figure 11.12

Heating & Air Conditioning Estimate Worksheet

Equipment	Type	Table Number	Quantity	Unit Cost	Total Costs	Calculations
Heat source Roof top	Multizone	D3030-206-3960	30,000 SF	$ 11.75	$ 352,500.00	
Pipe	Included					
Duct	Included					
Terminals						
Other						
Cold source	Included					
Pipe						
Duct						
Terminals						
Balance		RD3010-012	1 Allow	$ 3,000.00	$ 3,000.00	1/2% of HVAC for large jobs
Other						
SUBTOTAL					$ 355,500.00	
Quality complexity			7 %		$ 24,675.00	
TOTAL					$ 380,175.00	

Figure 11.13

ASSEMBLY NUMBER	DESCRIPTION	QTY	UNIT	TOTAL COST		COST PER S.F.
				UNIT	TOTAL	
DIVISION D	**SERVICES**					
D30	**Heating, Ventilating, and Air Conditioning**					
D3030-206-3960	Roof Top Multi-zone	30,000	S.F.	$ 11.75	$ 352,500	
	Balancing - Allowance				$ 3,000	
	Quality/Complexity (7%)				$ 24,675	
	Subtotal, Division D30, HVAC				$ 380,175	$ 12.67
D40	**Fire Protection**					

Figure 11.14

Chapter 12

Fire Protection

Fire Protection, Division D40 of UNIFORMAT II, has four specific purposes in the event of a fire:

- To provide occupants with a safe evacuation route or a place of refuge. In small buildings, evacuation is usually the minimum requirement, but in high-rise construction, a place of refuge is also necessary.
- To protect firefighters who enter the building by means of sufficient structural integrity to prevent building collapse.
- To protect adjacent property by providing adequate building separation and using appropriate exterior materials that reduce the risk of fire.
- To preserve the building itself.

To correctly interpret the national, regional, and local building codes that may apply to a structure, the building must first be classified. As a minimum standard, the following information must be established for each building:

- Fire zone
- Occupancy group
- Type of construction
- Number and height of floors
- Location on property
- Number of occupants
- Gross floor area

Building height, total area, and construction of exterior walls are limited by the fire zone, building use group, and distance from neighboring structures.

Model codes have established specific occupancy requirements. Buildings are broken down into broad groups as follows:

Group	Occupancy Type
A	Assembly
B	Business
E	Educational
F	Factory
H	Hazardous
I	Institutional
M	Mercantile
U	Miscellaneous
R	Residential
S	Storage

Buildings are further classified by the type of construction and degree of fire-resistance. Fire-resistance rates for major building elements vary from one to four hours. For example, the exterior bearing wall in a noncombustible, fire-resistive building may require four-hour construction, while the structural frame may require only three-hour construction.

Classes of Fires

There are four classes of fires, categorized by the type of material and equipment involved:

Class A: Ordinary materials, such as wood, paper, cloth, and rubber.
Class B: Flammable gases and liquids, such as natural gas, gasoline, and oil.
Class C: Electrical equipment.
Class D: Combustible metals such as sodium, potassium, and magnesium.

Fire extinguishing methods must be appropriate to the class of fire. Water, foam, soda acid, and other water-based extinguishing agents should never be used on Class C or D fires, for example. Systems with relatively low toxicity, such as FM-200, not IM-200, have been developed to safely extinguish Class B and C fires.

Standpipes

In buildings where water may be used as the extinguishing agent, standpipes are used to carry the water in large diameter pipes throughout the building. There are three standpipe systems that are recognized by the model codes: dry, wet, and combination standpipes.

A *dry standpipe* is a large, normally empty pipe that rises from the lowest floor to the roof. A Siamese connection is required at the street level to connect to the water supply. A dry standpipe is an extension of fire fighting equipment—once activated, hoses can be connected to extinguish a fire at any level. This system is not designed to be used by building occupants.

Buildings over four stories are required by model codes to have a *wet standpipe* system, which occupants can use to help suppress a fire until the fire department arrives. Wet standpipe systems also require a Siamese fitting at the street level and usually must be located so that every point of each floor is within 30 feet of the end of a 100-foot hose attached to a wet standpipe outlet.

Buildings higher than 150 feet must have a *combination standpipe* in every stairway or smoke-proof enclosure. This system replaces the wet and dry standpipes normally required.

Properly designed and maintained, standpipe systems are effective for extinguishing fires, especially in the upper stories of tall buildings, in the interior of large commercial or industrial facilities, or in other areas where construction features or access make using hoses time-consuming and hazardous.

There are three general classes of service for standpipes.

Class I: For use by fire department and personnel with special training only for heavy streams (2-1/2" hose connections).

Class II: For general use by building occupants until the fire department arrives (1-1/2" hose connector with hose).

Class III: For use by either the fire department or trained building occupants (2-1/2" and 1-1/2" hose connections or a 2-1/2" hose valve with an easily removable 2-1/2" by 1-1/2" adapter).

Sprinkler Systems

Automatic sprinkler systems have long been used as an effective means of controlling and extinguishing fires in their early stages. Because they are automatic, they do not require action by building occupants.

Building codes require automatic sprinkler systems in some building types. Even when not specifically mandated, these systems are highly recommended. The codes generally require automatic sprinklers in the following occupancies (there are exceptions, so check your local code for specifics):

- Basements and cellars, with the exception of residences and garages.
- Theater backstage areas, dressing rooms, workshops, and storage areas.
- Concealed areas above stairways in schools, hospitals, institutions such as prisons, and places of assembly such as theaters, stadiums, and schools.
- Most hazardous areas.
- All institutional occupancies, except prisons.
- Retail sales areas over 12,000 S.F./floor or over 24,000 S.F. gross.
- Places of assembly over 12,000 S.F.

There are a number of different types of sprinkler systems: wet pipe, dry pipe, preaction, deluge, and firecycle.

Wet Pipe Sprinkler System

A *wet pipe sprinkler system* has automatic sprinklers attached to a piping system that contains water and is connected to a water supply. When the sprinkler heads are opened by heat or fire, water immediately discharges. Benefits of the wet pipe system include quick response and a low initial cost.

Dry Pipe Sprinkler Systems

In a *dry pipe sprinkler system,* the water in the pipes has been replaced with compressed air. When the air is released by a sprinkler head opening, the dry pipe valve opens to allow water to flow to the open sprinkler head.

Preaction Sprinkler System

The *preaction sprinkler system* is a refinement of the dry pipe system. It uses automatic sprinklers connected to a piping system, which may or may not contain compressed air. To provide a second "look," a supplemental system of heat detectors with more sensitive characteristics is installed in the same areas as the sprinkler heads. When heat rises above the threshold of the detectors, a valve opens and water flows into the piping system from open heads. The advantage of this system is that when a sprinkler head is accidentally damaged, water does not flow as long as the heat detectors are not activated.

Deluge Sprinkler System

The *deluge system* is similar to the preaction system except that the sprinkler heads are always open. This system is normally used in areas with a high fire hazard.

Firecycle Sprinkler System

A *firecycle system* is fixed and uses water as the extinguishing agent. It is a time-delayed, recycling, preaction system that automatically shuts off the water when heat is reduced below the detector operating temperature. If the temperature is exceeded again, the water will automatically turn back on. The system senses a fire through a closed circuit electrical detector that controls water flow to the fire. A battery can supply up to 90 hours of emergency power. The piping remains dry until water is required, and is monitored with pressurized air. Should a leak occur, an alarm will sound, but water will not enter the system until heat is sensed by a firecycle detector.

Occupancy Hazards

Hazard levels for sprinklers have been established by building codes to determine the square feet of building floor area permitted for each sprinkler head. Since the model codes have varying rules regarding the installation of sprinkler systems, verification of your local code is essential before proceeding too far into project design. Depending on the occupancy classification, buildings will fall into one of three categories based on expected hazards: *light hazard*, *ordinary hazard*, or *extra hazard*. Figure 12.1 explains the three different classifications.

After the fire hazard classification has been determined, find the appropriate table for pricing. The pricing tables are set up by system type, fire hazard classification, and square footage that the system will cover. Standpipes and the other necessary components can be priced using the tables that follow. Figure 12.2 is a fire protection worksheet.

Assemblies Estimate: Three-Story Office Building

The example office building is in the light hazard classification and requires a wet pipe sprinkler system because of the potential for varied occupancies in the building.

Costs:
Wet Pipe Sprinkler System

(Figure 12.3, line #0620)	10,000 S.F. @ $1.69/S.F.	=	$16,900
(Figure 12.3, line #0740)	20,000 S.F. @ $1.36/S.F.	=	27,200
Standpipe			
(Figure 12.4, line #1540)	(12'/10') @ $4,125/floor	=	4,950
(Figure 12.4, line #1560)	(12'/10') @ $940/floor	=	2,256

Components:
Cabinet Assembly, incl. adapter, rack, hose & nozzle

(Figure 12.5, line #8400)	3 ea. @ $661 ea.	=	1,983

Siamese

(Figure 12.5, line #7200)	1 ea. @ $441 ea.	=	441

Alarm

(Figure 12.5, line #1050)	1 ea. @ $160.05 ea.	=	160

Adapter

(Figure 12.5, line #0100)	3 ea. @ $30.50 ea.	=	92

Fire Protection Sub-Total	$53,982
Quality/Complexity Allowance (10%)	$5,398
Fire Protection Total	$59,380
Cost per Square Foot	$1.98/S.F.

The costs for fire protection are shown in Figure 12.6.

Fire Protection Classification

System Classification

Rules for installation of sprinkler systems vary depending on the classification of occupancy falling into one of three categories as follows:

Light Hazard Occupancy

The protection area allotted per sprinkler should not exceed 200 S.F. with the maximum distance between lines and sprinklers on lines being 15'. The sprinklers do not need to be staggered. Branch lines should not exceed eight sprinklers on either side of a cross main. Each large area requiring more than 100 sprinklers and without a sub-dividing partition should be supplied by feed mains or risers sized for ordinary hazard occupancy.

Included in this group are:

Auditoriums	Museums
Churches	Nursing Homes
Clubs	Offices
Educational	Residential
Hospitals	Restaurants
Institutional	Schools
Libraries	Theaters
(except large stack rooms)	

Ordinary Hazard Occupancy

The protection area allotted per sprinkler shall not exceed 130 S.F. of noncombustible ceiling and 120 S.F. of combustible ceiling. The maximum allowable distance between sprinkler lines and sprinklers on line is 15'. Sprinklers shall be staggered if the distance between heads exceeds 12'. Branch lines should not exceed eight sprinklers on either side of a cross main.

Included in this group are:

Automotive garages	Electric generating stations
Bakeries	Feed mills
Beverage manufacturing	Grain elevators
Bleacheries	Ice manufacturing
Boiler houses	Laundries
Canneries	Machine shops
Cement plants	Mercantiles
Clothing factories	Paper mills
Cold storage warehouses	Printing and Publishing
Dairy products manufacturing	Shoe factories
Distilleries	Warehouses
Dry cleaning	Wood product assembly

Extra Hazard Occupancy

The protection area allotted per sprinkler shall not exceed 90 S.F. of noncombustible ceiling and 80 S.F. of combustible ceiling. The maximum allowable distance between lines and between sprinklers on lines is 12'. Sprinklers on alternate lines shall be staggered if the distance between sprinklers on lines exceeds 8'. Branch lines should not exceed six sprinklers on either side of a cross main.

Included in this group are:

Aircraft hangars	Paint shops
Chemical works	Shade cloth manufacturing
Explosives manufacturing	Solvent extracting
Linoleum manufacturing	Varnish works
Linseed oil mills	Volatile flammable
Oil refineries	liquid manufacturing & use

Figure 12.1

Fire Protection Estimate Worksheet

Type of Component	Hazard or Class	Table Number	Quantity		Unit Cost	Total Costs	Calculations
Sprinkler Wet pipe	Light	D4010-305-0820	10,000	SF	1.69	$ 16,900.00	
		D4020-305-0740	20,000	SF	1.36	$ 27,200.00	
Standpipe	III	D4020-310-1540	12/10	Floor	4125.00	$ 4,950.00	
		D4020-310-1560	24/10	Floor	940.00	$ 2,256.00	
Fire suppression							
Cabinets & Components		D4020-410-8400	3	EA	661.00	$ 1,983.00	
		D4020-410-7200	1	EA	441.00	$ 441.00	
		D4020-410-1650	1	EA	160.05	$ 160.05	
		D4020-410-0100	3	EA	30.50	$ 91.50	
SUBTOTAL						$ 53,981.55	
Quality complexity		RD4020-303	10%			$ 5,398.16	
TOTAL						$ 59,379.71	

Figure 12.2

D4010   Sprinklers

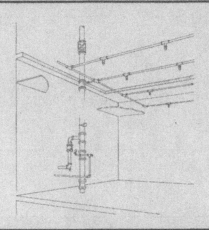

Wet Pipe System. A system employing automatic sprinklers attached to a piping system containing water and connected to a water supply so that water discharges immediately from sprinklers opened by heat from a fire.

All areas are assumed to be open.

System Components	QUANTITY	UNIT	COST EACH MAT.	COST EACH INST.	COST EACH TOTAL
SYSTEM D4010 305					
WET PIPE SPRINKLER, STEEL, BLACK, SCH. 40 PIPE					
LIGHT HAZARD, ONE FLOOR, 2000 S.F.					
Valve, gate, iron body, 125 lb, OS&Y, flanged, 4" diam	1.000	Ea.	285	187.50	472.50
Valve, swing check, bronze, 125 lb, regrinding disc, 2-1/2" pipe size	1.000	Ea.	147.75	37.50	185.25
Valve, angle, bronze, 150 lb, rising stem, threaded, 2" diam	1.000	Ea.	180.75	28.50	209.25
*Alarm valve, 2-1/2" pipe size	1.000	Ea.	615	187.50	802.50
Alarm, water motor, complete with gong	1.000	Ea.	137.25	78	215.25
Valve, swing check, w/balldrip CI with brass trim 4" pipe size	1.000	Ea.	116.25	187.50	303.75
Pipe, steel, black, schedule 40, 4" diam	10.000	L.F.	64.88	172.43	237.31
*Flow control valve, trim & gauges, 4" pipe size	1.000	Set	1,237.50	427.50	1,665
Fire alarm horn, electric	1.000	Ea.	30	45.75	75.75
Pipe, steel, black, schedule 40, threaded, cplg & hngr 10'OC, 2-1/2" diam	20.000	L.F.	75.75	225	300.75
Pipe, steel, black, schedule 40, threaded, cplg & hngr 10'OC, 2" diam	12.500	L.F.	29.81	109.69	139.50
Pipe, steel, black, schedule 40, threaded, cplg & hngr 10'OC, 1-1/4" diam	37.500	L.F.	63	236.25	299.25
Pipe steel, black, schedule 40, threaded cplg & hngr 10'OC, 1" diam	112.000	L.F.	160.44	659.40	819.84
Pipe Tee, malleable iron black, 150 lb threaded, 4" pipe size	2.000	Ea.	174	280.50	454.50
Pipe Tee, malleable iron black, 150 lb threaded, 2-1/2" pipe size	2.000	Ea.	48.75	124.50	173.25
Pipe Tee, malleable iron black, 150 lb threaded, 2" pipe size	1.000	Ea.	11.33	51	62.33
Pipe Tee, malleable iron black, 150 lb threaded, 1-1/4" pipe size	5.000	Ea.	26.63	200.63	227.26
Pipe Tee, malleable iron black, 150 lb threaded, 1" pipe size	4.000	Ea.	13.17	156	169.17
Pipe 90° elbow, malleable iron black, 150 lb threaded, 1" pipe size	6.000	Ea.	12.69	144	156.69
Sprinkler head, standard spray, brass 135°-286°F 1/2" NPT, 3/8" orifice	12.000	Ea.	79.20	312	391.20
Valve, gate, bronze, NRS, class 150, threaded, 1" pipe size	1.000	Ea.	27.38	16.50	43.88
*Standpipe connection, wall, single, flush w/plug & chain 2-1/2"x2-1/2"	1.000	Ea.	64.88	112.50	177.38
TOTAL			3,601.41	3,980.15	7,581.56
COST PER S.F.			1.80	1.99	3.79

*Not included in systems under 2000 S.F.

D4010 305	Wet Pipe Sprinkler Systems		COST PER S.F. MAT.	COST PER S.F. INST.	COST PER S.F. TOTAL
0520	Wet pipe sprinkler systems, steel, black, sch. 40 pipe				
0530	Light hazard, one floor, 500 S.F.		1.11	1.90	3.01
0560	1000 S.F.		1.72	1.97	3.69
0580	2000 S.F.	RD4010 -100	1.80	1.99	3.79
0600	5000 S.F.		.86	1.41	2.27
0620	10,000 S.F.	RD4020 -300	.53	1.16	1.69

Figure 12.3

D40 Fire Protection

D4010 Sprinklers

D4010 205	Wet Pipe Sprinkler Systems	COST PER S.F.		
		MAT.	INST.	TOTAL
0640	50,000 S.F.	.36	1.08	1.44
0660	Each additional floor, 500 S.F.	.54	1.61	2.15
0680	1000 S.F.	.51	1.50	2.01
0700	2000 S.F.	.47	1.36	1.83
0720	5000 S.F.	.33	1.15	1.48
0740	10,000 S.F.	.30	1.06	1.36
0760	50,000 S.F.	.26	.86	1.12
1000	Ordinary hazard, one floor, 500 S.F.	1.21	2.04	3.25
1020	1000 S.F.	1.70	1.95	3.65
1040	2000 S.F.	1.85	2.10	3.95
1060	5000 S.F.	.97	1.53	2.50
1080	10,000 S.F.	.66	1.55	2.21
1100	50,000 S.F.	.60	1.54	2.14
1140	Each additional floor, 500 S.F.	.64	1.82	2.46
1160	1000 S.F.	.49	1.49	1.98
1180	2000 S.F.	.54	1.53	2.07
1200	5000 S.F.	.54	1.43	1.97
1220	10,000 S.F.	.43	1.45	1.88
1240	50,000 S.F.	.46	1.36	1.82
1500	Extra hazard, one floor, 500 S.F.	3.23	3.14	6.37
1520	1000 S.F.	2.12	2.73	4.85
1540	2000 S.F.	1.96	2.78	4.74
1560	5000 S.F.	1.21	2.45	3.66
1580	10,000 S.F.	1.14	2.33	3.47
1600	50,000 S.F.	1.05	2.26	3.31
1660	Each additional floor, 500 S.F.	.83	2.24	3.07
1680	1000 S.F.	.83	2.12	2.95
1700	2000 S.F.	.72	2.14	2.86
1720	5000 S.F.	.60	1.88	2.48
1740	10,000 S.F.	.73	1.76	2.49
1760	50,000 S.F.	.68	1.66	2.34
2020	Grooved steel, black sch. 40 pipe, light hazard, one floor, 2000 S.F.	1.83	1.66	3.49
2060	10,000 S.F.	.70	1.02	1.72
2100	Each additional floor, 2000 S.F.	.52	1.09	1.61
2150	10,000 S.F.	.34	.87	1.21
2200	Ordinary hazard, one floor, 2000 S.F.	1.87	1.79	3.66
2250	10,000 S.F.	.65	1.31	1.96
2300	Each additional floor, 2000 S.F.	.56	1.22	1.78
2350	10,000 S.F.	.42	1.21	1.63
2400	Extra hazard, one floor, 2000 S.F.	2.01	2.26	4.27
2450	10,000 S.F.	.96	1.70	2.66
2500	Each additional floor, 2000 S.F.	.79	1.74	2.53
2550	10,000 S.F.	.66	1.52	2.18
3050	Grooved steel black sch. 10 pipe, light hazard, one floor, 2000 S.F.	1.81	1.65	3.46
3100	10,000 S.F.	.56	.96	1.52
3150	Each additional floor, 2000 S.F.	.50	1.08	1.58
3200	10,000 S.F.	.33	.86	1.19
3250	Ordinary hazard, one floor, 2000 S.F.	1.85	1.78	3.63
3300	10,000 S.F.	.64	1.27	1.91
3350	Each additional floor, 2000 S.F.	.54	1.21	1.75
3400	10,000 S.F.	.41	1.17	1.58
3450	Extra hazard, one floor, 2000 S.F.	1.99	2.26	4.25
3500	10,000 S.F.	.92	1.69	2.61
3550	Each additional floor, 2000 S.F.	.77	1.74	2.51
3600	10,000 S.F.	.65	1.51	2.16
4050	Copper tubing, type M, light hazard, one floor, 2000 S.F.	1.72	1.66	3.38
4100	10,000 S.F.	.55	.99	1.54
4150	Each additional floor, 2000 S.F.	.42	1.11	1.53

Figure 12.3 (cont.)

D4020 Standpipes

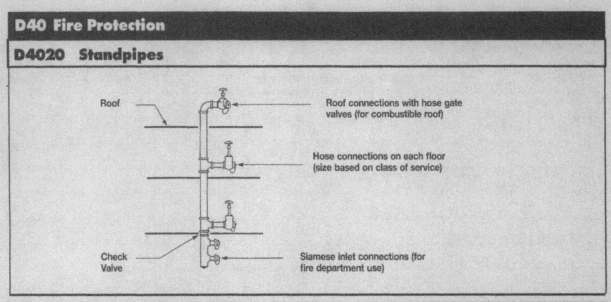

System Components	QUANTITY	UNIT	COST PER FLOOR		
			MAT.	INST.	TOTAL
SYSTEM D4020 310					
WET STANDPIPE RISER, CLASS I, STEEL, BLACK, SCH. 40 PIPE, 10' HEIGHT					
4" DIAMETER PIPE, ONE FLOOR					
Pipe, steel, black, schedule 40, threaded, 4" diam	20.000	L.F.	188	420	608
Pipe, Tee, malleable iron, black, 150 lb threaded, 4" pipe size	2.000	Ea.	232	374	606
Pipe, 90° elbow, malleable iron, black, 150 lb threaded 4" pipe size	1.000	Ea.	73.50	125	198.50
Pipe, nipple, steel, black, schedule 40, 2-1/2" pipe size x 3" long	2.000	Ea.	10.30	94	104.30
Fire valve, gate, 300 lb, brass w/handwheel, 2-1/2" pipe size	1.000	Ea.	104	59.50	163.50
Fire valve, pressure restricting, adj, rgh brs, 2-1/2" pipe size	1.000	Ea.	214	119	333
Valve, swing check, w/ball drip, CI w/brs ftngs, 4" pipe size	1.000	Ea.	155	250	405
Standpipe conn wall dble flush brs w/plugs & chains 2-1/2"x2-1/2"x4"	1.000	Ea.	291	150	441
Valve, swing check, bronze, 125 lb, regrinding disc, 2-1/2" pipe size	1.000	Ea.	197	50	247
Roof manifold, fire, w/valves & caps, horiz/vert brs 2-1/2"x2-1/2"x4"	1.000	Ea.	111	156	267
Fire, hydrolator, vent & drain, 2-1/2" pipe size	1.000	Ea.	45	34.50	79.50
Valve, gate, iron body 125 lb, OS&Y, threaded, 4" pipe size	1.000	Ea.	380	250	630
TOTAL			2,000.80	2,082	4,082.80

D4020 310	Wet Standpipe Risers, Class I		COST PER FLOOR		
			MAT.	INST.	TOTAL
0550	Wet standpipe risers, Class I, steel black sch. 40, 10' height				
0560	4" diameter pipe, one floor		2,000	2,075	4,075
0580	Additional floors		475	645	1,120
0600	6" diameter pipe, one floor		3,475	3,700	7,175
0620	Additional floors	RD4020 -300	850	1,050	1,900
0640	8" diameter pipe, one floor		5,150	4,475	9,625
0660	Additional floors		1,150	1,275	2,425
0680					

D4020 310	Wet Standpipe Risers, Class II		COST PER FLOOR		
			MAT.	INST.	TOTAL
1030	Wet standpipe risers, Class II, steel black sch. 40, 10' height				
1040	2" diameter pipe, one floor		845	745	1,590
1060	Additional floors		278	289	567
1080	2-1/2" diameter pipe, one floor		1,100	1,100	2,200
1100	Additional floors		310	335	645
1120					

Figure 12.4

D40 Fire Protection

D4020 Standpipes

D4020 310	Wet Standpipe Risers, Class III	COST PER FLOOR		
		MAT.	INST.	TOTAL
1530	Wet standpipe risers, Class III, steel black sch. 40, 10' height			
1540	4" diameter pipe, one floor	2,050	2,075	4,125
1560	Additional floors	400	540	940
1580	6" diameter pipe, one floor	3,525	3,700	7,225
1600	Additional floors	880	1,050	1,930
1620	8" diameter pipe, one floor	5,225	4,475	9,700
1640	Additional floors	1,175	1,275	2,450

Figure 12.4 (cont.)

D4020 410	Standpipe Equipment	COST EACH		
		MAT.	INST.	TOTAL
0100	Adapters, reducing, 1 piece, FxM, hexagon, cast brass, 2-1/2" x 1-1/2"	30.50		30.50
0200	Pin lug, 1-1/2" x 1"	9.70		9.70
0250	3" x 2-1/2"	38		38
0300	For polished chrome, add 75% mat.			
0400	Cabinets, D.S. glass in door, recessed, steel box, not equipped			
0500	Single extinguisher, steel door & frame	62	94	156
0550	Stainless steel door & frame	109	94	203
0600	Valve, 2-1/2" angle, steel door & frame	67.50	62.50	130
0650	Aluminum door & frame	90	62.50	152.50
0700	Stainless steel door & frame	118	62.50	180.50
0750	Hose rack assy, 2-1/2" x 1-1/2" valve & 100' hose, steel door & frame	137	125	262
0800	Aluminum door & frame	203	125	328
0850	Stainless steel door & frame	264	125	389
0900	Hose rack assy,& extinguisher,2-1/2"x1-1/2" valve & hose,steel door & frame	171	150	321
0950	Aluminum	259	150	409
1000	Stainless steel	305	150	455
1550	Compressor, air, dry pipe system, automatic, 200 gal., 1/3 H.P.	770	320	1,090
1600	520 gal., 1 H.P.	805	320	1,125
1650	Alarm, electric pressure switch (circuit closer)	144	16.05	160.05
2500	Couplings, hose, rocker lug, cast brass, 1-1/2"	28.50		28.50
2550	2-1/2"	37.50		37.50
3000	Escutcheon plate, for angle valves, polished brass, 1-1/2"	11.30		11.30
3050	2-1/2"	27		27
3500	Fire pump, electric, w/controller, fittings, relief valve			
3550	4" pump, 30 H.P., 500 G.P.M.	14,900	2,350	17,250
3600	5" pump, 40 H.P., 1000 G.P.M.	21,800	2,650	24,450
3650	5" pump, 100 H.P., 1000 G.P.M.	24,400	2,950	27,350
3700	For jockey pump system, add	2,675	375	3,050
5000	Hose, per linear foot, synthetic jacket, lined,			
5100	300 lb. test, 1-1/2" diameter	1.58	.29	1.87
5150	2-1/2" diameter	2.64	.34	2.98
5200	500 lb. test, 1-1/2" diameter	1.64	.29	1.93
5250	2-1/2" diameter	2.85	.34	3.19
5500	Nozzle, plain stream, polished brass, 1-1/2" x 10"	23.50		23.50
5550	2-1/2" x 15" x 13/16" or 1-1/2"	47.50		47.50
5600	Heavy duty combination adjustable fog and straight stream w/handle 1-1/2"	273		273
5650	"2-1/2" direct connection	390		390
6000	Rack, for 1-1/2" diameter hose 100 ft. long, steel	35.50	37.50	73
6050	Brass	55	37.50	92.50
6500	Reel, steel, for 50 ft. long 1-1/2" diameter hose	74	53.50	127.50
6550	For 75 ft. long 2-1/2" diameter hose	119	53.50	172.50
7050	Siamese, w/plugs & chains, polished brass, sidewalk, 4" x 2-1/2" x 2-1/2"	340	300	640
7100	6" x 2-1/2" x 2-1/2"	565	375	940
7200	Wall type, flush, 4" x 2-1/2" x 2-1/2"	291	150	441
7250	6" x 2-1/2" x 2-1/2"	395	163	558
7300	Projecting, 4" x 2-1/2" x 2-1/2"	266	150	416
7350	6" x 2-1/2" x 2-1/2"	435	163	598
7400	For chrome plate, add 15% mat.			
8000	Valves, angle, wheel handle, 300 Lb., rough brass, 1-1/2"	29	34.50	63.50
8050	2-1/2"	50.50	59.50	110
8100	Combination pressure restricting, 1-1/2"	46	34.50	80.50
8150	2-1/2"	98	59.50	157.50
8200	Pressure restricting, adjustable, satin brass, 1-1/2"	76	34.50	110.50
8250	2-1/2"	107	59.50	166.50
8300	Hydrolator, vent and drain, rough brass, 1-1/2"	45	34.50	79.50
8350	2-1/2"	45	34.50	79.50
8400	Cabinet assy, incls. adapter, rack, hose, and nozzle	435	226	661

Figure 12.5

ASSEMBLY NUMBER	DESCRIPTION	QTY	UNIT	TOTAL COST		COST PER S.F.
				UNIT	TOTAL	
DIVISION D	**SERVICES**					
D30	**Heating, Ventilating, and Air Conditioning**					
D3030-206-3960	Roof Top Multi-zone	30,000	S.F.	$ 11.75	$ 352,500	
	Balancing - Allowance				$ 3,000	
	Quality/Complexity (7%)				$ 24,675	
	Subtotal, Division D30, HVAC				$ 380,175	$ 12.67
D40	**Fire Protection**					
D4010-305-0620	Sprinkler System, Wet Pipe, Light Hazard, 1 Floor	10,000	S.F.	$ 1.69	$ 16,900	
D4010-305-0740	Sprinkler System, Wet Pipe, Light Hazard, Each Add. Floor.	20,000	S.F.	$ 1.36	$ 27,200	
D4020-310-1540	Standpipe Riser, 1 Floor	1.2	FLOOR	$ 4,125.00	$ 4,950	
D4020-310-1560	Standpipe Riser, Each Additional Floor	2.4	FLOOR	$ 940.00	$ 2,256	
D4020-410-8400	Cabinet Assembly, Including Adapter, Rack, Hose & Nozzle	3	Ea.	$ 661.00	$ 1,983	
D4020-410-7200	Siamese Connection	1	Ea.	$ 441.00	$ 441	
D4020-410-1650	Alarm	1	Ea.	$ 160.05	$ 160	
D4020-410-0100	Adaptor	3	Ea.	$ 30.50	$ 92	
	Fire Protection Subtotal				$ 53,982	
	Quality/Complexity Allowance (10%)				$ 5,398	
	Subtotal, Division D40, Fire Protection				$ 59,380	$ 1.98

Figure 12.6

Chapter 13

Electrical

The electrical assembly, UNIFORMAT II Division D50, is made up of several subassemblies that can be estimated in the early stages of a project even with limited information available. The building electrical load or total number of watts required must first be determined before sizing and estimating electrical components. The total amperes for the service are based on this calculation. The process involves simple math and requires a basic knowledge of the different components and where each fits into the overall system.

Budget Square Foot Costs

Energy codes set lighting limitations for many types of building occupancies. This wattage for lighting, combined with the power requirements for receptacles, air conditioning, elevators, and other systems, determines the total wattage. An approximation of all electrical systems of a building can be determined with minimal information. Because the electrical requirements for similar buildings are essentially the same, the costs per square foot are also the same. One bank or hospital will require approximately the same wattage as another bank or hospital. Use Figure 13.1 to obtain a rough approximation of electrical costs. The prices in this table are from actual projects that have been bid from 1991 to 2001 in the Northeastern part of the United States. Bid prices have been adjusted to January 1, 2001 price levels. The list of projects is by no means all-inclusive, but by carefully examining the various systems for a particular building type, certain cost relationships emerge. This table can be used to produce a budget square foot cost for the electrical portion of the job that is consistent with the amount of design information normally available at the conceptual estimate stage.

The following explanations apply when using the table in Figure 13.1.

Service & Distribution: The costs shown include the incoming primary feeder from the power company, the main building transformer, metering arrangement, switchboards, distribution panel boards, step-down transformers, and power and lighting panels. The construction types marked include the cost of the primary feeder and transformer. In all other projects, the primary feeder and transformer are provided by the power company.

Cost per S.F. for Electric Systems for Various Building Types

Type Construction	1. Service & Distrib.	2. Lighting	3. Devices	4. Equipment Connections	5. Basic Materials	6. Special Systems		
						Fire Alarm & Detection	Lightning Protection	Master TV Antenna
Apartment, luxury high rise	$1.41	$ .92	$ .70	$ .85	$2.22	$ .44		$.30
Apartment, low rise	.81	.76	.63	.72	1.30	.36		
Auditorium	1.78	4.66	.56	1.24	2.67	.56		
Bank, branch office	2.14	5.17	.94	1.25	2.53	1.61		
Bank, main office	1.61	2.81	.30	.53	2.74	.83		
Church	1.10	2.85	.36	.31	1.31	.83		
* College, science building	2.09	3.78	1.23	1.01	3.06	.73		
* College library	1.51	2.09	.27	.58	1.64	.83		
* College, physical education center	2.37	2.93	.38	.47	1.28	.47		
Department store	.79	2.02	.27	.84	2.20	.36		
* Dormitory, college	1.05	2.59	.28	.53	2.16	.59		.37
Drive-in donut shop	2.99	7.76	1.35	1.26	3.52	—		
Garage, commercial	.40	.96	.20	.39	.76	—		
* Hospital, general	5.77	4.00	1.54	.99	4.37	.51	$.15	
* Hospital, pediatric	5.05	5.96	1.33	3.66	8.10	.58		.45
* Hotel, airport	2.27	3.27	.28	.49	3.20	.48	.26	.41
Housing for the elderly	.64	.77	.37	.96	2.74	.58		.36
Manufacturing, food processing	1.44	4.09	.25	1.82	2.98	.36		
Manufacturing, apparel	.94	2.14	.31	.72	1.60	.32		
Manufacturing, tools	2.17	5.00	.29	.84	2.67	.37		
Medical clinic	.86	1.72	.47	1.30	2.10	.57		
Nursing home	1.49	3.24	.47	.38	2.67	.79		.30
Office Building	2.01	4.37	.25	.72	2.80	.42	.23	
Radio-TV studio	1.44	4.51	.70	1.32	3.29	.54		
Restaurant	5.36	4.28	.85	2.02	3.97	.32		
Retail Store	1.13	2.27	.28	.49	1.24	—		
School, elementary	1.90	4.04	.56	.49	3.33	.50		.22
School, junior high	1.16	3.37	.28	.89	2.64	.59		
* School, senior high	1.24	2.64	.49	1.17	2.94	.52		
Supermarket	1.28	2.28	.34	1.94	2.55	.25		
* Telephone Exchange	3.13	.96	.20	.84	1.74	.96		
Theater	2.47	3.12	.58	1.69	2.58	.71		
Town Hall	1.50	2.47	.58	.61	3.43	.47		
* U.S.Post Office	4.43	3.18	.59	.92	2.45	.47		
Warehouse, grocery	.82	1.37	.20	.49	1.83	.30		

*Includes cost of primary feeder and transformer. Cont'd. on next page.

Figure 13.1

226

Cost per S.F. for Electric Systems for Various Building Types (cont.)

Type Construction	6. Special Systems, (cont.)						
	Intercom Systems	Sound Systems	Closed Circuit TV	Snow Melting	Emergency Generator		Master Clock Sys
Apartment, luxury high rise	$ .67						
Apartment, low rise	.41						
Auditorium		$1.53	$ .71		$1.12		
Bank, branch office	.79		1.63			$1.37	
Bank, main office	.45		.35		.91	.75	$.32
Church	.56						
* College, science building	.58				1.15		.37
* College, library					.59		
* College, physical education center		.78					
Department store					.26		
* Dormitory, college	.77						
Drive-in donut shop							.15
Garage, commercial							.12
Hospital, general	.59		.25		1.56		
* Hospital, pediatric	4.05	.41	.45		.98		
* Hotel, airport	.58				.58		
Housing for the elderly	.71						
Manufacturing, food processing		.27			2.02		
Manufacturing apparel		.36					
Manufacturing, tools		.45		$.30			
Medical clinic							
Nursing home	1.34				.51		
Office Building		.23			.51	.25	.12
Radio-TV studio	.78				1.27		.56
Restaurant		.36					
Retail Store							
School, elementary		.25					.25
School, junior high		.66			.43		.45
* School, senior high	.54		.36		.59	.32	.33
Supermarket		.28			.54	.36	
* Telephone exchange					5.13	.19	
Theater		.53					
Town Hall							.25
* U.S. Post Office	.52			.11	.57		
Warehouse, grocery	.34						

*Includes cost of primary feeder and transformer. Cont'd. on next page.

Figure 13.1 (cont.)

Cost per S.F. for Total Electric Systems for Various Building Types (cont.)

Type Construction	Basic Description	Total Floor Area in Square Feet	Total Cost per Square Foot for Total Electric Systems
Apartment building, luxury high rise	All electric, 18 floors, 86 1 B.R., 34 2 B.R.	115,000	$ 7.41
Apartment building, low rise	All electric, 2 floors, 44 units, 1 & 2 B.R.	40,200	4.99
Auditorium	All electric, 1200 person capacity	28,000	14.83
Bank, branch office	All electric, 1 floor	2,700	17.43
Bank, main office	All electric, 8 floors	54,900	11.60
Church	All electric, incl. Sunday school	17,700	7.32
*College, science building	All electric, 3-1/2 floors, 47 rooms	27,500	14.00
*College, library	All electric	33,500	7.51
*College, physical education center	All electric	22,000	8.68
Department store	Gas heat, 1 floor	85,800	6.74
*Dormitory, college	All electric, 125 rooms	63,000	8.34
Drive-in donut shop	Gas heat, incl. parking area lighting	1,500	17.03
Garage, commercial	All electric	52,300	2.83
*Hospital, general	Steam heat, 4 story garage, 300 beds	540,000	19.73
*Hospital, pediatric	Steam heat, 6 stories	278,000	31.02
Hotel, airport	All electric, 625 guest rooms	536,000	11.82
Housing for the elderly	All electric, 7 floors, 100 1 B.R. units	67,000	7.13
Manufacturing, food processing	Electric heat, 1 floor	9,600	13.23
Manufacturing, apparel	Electric heat, 1 floor	28,000	6.39
Manufacturing, tools	Electric heat, 2 floors	42,000	12.09
Medical clinic	Electric heat, 2 floors	22,700	7.02
Nursing home	Gas heat, 3 floors, 60 beds	21,000	11.19
Office building	All electric, 15 floors	311,200	11.91
Radio-TV studio	Electric heat, 3 floors	54,000	14.41
Restaurant	All electric	2,900	17.16
Retail store	All electric	3,000	5.41
School, elementary	All electric, 1 floor	39,500	11.54
School, junior high	All electric, 1 floor	49,500	10.47
*School, senior high	All electric, 1 floor	158,300	11.14
Supermarket	Gas heat	30,600	9.82
*Telephone exchange	Gas heat, 300 KW emergency generator	24,800	13.15
Theater	Electric heat, twin cinema	14,000	11.68
Town Hall	All electric	20,000	9.31
*U.S. Post Office	All electric	495,000	13.24
Warehouse, grocery	All electric	96,400	5.35

*Includes cost of primary feeder and transformer.

Figure 13.1 (cont.)

Lighting: Costs include all interior fixtures for decor, illumination, exit, and emergency lighting. General exterior lighting fixtures are included, but those for area parking illumination are not unless specifically noted.

Devices: Costs include all outlet boxes, receptacles, switches for lighting control, dimmers, and cover plates.

Equipment Connections: Costs include all materials and equipment for making connections for heating, ventilating, and air conditioning. Food service and other motorized items or systems requiring connections are also included.

Basic Materials: This category includes all disconnect power switches that are not a part of the service equipment. Included are raceways for wires, pull boxes, junction boxes, supports, fittings, grounding materials, wireways, busways, and cable systems.

Special Systems: Costs for installed equipment are included for the particular assemblies shown.

Except in special cases, developing a preliminary cost by using cost per fixture or per receptacle is cumbersome and no more accurate. After the total wattage requirements are determined, simple formulas convert watts to amperes for the available voltages to size of service.

Estimating Procedure

To develop the building electrical loads, use the following guidelines:

1. Determine the building size and anticipated use.
2. Develop the total load in watts for:
   - Lighting
   - Receptacles
   - Air conditioning
   - Elevators
   - Other power requirements
3. Determine the voltage available from the utility company.
4. Determine the size of the building service using formulas.
5. Determine costs for service, panels, and feeders.
6. Determine costs for above subsystems using loads.

The conversion of watts to amperes is necessary to size the building service entrance. Once the number of watts required for the building is developed, ask the local utility company for the voltage available at or near the project site.

Power Requirements

Lighting	(Figure 13.2 and 13.3):		
3.0 watts/S.F.	30,000 S.F. × 3 W/S.F.	=	90,000 watts
Receptacles (Figure 13.2):			
2.0 watts/S.F.	30,000 S.F. × 2 W/S.F.	=	60,000 watts
HVAC (Figures 13.2 and 13.4):			
4.7 watts/S.F.	30,000 S.F. × 4.7 W/S.F.	=	141,000 watts
Misc. Motors & Power (Figure 13.2):			
1.2 watts/S.F.	30,000 S.F. × 1.2 W/S.F.	=	36,000 watts

Elevators:
Figure 13.5: 2 @ 3000 lb. and 100 F.P.M. = 30 hp
Figure 13.6: 30 hp, 460 volts = 40 amps

Formula:
Watts = 1.73 × v. × current × power factor × efficiency
Watts = 1.73 × 460 × 40 × 0.9 × 0.9 $\quad\quad\quad = \dfrac{25,784}{352,784 \text{ watts}}$

Nominal Watts Per S.F. for Electric Systems for Various Building Types

Type Construction	1. Lighting	2. Devices	3. HVAC	4. Misc.	5. Elevator	Total Watts
Apartment, luxury high rise	2	2.2	3	1		
Apartment, low rise	2	2	3	1		
Auditorium	2.5	1	3.3	.8		
Bank, branch office	3	2.1	5.7	1.4		
Bank, main office	2.5	1.5	5.7	1.4		
Church	1.8	.8	3.3	.8		
College, science building	3	3	5.3	1.3		
College, library	2.5	.8	5.7	1.4		
College, physical education center	2	1	4.5	1.1		
Department store	2.5	.9	4	1		
Dormitory, college	1.5	1.2	4	1		
Drive-in donut shop	3	4	6.8	1.7		
Garage, commercial	.5	.5	0	.5		
Hospital, general	2	4.5	5	1.3		
Hospital, pediatric	3	3.8	5	1.3		
Hotel, airport	2	1	5	1.3		
Housing for the elderly	2	1.2	4	1		
Manufacturing, food processing	3	1	4.5	1.1		
Manufacturing, apparel	2	1	4.5	1.1		
Manufacturing, tools	4	1	4.5	1.1		
Medical clinic	2.5	1.5	3.2	1		
Nursing home	2	1.6	4	1		
Office building, hi rise	3	2	4.7	1.2		
Office building, low rise	3	2	4.3	1.2		
Radio-TV studio	3.8	2.2	7.6	1.9		
Restaurant	2.5	2	6.8	1.7		
Retail store	2.5	.9	5.5	1.4		
School, elementary	3	1.9	5.3	1.3		
School, junior high	3	1.5	5.3	1.3		
School, senior high	2.3	1.7	5.3	1.3		
Supermarket	3	1	4	1		
Telephone exchange	1	.6	4.5	1.1		
Theater	2.5	1	3.3	.8		
Town Hall	2	1.9	5.3	1.3		
U.S. Post Office	3	2	5	1.3		
Warehouse, grocery	1	.6	0	.5		

Figure 13.2

Lighting Limit (Connected Load) for Listed Occupancies: New Building Proposed Energy Conservation Guideline

Type of Use	Maximum Watts per S.F.
Interior	
Category A: Classrooms, office areas, automotive mechanical areas, museums, conference rooms, drafting rooms, clerical areas, laboratories, merchandising areas, kitchens, examining rooms, book stacks, athletic facilities.	3.00
Category B: Auditoriums, waiting areas, spectator areas, restrooms, dining areas, transportation terminals, working corridors in prisons and hospitals, book storage areas, active inventory storage, hospital bedrooms, hotel and motel bedrooms, enclosed shopping mall concourse areas, stairways.	1.00
Category C: Corridors, lobbies, elevators, inactive storage areas.	0.50
Category D: Indoor parking.	0.25
Exterior	
Category E: Building perimeter: wall-wash, facade, canopy.	5.00 (per linear foot)
Category F: Outdoor parking.	0.10

Figure 13.3

Central Air Conditioning Watts per S.F., BTU's per Hour per S.F. of Floor Area and S.F. per Ton of Air Conditioning

Type Building	Watts per S.F.	BTUH per S.F.	S.F. per Ton	Type Building	Watts per S.F.	BTUH per S.F.	S.F. per Ton	Type Building	Watts per S.F.	BTUH per S.F.	S.F. per Ton
Apartments, Individual	3	26	450	Dormitory, Rooms	4.5	40	300	Libraries	5.7	50	240
Corridors	2.5	22	550	Corridors	3.4	30	400	Low Rise Office, Ext.	4.3	38	320
Auditoriums & Theaters	3.3	40	300/18*	Dress Shops	4.9	43	280	Interior	3.8	33	360
Banks	5.7	50	240	Drug Stores	9	80	150	Medical Centers	3.2	28	425
Barber Shops	5.5	48	250	Factories	4.5	40	300	Motels	3.2	28	425
Bars & Taverns	15	133	90	High Rise Off.-Ext. Rms.	5.2	46	263	Office (small suite)	4.9	43	280
Beauty Parlors	7.6	66	180	Interior Rooms	4.2	37	325	Post Office, Int. Office	4.9	42	285
Bowling Alleys	7.8	68	175	Hospitals, Core	4.9	43	280	Central Area	5.3	46	260
Churches	3.3	36	330/20*	Perimeter	5.3	46	260	Residences	2.3	20	600
Cocktail Lounges	7.8	68	175	Hotels, Guest Rooms	5	44	275	Restaurants	6.8	60	200
Computer Rooms	16	141	85	Public Spaces	6.2	55	220	Schools & Colleges	5.3	46	260
Dental Offices	6	52	230	Corridors	3.4	30	400	Shoe Stores	6.2	55	220
Dept. Stores, Basement	4	34	350	Industrial Plants, Offices	4.3	38	320	Shop'g. Ctrs., Sup. Mkts.	4	34	350
Main Floor	4.5	40	300	General Offices	4	34	350	Retail Stores	5.5	48	250
Upper Floor	3.4	30	400	Plant Areas	4.5	40	300	Specialty Shops	6.8	60	200

*Persons per ton

12,000 BTUH = 1 ton of air conditioning

Figure 13.4

Description: The table below shows approximate horsepower requirements for front opening elevators with 3-phase motors. Tabulations are for given car capacities at various travel speeds.

Watts = 1.73 × Volts × Current × Power Factor × Efficiency

$$\text{Horsepower} = \frac{\text{Volts} \times \text{Current} \times 1.73 \times \text{Power Factor}}{746 \text{ Watts}}$$

The power factor of electric motors varies from 80% to 90% in larger size motors. The efficiency likewise varies from 80% on a small motor to 90% on a large motor.

Horsepower Requirements For Elevators

Type	Maximum Travel Height	Travel Speeds In FPM	Capacity Of Cars In Lbs.								
			1200	1500	1800	2000	2500	3000	3500	4000	4500
Hydraulic	70 Ft.	70	10	15	15	15	20	20	20	25	30
		85	15	15	15	20	20	25	25	30	30
		100	15	15	20	20	25	30	30	40	40
		110	20	20	20	20	25	30	40	40	50
		125	20	20	20	25	30	40	40	50	50
		150	25	25	25	30	40	50	50	50	60
		175	25	30	30	40	50	50	60		
		200	30	30	40	40	50	60	60		
Geared Traction	300 Ft.	200				10	10	15	15		23
		350				15	15	23	23		36

Figure 13.5

Ampere Values Determined by Horsepower, Voltage and Phase Values

H.P.	Amperes					
	Single Phase		Three Phase			
	115V	230V	200 V	230V	460V	575V
1/6	4.4A	2.2A				
1/4	5.8	2.9				
1/3	7.2	3.6				
1/2	9.8	4.9	2.3A	2.0A	1.0A	0.8A
3/4	13.8	6.9	3.2	2.8	1.4	1.1
1	16	8	4.1	3.6	1.8	1.4
1-1/2	20	10	6.0	5.2	2.6	2.1
2	24	12	7.8	6.8	3.4	2.7
3	34	17	11.0	9.6	4.8	3.9
5			17.5	15.2	7.6	6.1
7-1/2			25.3	22	11	9
10			32.2	28	14	11
15			48.3	42	21	17
20			62.1	54	27	22
25			78.2	68	34	27
30			92.0	80	40	32
40			119.6	104	52	41
50			149.5	130	65	52
60			177	154	77	62
75			221	192	96	77
100			285	248	124	99
125			359	312	156	125
150			414	360	180	144
200			552	480	240	192

Figure 13.6

232

Voltage Available

Assume that the utility company has the following available at the site:

120/208 volt, 3 phase, 4 wire or 277/480 volt, 3 phase, 4 wire service.

Select the 277/480 volt service, as it allows smaller feeder to be used.

Size of Service

Formula: Amperes $= \dfrac{\text{watt}}{\text{volts} \times \text{power factor} \times 1.73}$

Amperes $= \dfrac{352{,}784}{480 \times 0.8 \times 1.73} = 531$ amps

Use 600 amp service.

Assemblies Estimate: Three-Story Office Building

Budget Method

Before going into a more detailed analysis of the assemblies costs, use Figure 13.1 to put together a budget estimate for the three-story office building.

Service & Distribution:	$2.01 × 30,000 S.F.	=	$ 60,300
Lighting:	$4.37 × 30,000 S.F.	=	131,100
Devices:	$0.25 × 30,000 S.F.	=	7,500
Equipment Connections:	$0.72 × 30,000 S.F.	=	21,600
Materials:	$2.80 × 30,000 S.F.	=	84,000
Fire Alarm:	$0.42 × 30,000 S.F.	=	12,600
Emergency Generator:	$0.51 × 30,000 S.F.	=	15,300
Total Budget Cost		=	$332,400
Budget Cost per Square Foot:	$332,400 / 30,000 S.F.	=	$11.08/S.F.

Assemblies Method

Using the above calculated loads, create an assemblies square foot estimate for the three story office building example:

Lighting:	(Figure 13.7, line #0280)		
	30,000 S.F. × $5.23	=	$156,900
Receptacles:	(Figure 13.8, line #0640)		
	30,000 S.F. × $3.00	=	90,000
HVAC:	(Figure 13.9, line #0280)		
	30,000 S.F. × $0.44	=	13,200
	(Interpolate 4.3 watts)		
Misc. Motors & Power:	(Figure 13.10, line #0320)		
	30,000 S.F. × $0.22	=	6,600
Elevators:			
2 @ 30 hp	(Figure 13.11, line #2200)		
	2 ea. × $2,450	=	4,900
Wall Switches:			
2/1000 S.F.	(Figure 13.12, line #0280)		
	30,000 S.F. × $0.28	=	8,400
Service 600 amp:	(Figure 13.13, line #0360)		
	1.25 × $7,425	=	9,281
	(See Figure 13.13, line #0570)		
Panels 600 amp:	(Figure 13.14, line #0240)		
	1.20 × $12,075	=	14,490
	(See Figure 13.14, line #0410)		
Feeders 600 amp:	(Figure 13.15, line #0360)		
	100 L.F. × $104.00	=	10,400

Fire Detection:	(Figure 13.1)				
	30,000 S.F. × $0.42		=	12,600	
Emergency	(Figure 13.1)				
Generator:	30,000 S.F. × $0.51		=	15,300	
Total Cost			=	$342,071	
Cost per Square Foot	$342,071/30,000 S.F.		=	$11.40/S.F.	

Cost Analysis

Compare the two cost estimates for the three-story office building:

Budget Cost per Square Foot: $11.08/S.F.
System Square Foot Cost: $11.40/S.F.

The cost difference is $0.32/S.F., or almost 3%. At the conceptual stage, is it worth spending the additional time to get a figure that is 3% more accurate?

Only the estimator can make that decision. For quick numbers, the budget cost is certainly attractive. However, the more detailed process

---

D50 Electrical

D5020 Lighting and Branch Wiring

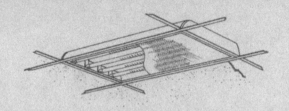

Type C. Recessed, mounted on grid ceiling suspension system, 2' x 4', four 40 watt lamps, acrylic prismatic diffusers.

5.3 watts per S.F. for 100 footcandles.
3 watts per S.F. for 57 footcandles.

System Components	QUANTITY	UNIT	COST PER S.F.		
			MAT.	INST.	TOTAL
SYSTEM D5020 210					
FLUORESCENT FIXTURES RECESS MOUNTED IN CEILING					
1 WATT PER S.F., 20 FC, 5 FIXTURES PER 1000 S.F.					
Steel intermediate conduit, (IMC) 1/2" diam	.128	L.F.	.18	.52	.70
Wire, 600 volt, type THW, copper, solid, #12	.003	C.L.F.	.02	.11	.13
Fluorescent fixture, recessed, 2'x4', four 40W, w/lens, for grid ceiling	.005	Ea.	.31	.44	.75
Steel outlet box 4" square	.005	Ea.	.04	.10	.14
Fixture whip, Greenfield w/#12 THHN wire	.005	Ea.	.01	.03	.04
TOTAL			.56	1.20	1.76

D5020 210	Fluorescent Fixtures (by Wattage)	COST PER S.F.		
		MAT.	INST.	TOTAL
0190	Fluorescent fixtures recess mounted in ceiling			
0200	1 watt per S.F., 20 FC, 5 fixtures per 1000 S.F.	.56	1.20	1.76
0240	2 watts per S.F., 40 FC, 10 fixtures per 1000 S.F.	1.11	2.37	3.48
0280	3 watts per S.F., 60 FC, 15 fixtures per 1000 S.F.	1.66	3.57	5.23
0320	4 watts per S.F., 80 FC, 20 fixtures per 1000 S.F.	2.21	4.72	6.93
0400	5 watts per S.F., 100 FC, 25 fixtures per 1000 S.F.	2.77	5.90	8.67

(RD5020-200)

Figure 13.7

based on specific information is more accurate. As long as the project being estimated is of a general nature—such as an office building—the budget cost per square foot method is reasonably accurate. Once the project becomes unique with unusual electrical components, the estimator should rely on the assemblies square foot method or a unit price estimate.

Figure 13.16 summarizes the costs of the project's electrical work.

D50 Electrical

D5020 Lighting and Branch Wiring

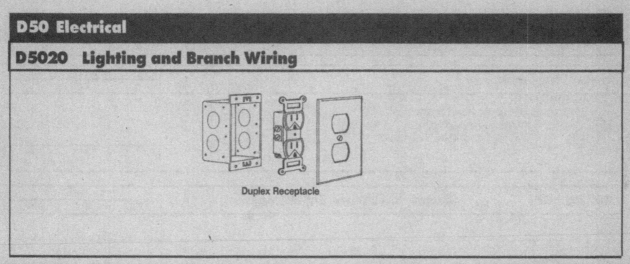

Duplex Receptacle

System Components			COST PER S.F.		
	QUANTITY	UNIT	MAT.	INST.	TOTAL
SYSTEM D5020 110					
RECEPTACLES INCL. PLATE, BOX, CONDUIT, WIRE & TRANS. WHEN REQUIRED					
2.5 PER 1000 S.F., .3 WATTS PER S.F.					
Steel intermediate conduit, (IMC) 1/2" diam	167.000	L.F.	.23	.68	.91
Wire 600V type THWN-THHN, copper solid #12	3.382	C.L.F.	.02	.13	.15
Wiring device, receptacle, duplex, 120V grounded, 15 amp	2.500	Ea.		.03	.03
Wall plate, 1 gang, brown plastic	2.500	Ea.		.01	.01
Steel outlet box 4" square	2.500	Ea.	.01	.05	.06
Steel outlet box 4" plaster rings	2.500	Ea.		.02	.02
TOTAL			.26	.92	1.18

D5020 110	Receptacle (by Wattage)		COST PER S.F.		
			MAT.	INST.	TOTAL
0190	Receptacles include plate, box, conduit, wire & transformer when required				
0200	2.5 per 1000 S.F., .3 watts per S.F.		.26	.92	1.18
0240	With transformer	RD5010 -110	.30	.97	1.27
0280	4 per 1000 S.F., .5 watts per S.F.		.30	1.07	1.37
0320	With transformer		.36	1.14	1.50
0360	5 per 1000 S.F., .6 watts per S.F.		.36	1.26	1.62
0400	With transformer		.44	1.35	1.79
0440	8 per 1000 S.F., .9 watts per S.F.		.37	1.38	1.75
0480	With transformer		.47	1.51	1.98
0520	10 per 1000 S.F., 1.2 watts per S.F.		.38	1.51	1.89
0560	With transformer		.55	1.72	2.27
0600	16.5 per 1000 S.F., 2.0 watts per S.F.		.46	1.90	2.36
0640	With transformer		.75	2.25	3
0680	20 per 1000 S.F., 2.4 watts per S.F.		.49	2.07	2.56
0720	With transformer		.83	2.48	3.31

Figure 13.8

D50 Electrical

D5020   Lighting and Branch Wiring

System D5020 140 includes all wiring and connections for central air conditioning units.

System Components	QUANTITY	UNIT	COST PER S.F.		
			MAT.	INST.	TOTAL
SYSTEM D5020 140					
CENTRAL AIR CONDITIONING POWER, 1 WATT					
Steel intermediate conduit, 1/2" diam.	.030	L.F.	.04	.12	.16
Wire 600V type THWN-THHN, copper solid #12	.001	C.L.F.	.01	.04	.05
TOTAL			.05	.16	.21

D5020 140	Central A. C. Power (by Wattage)	COST PER S.F.		
		MAT.	INST.	TOTAL
0200	Central air conditioning power, 1 watt	.05	.16	.21
0220	2 watts	.06	.19	.25
0240	3 watts	.07	.21	.28
0280	4 watts	.11	.27	.38
0320	6 watts	.18	.38	.56
0360	8 watts	.23	.40	.63
0400	10 watts	.30	.46	.76

Figure 13.9

D5020 Lighting and Branch Wiring

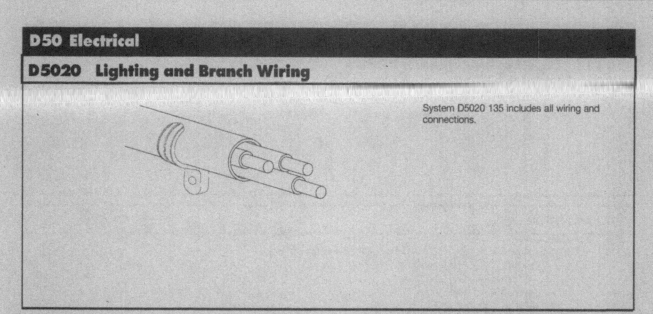

System D5020 135 includes all wiring and connections.

System Components			COST PER S.F.		
	QUANTITY	UNIT	MAT.	INST.	TOTAL
SYSTEM D5020 135					
MISCELLANEOUS POWER, TO .5 WATTS					
Steel intermediate conduit, (IMC) 1/2" diam	15.000	L.F.	.02	.06	.08
Wire 600V type THWN-THHN, copper solid #12	.325	C.L.F.		.01	.01
TOTAL			.02	.07	.09

D5020 135	Miscellaneous Power	COST PER S.F.		
		MAT.	INST.	TOTAL
0200	Miscellaneous power, to .5 watts	.02	.07	.09
0240	.8 watts	.03	.11	.14
0280	1 watt	.04	.13	.17
0320	1.2 watts	.05	.17	.22
0360	1.5 watts	.06	.19	.25
0400	1.8 watts	.07	.22	.29
0440	2 watts	.09	.27	.36
0480	2.5 watts	.10	.32	.42
0520	3 watts	.12	.38	.50

Figure 13.10

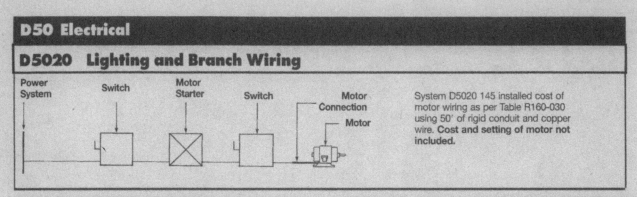

D50 Electrical

D5020 Lighting and Branch Wiring

System D5020 145 installed cost of motor wiring as per Table R160-030 using 50' of rigid conduit and copper wire. **Cost and setting of motor not included.**

System Components	QUANTITY	UNIT	COST EACH		
			MAT.	INST.	TOTAL
SYSTEM D5020 145					
MOTOR INSTALLATION, SINGLE PHASE, 115V, TO AND INCLUDING 1/3 HP MOTOR SIZE					
Wire 600V type THWN-THHN, copper solid #12	1.250	C.L.F.	8.06	46.25	54.31
Steel intermediate conduit, (IMC) 1/2" diam	50.000	L.F.	69	205	274
Magnetic FVNR, 115V, 1/3 HP, size 00 starter	1.000	Ea.	167	102	269
Safety switch, fused, heavy duty, 240V 2P 30 amp	1.000	Ea.	90	117	207
Safety switch, non fused, heavy duty, 600V, 3 phase, 30 A	1.000	Ea.	106	128	234
Flexible metallic conduit, Greenfield 1/2" diam	1.500	L.F.	.54	3.08	3.62
Connectors for flexible metallic conduit Greenfield 1/2" diam	1.000	Ea.	1.58	5.10	6.68
Coupling for Greenfield to conduit 1/2" diam flexible metalic conduit	1.000	Ea.	.83	8.20	9.03
Fuse cartridge nonrenewable, 250V 30 amp	1.000	Ea.	1.10	8.20	9.30
TOTAL			444.11	622.83	1,066.94

D5020 145	Motor Installation		COST EACH		
			MAT.	INST.	TOTAL
0200	Motor installation, single phase, 115V, to and including 1/3 HP motor size		445	625	1,070
0240	To and incl. 1 HP motor size		465	625	1,090
0280	To and incl. 2 HP motor size	RD5010 -170	500	665	1,165
0320	To and incl. 3 HP motor size		560	675	1,235
0360	230V, to and including 1 HP motor size		445	630	1,075
0400	To and incl. 2 HP motor size		470	630	1,100
0440	To and incl. 3 HP motor size		525	680	1,205
0520	Three phase, 200V, to and including 1-1/2 HP motor size		520	695	1,215
0560	To and incl. 3 HP motor size		555	755	1,310
0600	To and incl. 5 HP motor size		600	840	1,440
0640	To and incl. 7-1/2 HP motor size		615	860	1,475
0680	To and incl. 10 HP motor size		1,000	1,075	2,075
0720	To and incl. 15 HP motor size		1,325	1,200	2,525
0760	To and incl. 20 HP motor size		1,625	1,375	3,000
0800	To and incl. 25 HP motor size		1,650	1,375	3,025
0840	To and incl. 30 HP motor size		2,625	1,625	4,250
0880	To and incl. 40 HP motor size		3,175	1,925	5,100
0920	To and incl. 50 HP motor size		5,575	2,250	7,825
0960	To and incl. 60 HP motor size		5,725	2,375	8,100
1000	To and incl. 75 HP motor size		7,250	2,725	9,975
1040	To and incl. 100 HP motor size		15,000	3,200	18,200
1080	To and incl. 125 HP motor size		15,200	3,500	18,700
1120	To and incl. 150 HP motor size		18,400	4,125	22,525
1160	To and incl. 200 HP motor size		22,300	5,025	27,325
1240	230V, to and including 1-1/2 HP motor size		495	685	1,180
1280	To and incl. 3 HP motor size		535	745	1,280
1320	To and incl. 5 HP motor size		575	830	1,405
1360	To and incl. 7-1/2 HP motor size		575	830	1,405
1400	To and incl. 10 HP motor size		905	1,025	1,930
1440	To and incl. 15 HP motor size		1,025	1,100	2,125

Figure 13.11

D5020 145	Motor Installation	COST EACH		
		MAT.	INST.	TOTAL
1480	To and incl. 20 HP motor size	1,525	1,350	2,875
1520	To and incl. 25 HP motor size	1,625	1,375	3,000
1560	To and incl. 30 HP motor size	1,675	1,375	3,050
1600	To and incl. 40 HP motor size	3,125	1,875	5,000
1640	To and incl. 50 HP motor size	3,225	2,000	5,225
1680	To and incl. 60 HP motor size	5,575	2,250	7,825
1720	To and incl. 75 HP motor size	6,600	2,550	9,150
1760	To and incl. 100 HP motor size	7,550	2,850	10,400
1800	To and incl. 125 HP motor size	15,200	3,300	18,500
1840	To and incl. 150 HP motor size	16,300	3,750	20,050
1880	To and incl. 200 HP motor size	17,700	4,175	21,875
1960	460V, to and including 2 HP motor size	600	690	1,290
2000	To and incl. 5 HP motor size	635	755	1,390
2040	To and incl. 10 HP motor size	665	830	1,495
2080	To and incl. 15 HP motor size	890	950	1,840
2120	To and incl. 20 HP motor size	925	1,025	1,950
2160	To and incl. 25 HP motor size	1,000	1,075	2,075
2200	To and incl. 30 HP motor size	1,300	1,150	2,450
2240	To and incl. 40 HP motor size	1,625	1,250	2,875
2280	To and incl. 50 HP motor size	1,800	1,375	3,175
2320	To and incl. 60 HP motor size	2,775	1,625	4,400
2360	To and incl. 75 HP motor size	3,125	1,800	4,925
2400	To and incl. 100 HP motor size	3,400	2,000	5,400
2440	To and incl. 125 HP motor size	5,750	2,250	8,000
2480	To and incl. 150 HP motor size	7,075	2,525	9,600
2520	To and incl. 200 HP motor size	8,125	2,850	10,975
2600	575V, to and including 2 HP motor size	600	690	1,290
2640	To and incl. 5 HP motor size	635	755	1,390
2680	To and incl. 10 HP motor size	665	830	1,495
2720	To and incl. 20 HP motor size	890	950	1,840
2760	To and incl. 25 HP motor size	925	1,025	1,950
2800	To and incl. 30 HP motor size	1,300	1,150	2,450
2840	To and incl. 50 HP motor size	1,375	1,200	2,575
2880	To and incl. 60 HP motor size	2,725	1,600	4,325
2920	To and incl. 75 HP motor size	2,775	1,625	4,400
2960	To and incl. 100 HP motor size	3,125	1,800	4,925
3000	To and incl. 125 HP motor size	5,675	2,225	7,900
3040	To and incl. 150 HP motor size	5,750	2,250	8,000
3080	To and incl. 200 HP motor size	7,175	2,575	9,750

Figure 13.11 (cont.)

D5020 Lighting and Branch Wiring

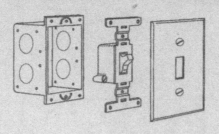

Description: Table D5020 130 includes the cost for switch, plate, box, conduit in slab or EMT exposed and copper wire. Add 20% for exposed conduit.

No power required for switches.

Federal energy guidelines recommend the maximum lighting area controlled per switch shall not exceed 1000 S.F. and that areas over 500 S.F. shall be so controlled that total illumination can be reduced by at least 50%.

System Components	QUANTITY	UNIT	COST PER S.F.		
			MAT.	INST.	TOTAL
SYSTEM D5020 130					
WALL SWITCHES, 1.0 PER 1000 S.F.					
Steel intermediate conduit, (IMC) 1/2" diam	22.000	L.F.	30.36	90.20	120.56
Wire 600V type THWN-THHN, copper solid #12	.420	C.L.F.	2.71	15.54	18.25
Toggle switch, single pole, 15 amp	1.000	Ea.	5.15	10.25	15.40
Wall plate, 1 gang, brown plastic	1.000	Ea.	.33	5.10	5.43
Steel outlet box 4" square	1.000	Ea.	2.26	20.50	22.76
Steel outlet box 4" plaster rings	1.000	Ea.	1.20	6.40	7.60
TOTAL			42.01	147.99	190
COST PER S.F.			.04	.15	.19

D5020 130	Wall Switch by Sq. Ft.	COST PER S.F.		
		MAT.	INST.	TOTAL
0200	Wall switches, 1.0 per 1000 S.F.	.04	.15	.19
0240	1.2 per 1000 S.F.	.04	.16	.20
0280	2.0 per 1000 S.F.	.05	.23	.28
0320	2.5 per 1000 S.F.	.08	.30	.38
0360	5.0 per 1000 S.F.	.18	.63	.81
0400	10.0 per 1000 S.F.	.35	1.28	1.63

Figure 13.12

D5010 Electrical Service/Distribution

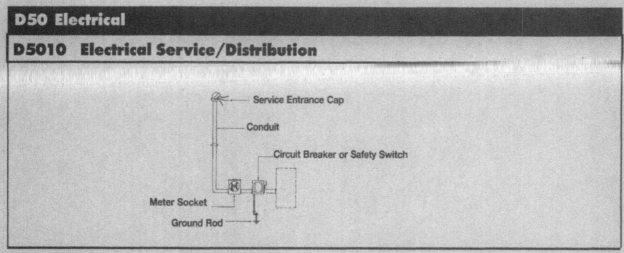

System Components			COST EACH		
	QUANTITY	UNIT	MAT.	INST.	TOTAL
SYSTEM D5010 120					
SERVICE INSTALLATION, INCLUDES BREAKERS, METERING, 20' CONDUIT & WIRE					
3 PHASE, 4 WIRE, 60 A					
Circuit breaker, enclosed (NEMA 1), 600 volt, 3 pole, 60 A	1.000	Ea.	415	146	561
Meter socket, single position, 4 terminal, 100 A	1.000	Ea.	30.50	128	158.50
Rigid galvanized steel conduit, 3/4", including fittings	20.000	L.F.	41	102	143
Wire, 600V type XHHW, copper stranded #6	.900	C.L.F.	29.70	56.70	86.40
Service entrance cap 3/4" diameter	1.000	Ea.	7.75	31.50	39.25
Conduit LB fitting with cover, 3/4" diameter	1.000	Ea.	10.75	31.50	42.25
Ground rod, copper clad, 8' long, 3/4" diameter	1.000	Ea.	33	77.50	110.50
Ground rod clamp, bronze, 3/4" diameter	1.000	Ea.	5.85	12.80	18.65
Ground wire, bare armored, #6-1 conductor	.200	C.L.F.	18.30	45.60	63.90
TOTAL			591.85	631.60	1,223.45

D5010 120	Electric Service, 3 Phase - 4 Wire		COST EACH		
			MAT.	INST.	TOTAL
0200	Service installation, includes breakers, metering, 20' conduit & wire				
0220	3 phase, 4 wire, 120/208 volts, 60 A		590	630	1,220
0240	100 A		735	760	1,495
0280	200 A		1,050	1,175	2,225
0320	400 A		2,300	2,150	4,450
0360	600 A	RD5010-110	4,525	2,900	7,425
0400	800 A		5,900	3,500	9,400
0440	1000 A		7,525	4,025	11,550
0480	1200 A		9,075	4,100	13,175
0520	1600 A		17,300	5,925	23,225
0560	2000 A		19,300	6,725	26,025
0570	Add 25% for 277/480 volt				
0580					
0610	1 phase, 3 wire, 120/240 volts, 100 A		390	685	1,075
0620	200 A		805	1,000	1,805

Figure 13.13

D50 Electrical

D5010 Electrical Service/Distribution

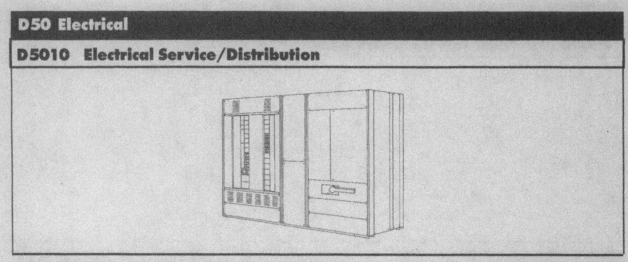

System Components			COST EACH		
	QUANTITY	UNIT	MAT.	INST.	TOTAL
SYSTEM D5010 240					
SWITCHGEAR INSTALLATION, INCL SWBD, PANELS & CIRC BREAKERS, 600 A					
Panelboard, NQOD 225A 4W 120/208V main CB, w/20A bkrs 42 circ	1.000	Ea.	1,850	1,475	3,325
Switchboard, alum. bus bars, 120/208V, 4 wire, 600V	1.000	Ea.	2,925	820	3,745
Distribution sect., alum. bus bar, 120/208 or 277/480 V, 4 wire, 600A	1.000	Ea.	1,350	820	2,170
Feeder section circuit breakers, KA frame, 70 to 225 A	3.000	Ea.	2,445	384	2,829
TOTAL			8,570	3,499	12,069

D5010 240	Switchgear		COST EACH		
			MAT.	INST.	TOTAL
0200	Switchgear inst., incl. swbd., panels & circ bkr, 400 A, 120/208volt		3,375	2,575	5,950
0240	600 A		8,575	3,500	12,075
0280	800 A	RD5010	10,700	4,950	15,650
0320	1200 A	-110	13,700	7,600	21,300
0360	1600 A		18,600	10,600	29,200
0400	2000 A		23,600	13,600	37,200
0410	Add 20% for 277/480 volt				

Figure 13.14

242

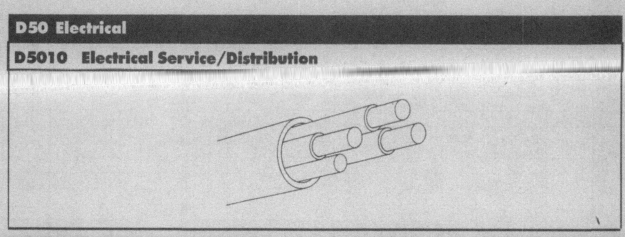

System Components				COST PER L.F.		
	QUANTITY	UNIT	MAT.	INST.	TOTAL
SYSTEM D5010 230					
FEEDERS, INCLUDING STEEL CONDUIT & WIRE, 60 A					
Rigid galvanized steel conduit, 3/4", including fittings	1.000	L.F.	2.05	5.10	7.15
Wire 600 volt, type XHHW copper stranded #6	.040	C.L.F.	1.32	2.52	3.84
TOTAL			3.37	7.62	10.99

D5010 230	Feeder Installation		COST PER L.F.		
			MAT.	INST.	TOTAL
0200	Feeder installation 600 V, including RGS conduit and XHHW wire, 60 A		3.37	7.60	10.97
0240	100 A		6.10	10.15	16.25
0280	200 A	RD5010	12.80	15.65	28.45
0320	400 A	-140	25.50	31.50	57
0360	600 A		53	51	104
0400	800 A		67	61	128
0440	1000 A		86	78	164
0480	1200 A		99.50	80	179.50
0520	1600 A		134	122	256
0560	2000 A		172	156	328
1200	Branch installation 600 V, including EMT conduit and THW wire, 15 A		.68	3.89	4.57
1240	20 A		.68	3.89	4.57
1280	30 A		1.05	4.79	5.84
1320	50 A		1.71	5.50	7.21
1360	65 A		2.05	5.85	7.90
1400	85 A		3.12	6.95	10.07
1440	100 A		3.90	7.35	11.25
1480	130 A		5.05	8.25	13.30
1520	150 A		6.05	9.45	15.50
1560	200 A		7.80	10.65	18.45

Figure 13.15

ASSEMBLY NUMBER	DESCRIPTION	QTY	UNIT	TOTAL COST		COST PER S.F.
				UNIT	TOTAL	
DIVISION D	**SERVICES**					
D30	**Heating, Ventilating, and Air Conditioning**					
D3030-206-3960	Roof Top Multi-zone	30,000	S.F.	$ 11.75	$ 352,500	
	Balancing - Allowance				$ 3,000	
	Quality/Complexity (7%)				$ 24,675	
	Subtotal, Division D30, HVAC				$ 380,175	$ 12.67
D40	**Fire Protection**					
D4010-305-0620	Sprinkler System, Wet Pipe, Light Hazard, 1 Floor	10,000	S.F.	$ 1.69	$ 16,900	
D4010-305-0740	Sprinkler System, Wet Pipe, Light Hazard, Each Add. Floor.	20,000	S.F.	$ 1.36	$ 27,200	
D4020-310-1540	Standpipe Riser, 1 Floor	1.2	FLOOR	$ 4,125.00	$ 4,950	
D4020-310-1560	Standpipe Riser, Each Additional Foor	2.4	FLOOR	$ 940.00	$ 2,256	
D4020-410-8400	Cabinet Assembly, Including Adapter, Rack, Hose & Nozzle	3	Ea.	$ 661.00	$ 1,983	
D4020-410-7200	Siamese Connection	1	Ea.	$ 441.00	$ 441	
D4020-410-1650	Alarm	1	Ea.	$ 160.05	$ 160	
D4020-410-0100	Adaptor	3	Ea.	$ 30.50	$ 92	
	Fire Protection Subtotal				$ 53,982	
	Quality/Complexity Allowance (10%)				$ 5,398	
	Subtotal, Division D40, Fire Protection				$ 59,380	$ 1.98
D50	**ELECTRICAL**					
D5020-210-0280	Lighting	30,000	S.F.	$ 5.23	$ 156,900	
D5020-110-0640	Receptacles	30,000	S.F.	$ 3.00	$ 90,000	
D5020-140-0280	HVAC	30,000	S.F.	$ 0.44	$ 13,200	
D5020-135-0320	Misc. Motors & Power	30,000	S.F.	$ 0.22	$ 6,600	
D5020-145-2200	Elevators, 2 @ 30 HP	2	Ea.	$ 2,450.00	$ 4,900	
D5020-130-0280	Wall Switches 2/1000 S.F.	30,000	S.F.	$ 0.28	$ 8,400	
D5010-120-0360	Service, 600 Amp	1.25	Ea.	$ 7,425.00	$ 9,281	
D5010-120-0570	Add for 277v/480v					
D5010-240-0240	Panels, 600	1.2	Ea.	$ 12,075.00	$ 14,490	
D5010-240-0410	Add for 277v/480v					
D5010-230-0360	Feeders, 600 Amp	100	L.F.	$ 104.00	$ 10,400	
	Fire Detection	30,000	S.F.	$ 0.42	$ 12,600	
	Emergency Generator	30,000	S.F.	$ 0.51	$ 15,300	
	Subtotal, Division D50, Electrical				$ 342,071	$ 11.40
	Total, Division D, Services				$ 980,597	$ 32.69

Figure 13.16

Chapter 14

Equipment, Furnishings, and Special Construction

Equipment and Furnishings

Building equipment and furnishings includes the specialized apparatus that transform a building shell into a functional space to meet a specific purpose. UNIFORMAT II Division E is divided into two major sections: Equipment and Furnishings. Equipment is further divided into Commercial Equipment, Institutional Equipment, Vehicular Equipment, and Other Equipment. Furnishings is differentiated between fixed and movable furnishings, including artwork, rugs and mats, and interior landscaping. These items are priced per each item, by the square foot, or by the linear foot.

Estimators should know which items will be provided within the construction budget, and which the owner will supply. Frequently equipment is purchased by the owner and installed by the contractor. Items to be considered include the following:

- Appliances
- Audio-visual equipment
- Bank equipment
- Dental equipment
- Detention equipment
- Display cases
- Ecclesiastical equipment
- Furnishings
- Kitchen equipment
- Laboratory equipment
- Loading dock equipment
- Medical equipment
- Parking control equipment
- Recreational equipment

Site equipment items, such as playground equipment, flagpoles, and site furnishings, are found in section G2040, Site Development.

Equipment and furnishing items in *Means Assemblies Cost Data* are general in nature and should be used for initial budgets only. Once more detailed information is known about the project, more specific costs from suppliers and manufacturers can be used to fine-tune the estimate.

Special Construction

UNIFORMAT II Division F, Special Construction, is divided into two major sections: Special Construction and Selective Building Demolition.

Special Construction is further broken down into Special Structures, Integrated Construction, Special Construction Systems, Special Facilities, and Special Controls and Instrumentation. Here we find pre-engineered structures, including greenhouses and metal storage buildings; special purpose rooms such as anaerobic chambers; and special construction such as radiation protection. Also included are aquatic facilities, ice rinks, kennels and animal shelters, and liquid and gas storage tanks. Special controls and instrumentation include building automation systems, recording instrumentation, and other special controls.

Selective Demolition includes the demolition of building components and the removal or encapsulation of hazardous building materials or components. This section can include a wide variety of tasks and is usually labor-intensive. The cost of removal or encapsulation of hazardous materials is particularly difficult to estimate, as it is usually governed by stringent local code requirements.

All items in this division are highly specific and not easily estimated using assemblies. There are no items in this division currently in the Means assembly database. It is recommended that items in this division be estimated using unit cost data.

Assemblies Estimate: Three-Story Office Building

Equipment and Furnishings may be included in a preliminary budget, or may be budgeted as a separate item to be furnished by the owner and installed by the contractor. The sample office building includes a safe.

Office Safe: 1 hour rating, 34″ × 20″ × 20″
(Figure 14.1, line #0300)   1 ea. @ $1,575   = $1,575

Figure 14.2 summarizes the costs for this division.

D5090   Other Electrical Systems

E1010 110	Security/Vault EACH	COST EACH		
		MAT.	INST.	TOTAL
0100	Bank equipment, drive up window, drawer & mike, no glazing, economy	4,025	955	4,980
0110	Deluxe	7,625	1,900	9,525
0120	Night depository, economy	6,800	955	7,755
0130	Deluxe	10,800	1,900	12,700
0140	Pneumatic tube systems, 2 station, standard	20,800	3,100	23,900
0150	Teller, automated, 24 hour, single unit	39,200	3,100	42,300
0160	Teller window, bullet proof glazing, 44" x 60"	2,600	575	3,175
0170	Pass through, painted steel, 72" x 40"	2,775	1,200	3,975
0180	Barber equipment, chair, hydraulic, economy	450	15.20	465.20
0190	Deluxe	2,975	23	2,998
0200	Console, including mirrors, deluxe	345	80.50	425.50
0210	Sink, hair washing basin	300	52	352
0220	Church equipment, altar, wood, custom, plain	1,825	261	2,086
0230	Granite, custom, deluxe	23,300	3,600	26,900
0300	Safe, office type, 1 hr. rating, 34" x 20" x 20"	1,575		1,575
0310	4 hr. rating, 62" x 33" x 20"	6,775		6,775
0320	Data storage, 4 hr. rating, 63" x 44" x 16"	9,375		9,375
0330	Jewelers, 63" x 44" x 16"	16,000		16,000
0340	Money, "B" label, 9" x 14" x 14"	490		490
0350	Tool and torch resistive, 24" x 24" x 20"	7,000	143	7,143
0500	Security gates-scissors type, painted steel, single, 6' high, 5-1/2' wide	125	238	363
0510	Double gate, 7-1/2' high, 14' wide	445	475	920

E1010 510	Mercantile Equipment, EACH	COST EACH		
		MAT.	INST.	TOTAL
0100	Checkout counter, single belt	2,175	57	2,232
0110	Double belt, power take-away	3,750	63.50	3,813.50
0200	Display cases, freestanding, glass and aluminum, 3'-6" x 3' x 1'-0" deep	455	91	546
0210	5'-10" x 4' x 1'-6" deep	1,325	122	1,447
0220	Wall mounted, glass and aluminum, 3' x 4' x 1'-4" deep	1,100	146	1,246
0230	16' x 4' x 1'-4" deep	3,200	485	3,685
0240	Table exhibit, flat, 3' x 4' x 2'-0" wide	785	146	931
0250	Sloping, 3' x 4' x 3'-0" wide	1,150	243	1,393
0300	Refrigerated food cases, dairy, multi-deck, rear sliding doors, 6 ft. long	8,875	253	9,128
0310	Delicatessen case, service type, 12 ft. long	5,225	194	5,419
0320	Frozen food, chest type, 12 ft. long	5,250	230	5,480
0330	Glass door, reach-in, 5 door	10,100	253	10,353
0340	Meat case, 12 ft. long, single deck	4,325	230	4,555
0350	Multi-deck	7,450	245	7,695
0360	Produce case, 12 ft. long, single deck	5,725	230	5,955
0370	Multi-deck	6,375	245	6,620

E1010 610	Laundry/Dry Cleaning, EACH	COST EACH		
		MAT.	INST.	TOTAL
0100	Laundry equipment, dryers, gas fired, residential, 16 lb. capacity	570	139	709
0110	Commercial, 30 lb. capacity, single	2,625	139	2,764
0120	Dry cleaners, electric, 20 lb. capacity	30,700	4,125	34,825
0130	30 lb. capacity	42,100	5,500	47,600
0140	Ironers, commercial, 120" with canopy, 8 roll	150,000	11,800	161,800
0150	Institutional, 110", single roll	26,600	2,050	28,650
0160	Washers, residential, 4 cycle	655	139	794
0170	Commercial, coin operated, deluxe	2,750	139	2,889

Figure 14.1

| ASSEMBLY NUMBER | DESCRIPTION | QTY | UNIT | TOTAL COST | | COST PER S.F. |
				UNIT	TOTAL	
DIVISION E	**EQUIPMENT & FURNISHINGS**					
E1010-110-0300	Office Safe, 1 hr Rating, 34" x 20" x 20"	1	EA	$ 1,575.00	$ 1,575	
	Subtotal, Division E, Equipment & Furnishings				$ 1,575	$ 0.05
DIVISION G	**SITEWORK**					

Figure 14.2

Chapter 15

Building Sitework

Building Sitework, UNIFORMAT II Division G, is subdivided into several major sections: Site Preparation, Improvements, Mechanical Utilities, Electrical Utilities, and Other Site Construction.

Site Preparation

Site Preparation, G10, includes site clearing, grubbing, and tree removal and thinning. It also includes demolition and/or relocation of buildings, site components, and utilities; removal of hazardous waste materials and contaminated soil; and site earthwork (not including building excavation).

Site Improvements

Site Improvements, G20, includes the construction of roadways, parking lots and pedestrian paving, fences and gates, retaining walls, site signage, site furnishings, playing fields, and landscaping. It is unnecessary to develop specific figures in the conceptual phases of a project for these items. An estimate of the area involved is enough to produce a budget or allowance for site improvements. A unit price cost data source, such as *Means Building Construction Cost Data*, can provide the necessary information for most site improvements.

Site Mechanical and Electrical Utilities

Site Mechanical Utilities, G30, consist of water supply; sanitary sewer work; storm sewers; and heating, cooling, and fuel distribution. Site Electrical Utilities, G40, include electrical distribution, site lighting, and site communications and security. These items are easily quantified and priced using assemblies.

Other Site Construction, G90, consists of service and pedestrian tunnels, prefabricated service tunnels, and trench boxes.

The following sections detail some of the components of the previous UNIFORMAT II divisions.

Earthwork

Earthwork includes bulk excavation, trench excavation, and backfill. (Bulk excavation and backfill, as they relate to the building itself, have been discussed in Chapter 4.) As with site improvement, it is better to price these items using a unit price estimate if the work is of significant value in the project. Sheeting, dewatering, drilling and blasting, traffic control, equipment mobilization and demobilization, and any unusual factors pertaining to the site must also be accounted for in the earthwork estimate using unit cost data.

Trenching

Trenching can be a significant cost on many projects and may vary depending on the type of soil, width and depth of the trench, and the slope required. Figure 15.1 is representative of the type of pricing information available for trenching systems, which are shown on a cost per linear foot basis. The prices include excavation as well as backfill material that is placed and compacted for various depths and trench bottom widths. The trench side slopes represent sound construction practice for most soils. (*Note: Trenching for the building foundation can be found in Division A1010, Foundations.*)

Utilities

Site utilities and drainage account for the costs of pipe bedding, manholes, and catch basins. Drainage and utilities require piping and related fittings to carry services to and from the building and site. If a site has been selected, and an approximate location of the building has been determined, the estimator can approximate the type and location of the required site utilities. Electrical cables and ducts, though installed in the sitework area, are more properly priced in the electrical division.

Pipe Bedding Systems

Pipe bedding is used to provide a stable substrate in the trench bottom to support and surround utility piping. The assemblies shown in Figure 15.2 are for various pipe diameters. Compacted sand is used for the bedding material and fill to 12" over the pipe. No backfill is included in the price. Various side slopes are shown to accommodate different soil conditions. Prices for the pipe are not included and must be added.

Manhole and Catch Basin Systems

Included in Figure 15.3 are excavation with a backhoe, formed concrete footings, concrete or masonry walls, aluminum steps, and compacted backfill. Inside diameters range from 4' to 6', with depths ranging from 4' to 14'. The construction materials used are concrete, concrete block, pre-cast concrete, or brick.

Pavements

If a site plan is not available, approximate the amount of roadway and parking area required using the anticipated number of occupants and local zoning requirements. Parking lots require between 300 and 400 S.F. per car, depending upon engineering layout and design. Once the total square foot areas of roadway and the number of parking spaces have been determined, a reliable estimate can be prepared using assemblies. The roadway and parking assemblies include quantities and prices for typical designs with gravel base and asphaltic concrete pavement. Different climate, soil conditions, material availability, and the owner's requirements can be adapted to the parameters shown.

Roadway Systems

The roadway prices in Figure 15.4 include 8" of compacted bank-run gravel, fine grading with a grader and roller, and a bituminous wearing course. The roadway systems are shown on a cost-per-linear-foot basis for the thickness and width indicated.

Parking Lot Systems

Figure 15.5 includes prices for various parking lot assemblies and layouts. These include compacted bank-run gravel, fine grading with a grader and roller, bituminous concrete wearing course, final stall design, and layout with white striping paint. All of the assemblies

G1030 Site Earthwork

Trenching Systems are shown on a cost per linear foot basis. The systems include: excavation; backfill and removal of spoil; and compaction for various depths and trench bottom widths. The backfill has been reduced to accommodate a pipe of suitable diameter and bedding.

The slope for trench sides varies from 0:1 to 2:1.

The Expanded System Listing shows Trenching Systems that range from 2' to 12' in width. Depths range from 2' to 25'.

System Components			COST PER L.F.		
	QUANTITY	UNIT	EQUIP.	LABOR	TOTAL
SYSTEM G1030 805					
TRENCHING, BACKHOE, 0 TO 1 SLOPE, 2' WIDE, 2' DP, 3/8 C.Y. BUCKET					
Excavation, trench, hyd. backhoe, track mtd., 3/8 C.Y. bucket	.174	C.Y.	.23	.76	.99
Backfill and load spoil, from stockpile	.174	C.Y.	.11	.22	.33
Compaction by rammer tamper, 8" lifts, 4 passes	.014	C.Y.	.01	.03	.04
Remove excess spoil, 6 C.Y. dump truck, 2 mile roundtrip	.160	C.Y.	.65	.47	1.12
Total			1	1.48	2.48

G1030 805	Trenching	COST PER L.F.		
		EQUIP.	LABOR	TOTAL
1310	Trenching, backhoe, 0 to 1 slope, 2' wide, 2' deep, 3/8 C.Y. bucket	1	1.48	2.48
1320	3' deep, 3/8 C.Y. bucket	1.22	2.16	3.38
1330	4' deep, 3/8 C.Y. bucket	1.43	2.86	4.29
1340	6' deep, 3/8 C.Y. bucket	1.84	3.64	5.48
1350	8' deep, 1/2 C.Y. bucket	2.09	4.62	6.71
1360	10' deep, 1 C.Y. bucket	3.13	5.55	8.68
1400	4' wide, 2' deep, 3/8 C.Y. bucket	2.10	2.99	5.09
1410	3' deep, 3/8 C.Y. bucket	2.96	4.44	7.40
1420	4' deep, 1/2 C.Y. bucket	2.84	4.45	7.29
1430	6' deep, 1/2 C.Y. bucket	3.67	6.75	10.42
1440	8' deep, 1/2 C.Y. bucket	5.60	8.85	14.45
1450	10' deep, 1 C.Y. bucket	6.60	10.95	17.55
1460	12' deep, 1 C.Y. bucket	8.25	14.05	22.30
1470	15' deep, 1-1/2 C.Y. bucket	7.25	12	19.25
1480	18' deep, 2-1/2 C.Y. bucket	9.85	17.15	27
1520	6' wide, 6' deep, 5/8 C.Y. bucket	7.40	9.50	16.90
1530	8' deep, 3/4 C.Y. bucket	8.55	11.45	20
1540	10' deep, 1 C.Y. bucket	9.80	13.80	23.60
1550	12' deep, 1-1/4 C.Y. bucket	8.70	10.85	19.55
1560	16' deep, 2 C.Y. bucket	14.20	12.35	26.55
1570	20' deep, 3-1/2 C.Y. bucket	15.15	22	37.15
1580	24' deep, 3-1/2 C.Y. bucket	24	37	61
1640	8' wide, 12' deep, 1-1/4 C.Y. bucket	13.30	14.85	28.15
1650	15' deep, 1-1/2 C.Y. bucket	15.40	18.25	33.65
1660	18' deep, 2-1/2 C.Y. bucket	22	18.45	40.45
1680	24' deep, 3-1/2 C.Y. bucket	33.50	49.50	83
1730	10' wide, 20' deep, 3-1/2 C.Y. bucket	28	37.50	65.50
1740	24' deep, 3-1/2 C.Y. bucket	42.50	62	104.50
1780	12' wide, 20' deep, 3-1/2 C.Y. bucket	34	45	79
1790	25' deep, bucket	57	83	140
1800	1/2 to 1 slope, 2' wide, 2' deep, 3/8 C.Y. bucket	1.34	1.90	3.24
1810	3' deep, 3/8 C.Y. bucket	1.84	3.14	4.98
1820	4' deep, 3/8 C.Y. bucket	2.31	4.58	6.89
1840	6' deep, 3/8 C.Y. bucket	3.44	7.05	10.49

RG1010 -020

Figure 15.1

G1030   Site Earthwork

The Pipe Bedding System is shown for various pipe diameters. Compacted bank sand is used for pipe bedding and to fill 12" over the pipe. No backfill is included. Various side slopes are shown to accommodate different soil conditions. Pipe sizes vary from 6" to 84" diameter.

System Components	QUANTITY	UNIT	COST PER L.F.		
			MAT.	INST.	TOTAL
SYSTEM G1030 815					
PIPE BEDDING, SIDE SLOPE 0 TO 1, 1' WIDE, PIPE SIZE 6" DIAMETER					
Borrow, bank sand, 2 mile haul, machine spread	.067	C.Y.	.26	.33	.59
Compaction, vibrating plate	.067	C.Y.		.12	.12
TOTAL			.26	.45	.71

G1030 815	Pipe Bedding	COST PER L.F.		
		MAT.	INST.	TOTAL
1440	Pipe bedding, side slope 0 to 1, 1' wide, pipe size 6" diameter	.26	.45	.71
1460	2' wide, pipe size 8" diameter	.58	.99	1.57
1480	Pipe size 10" diameter	.59	1.02	1.61
1500	Pipe size 12" diameter	.61	1.06	1.67
1520	3' wide, pipe size 14" diameter	1	1.73	2.73
1540	Pipe size 15" diameter	1.02	1.75	2.77
1560	Pipe size 16" diameter	1.03	1.77	2.80
1580	Pipe size 18" diameter	1.05	1.81	2.86
1600	4' wide, pipe size 20" diameter	1.52	2.61	4.13
1620	Pipe size 21" diameter	1.53	2.64	4.17
1640	Pipe size 24" diameter	1.57	2.70	4.27
1660	Pipe size 30" diameter	1.61	2.77	4.38
1680	6' wide, pipe size 32" diameter	2.81	4.85	7.66
1700	Pipe size 36" diameter	2.89	4.98	7.87
1720	7' wide, pipe size 48" diameter	3.76	6.50	10.26
1740	8' wide, pipe size 60" diameter	4.69	8.05	12.74
1760	10' wide, pipe size 72" diameter	6.75	11.65	18.40
1780	12' wide, pipe size 84" diameter	9.20	15.85	25.05
2140	Side slope 1/2 to 1, 1' wide, pipe size 6" diameter	.55	.95	1.50
2160	2' wide, pipe size 8" diameter	.91	1.57	2.48
2180	Pipe size 10" diameter	.99	1.70	2.69
2200	Pipe size 12" diameter	1.06	1.84	2.90
2220	3' wide, pipe size 14" diameter	1.52	2.62	4.14
2240	Pipe size 15" diameter	1.56	2.68	4.24
2260	Pipe size 16" diameter	1.61	2.77	4.38
2280	Pipe size 18" diameter	1.70	2.94	4.64
2300	4' wide, pipe size 20" diameter	2.24	3.87	6.11
2320	Pipe size 21" diameter	2.30	3.96	6.26
2340	Pipe size 24" diameter	2.46	4.25	6.71
2360	Pipe size 30" diameter	2.76	4.76	7.52
2380	6' wide, pipe size 32" diameter	4.07	7	11.07
2400	Pipe size 36" diameter	4.36	7.50	11.86
2420	7' wide, pipe size 48" diameter	5.95	10.25	16.20
2440	8' wide, pipe size 60" diameter	7.75	13.35	21.10
2460	10' wide, pipe size 72" diameter	10.85	18.70	29.55
2480	12' wide, pipe size 84" diameter	14.45	25	39.45
2620	Side slope 1 to 1, 1' wide, pipe size 6" diameter	.84	1.45	2.29
2640	2' wide, pipe size 8" diameter	1.26	2.18	3.44

Figure 15.2

G3030 Storm Sewer

Manhole   Catch Basin

The Manhole and Catch Basin System includes: excavation with a backhoe; a formed concrete footing; frame and cover; cast iron steps and compacted backfill.

The Expanded System Listing shows manholes that have a 4', 5' and 6' inside diameter riser. Depths range from 4' to 14'. Construction material shown is either concrete, concrete block, precast concrete, or brick.

System Components	QUANTITY	UNIT	COST PER EACH		
			MAT.	INST.	TOTAL
SYSTEM G3030 210					
MANHOLE/CATCH BASIN, BRICK, 4' I.D. RISER, 4' DEEP					
Excavation, hydraulic backhoe, 3/8 C.Y. bucket	14.815	C.Y.		75.26	75.26
Trim sides and bottom of excavation	64.000	S.F.		40.32	40.32
Forms in place, manhole base, 4 uses	20.000	SFCA	13.20	66.60	79.80
Reinforcing in place footings, #4 to #7	.019	Ton	11.02	16.06	27.08
Concrete, 3000 psi		C.Y.	65.68		65.68
Place and vibrate concrete, footing, direct chute		C.Y.		31.47	31.47
Catch basin or MH, brick, 4' ID, 4' deep	1.000	Ea.	298	650	948
Catch basin or MH steps; heavy galvanized cast iron	1.000	Ea.	10.40	9.15	19.55
Catch basin or MH frame and cover	1.000	Ea.	174	143.50	317.50
Fill, granular		C.Y.	108.17		108.17
Backfill, spread with wheeled front end loader		C.Y.		19.04	19.04
Backfill compaction, 12" lifts, air tamp	12.954	C.Y.		74.88	74.88
TOTAL			680.47	1,126.28	1,806.75

G3030 210	Manholes & Catch Basins	COST PER EACH		
		MAT.	INST.	TOTAL
1920	Manhole/catch basin, brick, 4' I.D. riser, 4' deep	680	1,125	1,805
1940	6' deep	915	1,575	2,490
1960	8' deep	1,175	2,150	3,325
1980	10' deep	1,500	2,650	4,150
3000	12' deep	1,925	2,900	4,825
3020	14' deep	2,425	4,050	6,475
3200	Block, 4' I.D. riser, 4' deep	630	910	1,540
3220	6' deep	825	1,300	2,125
3240	8' deep	1,050	1,775	2,825
3260	10' deep	1,225	2,175	3,400
3280	12' deep	1,525	2,800	4,325
3300	14' deep	1,850	3,400	5,250
4620	Concrete, cast-in-place, 4' I.D. riser, 4' deep	760	1,600	2,360
4640	6' deep	1,050	2,125	3,175
4660	8' deep	1,425	3,075	4,500
4680	10' deep	1,725	3,800	5,525
4700	12' deep	2,100	4,725	6,825
4720	14' deep	2,550	5,650	8,200
5820	Concrete, precast, 4' I.D. riser, 4' deep	715	835	1,550
5840	6' deep	960	1,125	2,085

Figure 15.3

G2010 Roadways

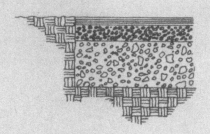

The Roadway System includes surveying; 8" of compacted bank-run gravel; fine grading with a grader and roller; and bituminous wearing course. Roadway Systems are shown on a cost per linear foot basis.

The Expanded System Listing shows Roadway Systems with widths that vary from 20' to 34'; for various pavement thicknesses.

System Components	QUANTITY	UNIT	COST PER L.F.		
			MAT.	INST.	TOTAL
SYSTEM G2010 210					
ROADWAYS, BITUMINOUS CONC. PAVING, 2-1/2" THICK, 20' WIDE					
4 man surveying crew	.010	Ea.		15	15
Bank gravel, 2 mi. haul, place and spread by dozer	.494	C.Y.	7.39	2.44	9.83
Compaction, granular material to 98%	.494	C.Y.		2.35	2.35
Grading, fine grade, 3 passes with grader plus rolling	2.220	S.Y.		7.80	7.80
Bituminous, paving, wearing course 2" - 2-1/2"	2.220	S.Y.	9.48	2.37	11.85
Curbs, granite, split face, straight, 5" x 16"	2.000	L.F.	20	10.14	30.14
Painting lines, reflectorized, 4" wide	1.000	L.F.	.12	.12	.24
TOTAL			36.99	40.22	77.21

G2010 210	Roadway Pavement	COST PER L.F.		
		MAT.	INST.	TOTAL
1500	Roadways, bituminous conc. paving, 2-1/2" thick, 20' wide	37	40	77
1520	24' wide	40.50	41.50	82
1540	26' wide	42	43	85
1560	28' wide	43.50	44.50	88
1580	30' wide	45.50	46	91.50
1600	32' wide	47	47	94
1620	34' wide	49	48.50	97.50
1800	3" thick, 20' wide	39	39	78
1820	24' wide	42.50	42	84.50
1840	26' wide	44.50	43.50	88
1860	28' wide	46.50	44.50	91
1880	30' wide	48	46	94
1900	32' wide	50	47.50	97.50
1920	34' wide	52	48.50	100.50
2100	4" thick, 20' wide	41.50	39	80.50
2120	24' wide	46	42	88
2140	26' wide	48	43.50	91.50
2160	28' wide	50.50	44.50	95
2180	30' wide	52.50	46	98.50
2200	32' wide	54.50	47.50	102

RG2010 -100

Figure 15.4

shown are on a cost-per-car basis and include travel areas, ingress, and egress. Prices vary according to the thickness of the different materials and the parking angle. The three parking arrangements have different efficiencies and installed costs. The 90-degree arrangement is the most efficient and least expensive.

Assemblies Estimate: Three-Story Office Building

The following sitework is included in the estimate for the sample office building.

Site Preparation, Clear & Grub			
(Figure 15.6)	1 Acre × $7,100	=	$7,100
Trenching for Utility Piping			
(Figure 15.1, line #1330)	300 L.F. @ $4.29	=	1,287
Pipe bedding, 2' wide trench, pipe < 8" dia.			
(Figure 15.2, line #1460)	300 L.F. @ $1.57	=	471
Drainage/Sewage Piping, Concrete, non-reinforced, 6"			
(Figure 15.7, line #4150)	100 L.F. @ $11.38	=	1,138
Drainage/Sewage Piping, Concrete, non-reinforced, 8"			
(Figure 15.7, line #4160)	160 L.F. @ $13.16	=	2,106
Gas Service Piping, Steel, Sched. 40, 3"			
(Figure 15.8, line #3110)	25 L.F. @ $12.60	=	315
Water Distribution Piping, copper, Type K, 3"			
(Figure 15.9, line #4110)	25 L.F. @ $15.10	=	378
Water Distribution Piping, copper, Type K, 4"			
(Figure 15.9, line #4130)	25 L.F. @ $22.70	=	568
Manhole/Catchbasin, Concrete, C.I.P., 4' ID Riser, 4' Deep			
(Figure 15.3, line #4620)	2 ea. @ $2,360	=	4,720
Manhole/Catchbasin, Concrete, C.I.P., 4' ID Riser, 6' Deep			
(Figure 15.3, line #4640)	1 ea. @ $3,175	=	3,175
Roadway, Bit. Conc. Paving, 3" Thick, 20 Wide			
(Figure 15.4, line #1800)	260 L.F. @ $78.00	=	20,280
Parking Lot, 60 Cars			
(Figure 15.5, line #1520)	60 cars × $571	=	34,260
(23,200 S.F./390 S.F./car = 60 cars)			
Site Improvements (grass seed, shrubs, trees)			
(Allowance)	6,800 S.F. × $2.00/ S.F.	=	13,600
Total Cost			$89,398
Cost per Square Foot: $89,398/30,000		=	$2.98/S.F.

Figure 15.10 summarizes the costs for this division.

G20 Site Improvements

G2020 Parking Lots

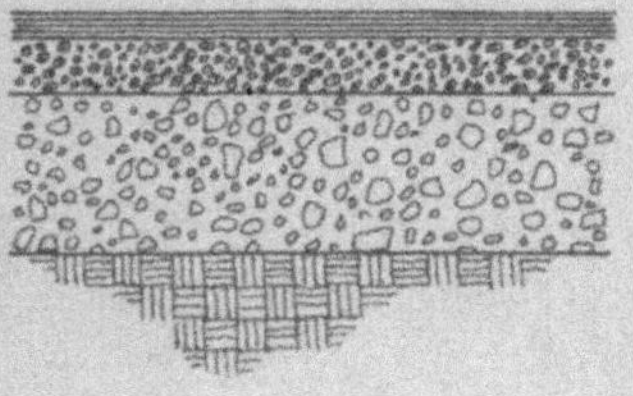

The Parking Lot System includes: compacted bank-run gravel; fine grading with a grader and roller; and bituminous concrete wearing course. All Parking Lot systems are on a cost per car basis. There are three basic types of systems: 90° angle, 60° angle, and 45° angle. The gravel base is compacted to 98%. Final stall design and lay-out of the parking lot with precast bumpers, sealcoating and white paint is also included.

The Expanded System Listing shows the three basic parking lot types with various depths of both gravel base and wearing course. The gravel base depths range from 6" to 10". The bituminous paving wearing course varies from a depth of 3" to 6".

System Components	QUANTITY	UNIT	COST PER CAR MAT.	COST PER CAR INST.	COST PER CAR TOTAL
SYSTEM G2020 210					
PARKING LOT, 90° ANGLE PARKING, 3" BITUMINOUS PAVING, 6" GRAVEL BASE					
Surveying crew for layout, 4 man crew	.020	Day		30	30
Borrow, bank run gravel, haul 2 mi., spread w/dozer, no compaction	5.556	C.Y.	83.06	27.50	110.56
Grading, fine grade 3 passes with motor grader	33.333	S.Y.		49.13	49.13
Compact w/ vibrating plate, 8" lifts, granular mat'l. to 98%	5.556	C.Y.		26.45	26.45
Bituminous paving, 3" thick	33.333	S.Y.	170	40	210
Seal coating, petroleum resistant under 1,000 S.Y.	33.333	S.Y.	19	28	47
Painting lines on pavement, parking stall, white	1.000	Ea.	2.86	5.13	7.99
Precast concrete parking bar, 6" x 10" x 6'-0"	1.000	Ea.	31	12.15	43.15
TOTAL			305.92	218.36	524.28

G2020 210	Parking Lots	COST PER CAR MAT.	COST PER CAR INST.	COST PER CAR TOTAL
1500	Parking lot, 90° angle parking, 3" bituminous paving, 6" gravel base	305	219	524
1520	8" gravel base	335	236	571
1540	10" gravel base RG2020 -500	360	255	615
1560	4" bituminous paving, 6" gravel base	350	217	567
1580	8" gravel base	375	236	611
1600	10" gravel base	405	253	658
1620	6" bituminous paving, 6" gravel base	450	226	676
1640	8" gravel base	475	244	719
1661	10" gravel base	505	263	768
1800	60° angle parking, 3" bituminous paving, 6" gravel base	350	239	589
1820	8" gravel base	385	260	645
1840	10" gravel base	415	281	696
1860	4" bituminous paving, 6" gravel base	400	237	637
1880	8" gravel base	435	258	693
1900	10" gravel base	465	280	745
1920	6" bituminous paving, 6" gravel base	520	248	768
1940	8" gravel base	550	269	819
1961	10" gravel base	585	290	875
2200	45° angle parking, 3" bituminous paving, 6" gravel base	395	259	654
2220	8" gravel base	435	283	718
2240	10" gravel base	470	305	775
2260	4" bituminous paving, 6" gravel base	455	257	712
2280	8" gravel base	490	281	771
2300	10" gravel base	530	305	835
2320	6" bituminous paving, 6" gravel base	590	270	860
2340	8" gravel base	625	294	919
2361	10" gravel base	660	470	1,130

Figure 15.5

Site Preparation

Description	Unit	Total Costs
Clear and Grub Average	Acre	$7,100
Bulk Excavation with Front End Loader	C.Y.	1.51
Spread and Compact Dumped Material	C.Y.	1.88
Fine Grade and Seed	S.Y.	2.18

Figure 15.6

G3020 Sanitary Sewer

G3020 110	Drainage & Sewage Piping	COST PER L.F.		
		MAT.	INST.	TOTAL
2000	Piping, excavation & backfill excluded, PVC, plain			
2130	4" diameter	1.55	2.59	4.14
2150	6" diameter	2.65	2.78	5.43
2160	8" diameter	4.84	2.90	7.74
2900	Box culvert, precast, 8' long			
3000	6' x 3'	164	23.50	187.50
3020	6' x 7'	253	26.50	279.50
3040	8' x 3'	186	25	211
3060	8' x 8'	310	33	343
3080	10' x 3'	400	30	430
3100	10' x 8'	385	41	426
3120	12' x 3'	315	33	348
3140	12' x 8'	550	49.50	599.50
4000	Concrete, nonreinforced			
4150	6" diameter	3.83	7.55	11.38
4160	8" diameter	4.21	8.95	13.16
4170	10" diameter	4.67	9.25	13.92
4180	12" diameter	5.75	10	15.75
4200	15" diameter	6.70	11.10	17.80
4220	18" diameter	8.25	13.95	22.20
4250	24" diameter	12.45	20	32.45
4400	Reinforced, no gasket			
4580	12" diameter	10.75	9.55	20.30
4600	15" diameter	13.30	11.45	24.75
4620	18" diameter	13.40	15.45	28.85
4650	24" diameter	19.60	20	39.60
4670	30" diameter	27	32.50	59.50
4680	36" diameter	38.50	39.50	78
4690	42" diameter	52.50	44	96.50
4700	48" diameter	65.50	49	114.50
4720	60" diameter	105	65	170
4730	72" diameter	148	78.50	226.50
4740	84" diameter	269	98	367
4800	With gasket			
4980	12" diameter	11.90	7.85	19.75
5000	15" diameter	14.25	8.25	22.50
5020	18" diameter	17.85	8.70	26.55
5050	24" diameter	27	9.70	36.70
5070	30" diameter	35.50	32.50	68
5080	36" diameter	53.50	39.50	93
5090	42" diameter	70	39.50	109.50
5100	48" diameter	87.50	49	136.50
5120	60" diameter	118	63	181
5130	72" diameter	226	78.50	304.50
5140	84" diameter	325	131	456
5700	Corrugated metal, alum. or galv. bit. coated			
5760	8" diameter	6.45	6.05	12.50
5770	10" diameter	7.70	7.70	15.40
5780	12" diameter	9.30	9.55	18.85
5800	15" diameter	11.25	10	21.25
5820	18" diameter	14.50	10.55	25.05
5850	24" diameter	17.75	12.50	30.25
5870	30" diameter	23	23.50	46.50
5880	36" diameter	34	23.50	57.50
5900	48" diameter	52	28.50	80.50
5920	60" diameter	67	41.50	108.50
5930	72" diameter	102	69.50	171.50
6000	Plain			

Figure 15.7

G30 Site Mechanical Utilities

G3060 Fuel Distribution

G3060 110	Gas Service Piping		COST PER L.F.		
			MAT.	INST.	TOTAL
2000	Piping, excavation & backfill excluded, polyethylene				
2070	1-1/4" diam, SDR 10		.71	2.47	3.18
2090	2" diam, SDR 11		.87	2.76	3.63
2110	3" diam, SDR 11.5		1.85	3.30	5.15
2130	4" diam, SDR 11		4.22	6.25	10.47
2150	6" diam, SDR 21		13.15	6.70	19.85
2160	8" diam, SDR 21		17.95	8.05	26
3000	Steel, schedule 40, plain end, tarred & wrapped				
3060	1" diameter		2.33	5.65	7.98
3090	2" diameter		3.66	6.05	9.71
3110	3" diameter		6.05	6.55	12.60
3130	4" diameter		7.90	10.35	18.25
3140	5" diameter		11.45	12	23.45
3150	6" diameter		14	14.70	28.70
3160	8" diameter		22	18.90	40.90
3170	10" diameter		39.50	26.50	66
3180	12" diameter		56	33	89
3190	14" diameter		56	35	91
3200	16" diameter		65	37.50	102.50
3210	18" diameter		83.50	40.50	124
3220	20" diameter		130	44	174
3230	24" diameter		149	53	202

Figure 15.8

259

G30 Site Mechanical Utilities

G3010 Water Supply

G3010 110	Water Distribution Piping	COST PER L.F.		
		MAT.	INST.	TOTAL
2000	Piping, excav. & backfill excl., ductile iron class 250, mech. joint			
2130	4" diameter	9.65	6.75	16.40
2150	6" diameter	10	7.70	17.70
2160	8" diameter	12.85	9	21.85
2170	10" diameter	16.65	10.80	27.45
2180	12" diameter	21.50	18.25	39.75
2210	16" diameter	28.50	28.50	57
2220	18" diameter	37.50	31	68.50
3000	Tyton joint			
3130	4" diameter	7.10	6.25	13.35
3150	6" diameter	8.15	7.20	15.35
3160	8" diameter	11.55	8.45	20
3170	10" diameter	17.55	9.90	27.45
3180	12" diameter	18.50	12.45	30.95
3210	16" diameter	28.50	25.50	54
3220	18" diameter	32	30.50	62.50
3230	20" diameter	35	32.50	67.50
3250	24" diameter	45	33	78
4000	Copper tubing, type K			
4050	3/4" diameter	1.45	1.87	3.32
4060	1" diameter	1.85	2.34	4.19
4080	1-1/2" diameter	2.97	2.83	5.80
4090	2" diameter	4.74	3.26	8
4110	3" diameter	9.50	5.60	15.10
4130	4" diameter	14.80	7.90	22.70
4150	6" diameter	46.50	12.25	58.75
5000	Polyvinyl chloride class 160, S.D.R. 26			
5130	1-1/2" diameter	.65	3.24	3.89
5150	2" diameter	.98	3.89	4.87
5160	4" diameter	3.40	4.86	8.26
5170	6" diameter	7.35	5.40	12.75
5180	8" diameter	12.40	8.25	20.65
6000	Polyethylene 160 psi, S.D.R. 7			
6050	3/4" diameter	.23	1.85	2.08
6060	1" diameter	.39	2	2.39
6080	1-1/2" diameter	.89	2.16	3.05
6090	2" diameter	1.52	2.66	4.18

Figure 15.9

260

ASSEMBLY NUMBER	DESCRIPTION	QTY	UNIT	TOTAL COST UNIT	TOTAL COST TOTAL	COST PER S.F.
DIVISION E	**EQUIPMENT & FURNISHINGS**					
E1010-110-0300	Office Safe, 1 hr Rating, 34" x 20" x 20"	1	EA	$ 1,575.00	$ 1,575	
	Subtotal, Division E, Equipment & Furnishings				$ 1,575	$ 0.05
DIVISION G	**SITEWORK**					
RG1010-011	Clear & Grub	1	ACRE	$ 7,100.00	$ 7,100	
G1030-805-1330	Utility trenching, backhoe, 0 to 1 slope, 4' deep	300	L.F.	$ 4.29	$ 1,287	
G1030-815-1460	Pipe bedding, 2' wide trench, pipe <8" dia.	300	L.F.	$ 1.57	$ 471	
G3020-110-4150	Drainage/sewage piping, conc., non-reinf., 6" dia.	100	L.F	$ 11.38	$ 1,138	
G3020-110-4160	Drainage/sewage piping, conc., non-reinf., 8" dia.	160	L.F.	$ 13.16	$ 2,106	
G3060-110-3110	Gas service piping, steel, sched. 40, PE, 3" dia.	25	L.F.	$ 12.60	$ 315	
G3010-110-4110	Water distribution piping, copper, type K, 3" dia.	25	L.F.	$ 15.10	$ 378	
G3010-110-4130	Water distribution piping, copper, type K, 4" dia.	25	L.F.	$ 22.70	$ 568	
G3030-210-4620	Manhole/catchbasin, concrete, CIP, 4' ID riser, 4' deep	2	Ea.	$ 2,360.00	$ 4,720	
G3030-210-4640	Manhole/catchbasin, concrete, CIP, 4' ID riser, 6' deep	1	Ea.	$ 3,175.00	$ 3,175	
G2010-210-1800	Roadway, Bit. Conc. Paving, 3" thick, 20' wide	260	L.F.	$ 78.00	$ 20,280	
G2020-210-1520	Parking Lot	60	CAR	$ 571.00	$ 34,260	
Allowance	Site Improvements (Grass, Trees, etc.)	6,800	S.F.	$ 2.00	$ 13,600	
	SUBTOTAL, DIVISION G, SITEWORK				$ 89,398	$ 2.98

Figure 15.10

Chapter 16

Estimate Summary

The assemblies costs in *Means Assemblies Cost Data* include overhead and profit for the installing contractor. If the project estimate is to reflect the actual price charged by a general contractor for the work, then general conditions, overhead, and profit costs must be added to the estimate. The usual practice is to add these costs at the end of the estimate after all items have been totaled and the project's scope and duration have been determined. The following discussion defines overhead and profit items for both the installing contractor and the general contractor.

The Installing Contractor's Overhead & Profit

Means Assemblies Cost Data includes the installing contractor's overhead and profit for each item. The column, "Total Overhead & Profit," in Figure 16.1 shows the total percentage and cost added to each trade for the installing contractor's overhead and profit. These costs are developed as follows.

Labor Base Rate

The *base rate* is the figure on which the actual payroll is calculated. The following deductions are made each week (with the net balance going to the employee): standard deductions such as taxes, Social Security, saving plans, hospitalization, and any voluntary employee-paid deductions.

Base Rate Including Fringe Benefits

The *base rate including fringe benefits* is the sum of the base rate plus all employer-paid fringe benefits, which might include vacation pay, health and welfare insurance, pension, apprentice training, and industry advancement funds. Figure 15.1 lists the 40 construction trades plus the skilled workers' and helpers' average for 30 major United States cities. The table is based on union labor rates for January 1, 2001 and reflects the national average.

Billing Rate

When adjusted to include all direct and indirect expenses of the installing contractor, the base rate including fringe benefits is called the *billing rate*. This rate includes the following direct and indirect expenses.

Installing Contractor's Overhead & Profit

Below are the **average** installing contractor's percentage mark-ups applied to base labor rates to arrive at typical billing rates.

Column A: Labor rates are based on union wages averaged for 30 major U.S. cities. Base rates including fringe benefits are listed hourly and daily. These figures are the sum of the wage rate and employer-paid fringe benefits such as vacation pay, employer-paid health and welfare costs, pension costs, plus appropriate training and industry advancement funds costs.

Column B: Workers' Compensation rates are the national average of state rates established for each trade.

Column C: Column C lists average fixed overhead figures for all trades. Included are Federal and State Unemployment costs set at 7.0%; Social Security Taxes (FICA) set at 7.65%; Builder's Risk Insurance costs set at 0.34%; and Public Liability costs set at 1.55%. All the percentages except those for Social Security Taxes vary from state to state as well as from company to company.

Columns D and E: Percentages in Columns D and E are based on the presumption that the installing contractor has annual billing of $1,500,000 and up. Overhead percentages may increase with smaller annual billing. The overhead percentages for any given contractor may vary greatly and depend on a number of factors, such as the contractor's annual volume, engineering and logistical support costs, and staff requirements. The figures for overhead and profit will also vary depending on the type of job, the job location, and the prevailing economic conditions. All factors should be examined very carefully for each job.

Column F: Column F lists the total of Columns B, C, D, and E.

Column G: Column G is Column A (hourly base labor rate) multiplied by the percentage in Column F (O&P percentage).

Column H: Column H is the total of Column A (hourly base labor rate) plus Column G (Total O&P).

Column I: Column I is Column H multiplied by eight hours.

		A		B	C	D	E	F		H	I
		Base Rate Incl. Fringes		Workers' Comp. Ins.	Average Fixed Overhead	Overhead	Profit	Total Overhead & Profit		Rate with O & P	
Abbr.	Trade	Hourly	Daily					%	Amount	Hourly	Daily
Skwk	Skilled Workers Average (35 trades)	$29.85	$238.80	17.5%	16.5%	13.0%	10%	57.0%	$17.00	$46.85	$374.80
	Helpers Average (5 trades)	22.15	177.20	19.3		11.0		56.8	12.55	34.70	277.60
	Foreman Average, Inside ($.50 over trade)	30.35	242.80	17.5		13.0		57.0	17.30	47.65	381.20
	Foreman Average, Outside ($2.00 over trade)	31.85	254.80	17.5		13.0		57.0	18.15	50.00	400.00
Clab	Common Building Laborers	22.85	182.80	19.0		11.0		56.5	12.90	35.75	286.00
Asbe	Asbestos/Insulation Workers/Pipe Coverers	32.15	257.20	17.7		16.0		60.2	19.35	51.50	412.00
Boil	Boilermakers	34.55	276.40	15.9		16.0		58.4	20.20	54.75	438.00
Bric	Bricklayers	29.60	236.80	17.0		11.0		54.5	16.15	45.75	366.00
Brhe	Bricklayer Helpers	22.95	183.60	17.0		11.0		54.5	12.50	35.45	283.60
Carp	Carpenters	29.15	233.20	19.0		11.0		56.5	16.45	45.60	364.80
Cefi	Cement Finishers	27.95	223.60	11.1		11.0		48.6	13.60	41.55	332.40
Elec	Electricians	34.30	274.40	6.8		16.0		49.3	16.90	51.20	409.60
Elev	Elevator Constructors	35.65	285.20	8.4		16.0		50.9	18.15	53.80	430.40
Eqhv	Equipment Operators, Crane or Shovel	31.15	249.20	10.9		14.0		51.4	16.00	47.15	377.20
Eqmd	Equipment Operators, Medium Equipment	30.35	242.80	10.9		14.0		51.4	15.60	45.95	367.60
Eqlt	Equipment Operators, Light Equipment	28.75	230.00	10.9		14.0		51.4	14.80	43.55	348.40
Eqol	Equipment Operators, Oilers	25.75	206.00	10.9		14.0		51.4	13.25	39.00	312.00
Eqmm	Equipment Operators, Master Mechanics	31.80	254.40	10.9		14.0		51.4	16.35	48.15	385.20
Glaz	Glaziers	28.55	228.40	14.1		11.0		51.6	14.75	43.30	346.40
Lath	Lathers	28.20	225.60	11.8		11.0		49.3	13.90	42.10	336.80
Marb	Marble Setters	29.20	233.60	17.0		11.0		54.5	15.90	45.10	360.80
Mill	Millwrights	30.55	244.40	11.2		11.0		48.7	14.90	45.45	363.60
Mstz	Mosaic & Terrazzo Workers	28.30	226.40	10.3		11.0		47.8	13.55	41.85	334.80
Pord	Painters, Ordinary	26.10	208.80	14.8		11.0		52.3	13.65	39.75	318.00
Psst	Painters, Structural Steel	27.15	217.20	50.0		11.0		87.5	23.75	50.90	407.20
Pape	Paper Hangers	26.25	210.00	14.8		11.0		52.3	13.75	40.00	320.00
Pile	Pile Drivers	28.80	230.40	27.7		16.0		70.2	20.20	49.00	392.00
Plas	Plasterers	27.30	218.40	15.7		11.0		53.2	14.50	41.80	334.40
Plah	Plasterer Helpers	23.25	186.00	15.7		11.0		53.2	12.35	35.60	284.80
Plum	Plumbers	34.45	275.60	8.5		16.0		51.0	17.55	52.00	416.00
Rodm	Rodmen (Reinforcing)	32.75	262.00	29.2		14.0		69.7	22.85	55.60	444.80
Rofc	Roofers, Composition	25.70	205.60	34.3		11.0		71.8	18.45	44.15	353.20
Rots	Roofers, Tile & Slate	26.00	208.00	34.3		11.0		71.8	18.65	44.65	357.20
Rohe	Roofers, Helpers (Composition)	19.10	152.80	34.3		11.0		71.8	13.70	32.80	262.40
Shee	Sheet Metal Workers	33.55	268.40	11.9		16.0		54.4	18.25	51.80	414.40
Spri	Sprinkler Installers	34.45	275.60	8.8		16.0		51.3	17.65	52.10	416.80
Stpi	Steamfitters or Pipefitters	34.85	278.80	8.5		16.0		51.0	17.75	52.60	420.80
Ston	Stone Masons	29.75	238.00	17.0		11.0		54.5	16.20	45.95	367.60
Sswk	Structural Steel Workers	32.85	262.80	40.9		14.0		81.4	26.75	59.60	476.80
Tilf	Tile Layers	28.15	225.20	10.3		11.0		47.8	13.45	41.60	332.80
Tilh	Tile Layers Helpers	22.70	181.60	10.3		11.0		47.8	10.85	33.55	268.40
Trlt	Truck Drivers, Light	23.40	187.20	15.3		11.0		52.8	12.35	35.75	286.00
Trhv	Truck Drivers, Heavy	23.95	191.60	15.3		11.0		52.8	12.65	36.60	292.80
Sswl	Welders, Structural Steel	32.85	262.80	40.9		14.0		81.4	26.75	59.60	476.80
Wrck	*Wrecking	22.85	182.80	42.2		11.0		79.7	18.20	41.05	328.40

*Not included in averages

Figure 16.1

Workers' Compensation and Employer's Liability Insurance

Workers' compensation and employer's liability insurance rates vary from state to state and from one trade to another based on the relative safety record for that trade during the previous year. Rates also vary from company to company based on experience rating.

Average Fixed Overhead

Average fixed overhead includes the costs paid by the employer for U.S. and state unemployment, Social Security, Medicare, builder's risk, and public liability.

Overhead

Contractors have business expenses, or overhead, that must be paid whether or not they are working. These may not contribute directly to the actual construction activities but are, nevertheless, necessary in order to get the job done. The cost of overhead can represent a considerable portion of a contractor's annual business volume. The range used in *Means Assemblies Cost Data* is representative for contractors whose projects cost $500,000 and more. Overhead ranges from 11% to 16% and varies with the type of contractor and annual volume of business.

Profit and Contingency

The percentage in the profit column of Figure 16.1 represents the fee added by the installing contractor for a return on the investment, plus an allowance to cover the project's risks. Profit varies for each job depending on several factors, including the economic outlook for future work, the contractor's current need for work, the time of year and weather, the amount of work to be subcontracted, the number of competing bidders, and the estimated risk involved in the job. For estimating purposes, *Means Assemblies Cost Data* builds in an allowance of 10% as a reasonable profit factor. Contractors, however, usually set their own profit based on their own unique situations.

General Contractor's Mark-ups

When all construction items have been considered and totaled, account for all other items that will affect the cost of the project as a whole. Add an allowance for sales taxes, general conditions, overhead, and profit to the prices in the assemblies estimate. These items are typically added as percentage mark-ups to the total project cost in the estimate summary. Figure 16.2 is a final summary form.

Sales Tax

Where applicable, add sales tax on materials, labor, and rental equipment. Figure 16.3 lists state sales taxes, but does not include any additional taxes levied by individual cities and counties. To derive the cost of sales tax for an assemblies square foot estimate, it is reasonable to assume that 50% of the subtotal cost represents the material costs, and 50% is the labor cost. Multiply the appropriate cost by the sales tax percentage to derive the sales tax.

General Conditions

General Conditions refers to those cost items provided by the general contractor that affect the operation of the entire project. The general contractor's costs include those for the field office, site supervision, site security, temporary facilities and utilities, tools and minor equipment, bonds, testing costs, and so forth. These are itemized and priced

Preliminary Estimate Cost Summary

PROJECT	TOTAL AREA	SHEET NO.
LOCATION	TOTAL VOLUME	ESTIMATE NO.
ARCHITECT	COST PER S.F.	DATE
OWNER	COST PER C.F.	NO OF STORIES
QUANTITIES BY	EXTENSIONS BY	CHECKED BY

DIV	DESCRIPTION	SUBTOTAL COST	COST/S.F.	PERCENTAGE
A	SUBSTRUCTURE			
B10	SHELL: SUPERSTRUCTURE			
B20	SHELL: EXTERIOR CLOSURE			
B30	SHELL: ROOFING			
C	INTERIOR CONSTRUCTION			
D10	SERVICES: CONVEYING			
D20	SERVICES: PLUMBING			
D30	SERVICES: HVAC			
D40	SERVICES: FIRE PROTECTION			
D50	SERVICES: ELECTRICAL			
E	EQUIPMENT & FURNISHINGS			
F	SPECIAL CONSTRUCTION			
G	SITEWORK			

BUILDING SUBTOTAL $ _____ -

Sales Tax ___ % x Subtotal /2 $ _____ -

General Conditions ___ % x Subtotal $ _____ -

         Subtotal "A" $ _____ -

Overhead ___ % x Subtotal "A" $ _____ -

         Subtotal "B" $ _____ -

Profit ___ % x Subtotal "B" $ _____ -

         Subtotal "C" $ _____ -

Location Factor ____ % x Subtotal "C"   Localized Cost $ _____ -
     (Boston, MA)

Architects Fee ___ x Localized Cost = $ _____ -
Contingency ____ x Localized Cost = $ _____ -

         Project Total Cost $ -

Square Foot Cost $ ___ / S.F. =   **S.F. Cost** $ -
Cubic Foot Cost $ ___ / C.F. =   **C.F. Cost** $ -

Figure 16.2

266

individually in a unit price estimate. In a square foot assemblies estimate, a percentage mark-up is traditionally used in the estimate summary to provide for these costs. Figure 16.4 lists factors that may be used for these project overhead items.

Office Overhead

There are certain indirect expense items that are necessary to operate a company, attract business, and bid work. The general contractor must pass these indirect costs for main office expenses on as a percentage mark-up on all projects. The percentage may decline with increased annual volume, since overhead costs do not appreciably increase when the annual volume of work increases. Typical main office expenses range from 2% to 20% of project costs, with a median of 7.2% of the total volume. This equals 7.7% of the job direct costs.

Sales Tax by State

State sales tax on materials is tabulated below (5 states have no sales tax). Many states allow local jurisdictions, such as a county or city, to levy additional sales tax.

Some projects may be sales tax exempt, particularly those constructed with public funds.

State	Tax (%)	State	Tax (%)	State	Tax (%)	State	Tax (%)
Alabama	4	Illinois	6.25	Montana	0	Rhode Island	7
Alaska	0	Indiana	5	Nebraska	4.5	South Carolina	5
Arizona	5	Iowa	5	Nevada	6.875	South Dakota	4
Arkansas	4.625	Kansas	4.9	New Hampshire	0	Tennessee	6
California	7.25	Kentucky	6	New Jersey	6	Texas	6.25
Colorado	3	Louisiana	4	New Mexico	5	Utah	4.75
Connecticut	6	Maine	6	New York	4	Vermont	5
Delaware	0	Maryland	5	North Carolina	4	Virginia	4.5
District of Columbia	5.75	Massachusetts	5	North Dakota	5	Washington	6.5
Florida	6	Michigan	6	Ohio	5	West Virginia	6
Georgia	4	Minnesota	6.5	Oklahoma	4.5	Wisconsin	5
Hawaii	4	Mississippi	7	Oregon	0	Wyoming	4
Idaho	5	Missouri	4.225	Pennsylvania	6	Average	4.71 %

Figure 16.3

General Contractor's Overhead

The table below shows a contractor's overhead as a percentage of direct cost in two ways. The figures on the right are for the overhead, markup based on both material and labor. The figures on the left are based on the entire overhead applied only to the labor. This figure would be used if the owner supplied the materials or if a contract is for labor only.

Items of General Contractor's Indirect Costs	% of Direct Costs	
	As a Markup of Labor Only	As a Markup of Both Material and Labor
Field Supervision	6.0%	2.9%
Main Office Expense (see details below)	16.2	7.7
Tools and Minor Equipment	1.0	0.5
Workers' Compensation & Employers' Liability. See H1020-203	17.5	8.3
Field Office, Sheds, Photos, Etc.	1.5	0.7
Performance and Payment Bond, 0.7% to 1.5%. See H1020-302	2.3	1.1
Unemployment Tax See RH1020-300 (Combined Federal and State)	7.0	3.3
Social Security and Medicare, See RH1020-300	7.7	3.7
Sales Tax — add if applicable 42/80 x % as markup of total direct costs including both material and labor. See RH1030-100		
Sub Total	59.2%	28.2%
*Builder's Risk Insurance ranges from 0.141% to 0.586%. See H1010-301	0.6	0.3
*Public Liability Insurance	3.2	1.5
Grand Total	63.0%	30.0%

*Paid by Owner or Contractor

Figure 16.4

Figure 16.5 shows approximate percentages for some of the items usually included in a general/prime contractor's main office overhead. These percentages may vary with different accounting procedures.

Profit Margin

The profit assumed in *Means Assemblies Cost Data* is 10% on material, labor, and equipment. Since this is the profit margin for the installing contractor, an additional percentage must be included to cover the profit of the general or prime contractor. The general contractor's profit depends on many factors, including economic conditions and the risks involved in the project. Profit can range from negative numbers to 25-35%, depending on specific job conditions.

Location Factors

Publications such as *Means Assemblies Cost Data* contain building cost information based on a national average of major cities for each year. Estimates derived from this data must be "localized" so that the costs reflect, as closely as possible, those in the area where the project is to be constructed. This can be done using the City Cost Indexes published in each edition of *Means Assemblies Cost Data*. (See Figure 16.6, a sample City Cost Index page.) The indexes are calculated for over 300 cities in the United States and Canada. *(Use of Canadian indexes results in project costs in Canadian dollars.)* Both material and installation factors are shown. January 1, 2001 is the reference point for the 30 major city average. *(Note: The index figures in Figure 16.6 are based on prices shown in* Means Assemblies Cost Data *being equal to a reference of 100.0.)*

There are certain factors that the indexes cannot take into account, including variations such as:

- Productivity
- Labor availability
- Contractor management efficiency
- Competitive conditions
- Automation
- Restrictive union practices
- Clients' unique requirements

Main Office Expense

A General Contractor's main office expense consists of many items not detailed in the front portion of the book. The percentage of main office expense declines with increased annual volume of the contractor. Typical main office expense ranges from 2% to 20% with the median about 7.2% of total volume. This equals about 7.7% of direct costs. The following are approximate percentages of total overhead for different items usually included in a General Contractor's main office overhead. With different accounting procedures, these percentages may vary.

Item	Typical Range			Average
Managers', clerical and estimators' salaries	40 %	to	55 %	48%
Profit sharing, pension and bonus plans	2	to	20	12
Insurance	5	to	8	6
Estimating and project management (not including salaries)	5	to	9	7
Legal, accounting and data processing	0.5	to	5	3
Automobile and light truck expense	2	to	8	5
Depreciation of overhead capital expenditures	2	to	6	4
Maintenance of office equipment	0.1	to	1.5	1
Office rental	3	to	5	4
Utilities including phone and light	1	to	3	2
Miscellaneous	5	to	15	8
Total				100%

Figure 16.5

Design and Engineering
Architectural engineering fees are typical percentage fees for design services. Some fees may vary significantly from those listed in Figure 2.4 because of economic conditions and the scope of work. The architectural fees include the engineering fees in Figures 16.7 and 16.8.

\multicolumn{4}{c}{City: Boston}
Div. No.	System	Mat.	Inst.	Total
---	---	---	---	---
A	Substructure	104.1	129.4	120.2
B10	Superstructure	102.0	129.2	113.8
B20	Exterior Closure	116.5	139.5	127.5
B30	Roofing	100.0	136.4	116.1
C	Interior Construction	101.0	135.8	115.2
D10	Conveying	100.0	125.9	107.3
D20-40	Mechanical	99.8	122.0	110.1
D50	Electrical	99.1	125.1	116.9
E	Special Construction	100.0	129.8	101.9
G	Sitework	93.7	110.2	106.0
A-G	Weighted Average	102.6	128.0	115.0

Figure 16.6

Engineering Fees
Typical **Structural Engineering Fees** based on type of construction and total project size. These fees are included in Architectural Fees.

Type of Construction	\multicolumn{4}{c}{Total Project Size (in thousands of dollars)}			
	$500	$500-$1,000	$1,000-$5,000	Over $5000
Industrial buildings, factories & warehouses	Technical payroll times 2.0 to 2.5	1.60%	1.25%	1.00%
Hotels, apartments, offices, dormitories, hospitals, public buildings, food stores		2.00%	1.70%	1.20%
Museums, banks, churches and cathedrals		2.00%	1.75%	1.25%
Thin shells, prestressed concrete, earthquake resistive		2.00%	1.75%	1.50%
Parking ramps, auditoriums, stadiums, convention halls, hangars & boiler houses		2.50%	2.00%	1.75%
Special buildings, major alterations, underpinning & future expansion	▼	Add to above 0.5%	Add to above 0.5%	Add to above 0.5%

For complex reinforced concrete or unusually complicated structures, add 20% to 50%.

Figure 16.7

Mechanical and Electrical Fees
Typical **Mechanical and Electrical Engineering Fees** based on the size of the subcontract. These fees are included in Architectural Fees.

Type of Construction	\multicolumn{8}{c}{Subcontract Size}							
	$25,000	$50,000	$100,000	$225,000	$350,000	$500,000	$750,000	$1,000,000
Simple structures	6.4%	5.7%	4.8%	4.5%	4.4%	4.3%	4.2%	4.1%
Intermediate structures	8.0	7.3	6.5	5.6	5.1	5.0	4.9	4.8
Complex structures	12.0	9.0	9.0	8.0	7.5	7.5	7.0	7.0

For renovations, add 15% to 25% to applicable fee.

Figure 16.8

Fees may be interpolated horizontally and vertically. Adjust the various portions of the same project that require different fees proportionately. For alterations, add 50% to the fee for the first $500,000 of the estimated project cost, and 25% for over $500,000.

Contingencies

An allowance for contingencies provides for unforeseen conditions, construction difficulties, or oversights during the estimating process. Since risk should diminish as more detailed information becomes available, different factors should be used at various stages of design completion. The following percentages are guidelines.

Conceptual Stage, add:	15–20%
Preliminary Drawings, add:	10–15%
Working Drawings, 60% Design Complete, add:	7–10%
Final Working Drawings, 100% Checked Finals, add:	2–7%
Field Contingencies, add:	0–3%

Also consider inflationary price trends and material and labor shortages that may occur during the course of the job. Cost escalation factors depend on both economic conditions and the anticipated time between the estimate and actual construction. It is customary to include a cost escalation factor from the date of the estimate to the date of the midpoint of construction. In the final summary, contingencies are a matter of estimating judgment.

Assemblies Estimate: Three-Story Office Building

Location Factor

Using the index figures in Figure 16.6, adjust the developed cost for the sample three-story office building for a construction site in Boston, Massachusetts. Since the costs used when developing the estimate were from the "Total" Columns, use the Total index factor. (See Figure 16.9.)

Division Number	System	Estimated Cost	Boston Index	Adjusted Cost
A	Substructure	$ 75,819	120.2	$ 91,134
B10	Superstructure	293,896	113.8	334,454
B20	Exterior Closure	363,262	127.5	463,159
B30	Roofing	36,095	116.1	41,906
C	Interiors	444,011	115.2	511,501
D10	Conveying Systems	107,700	107.3	115,562
D20-40	Mechanical	530,826	110.1	584,439
D50	Electrical	342,071	116.9	399,881
E	Equipment & Furnishings	1,575	101.9	1,605
G	Building Sitework	89,398	106.0	94,762
	Subtotal	$2,284,653		$2,638,403
	Cost per Square Foot	$76.16/S.F.		$87.95/S.F.

Figure 16.9

If the "Weighted Average" for Divisions A-G had been used, the results would be:

$2,284,653 × 115.0% = $2,627,351, or a difference of $342,698
The new square foot cost ($2,627,351/30,000 S.F.) = $87.58

Still another way to use the indexes is to accumulate separate material and installation prices, then apply the appropriate factors for each item.

Percentage Mark-ups

The example three-story office building has the following percentages added for general conditions (see Figure 16.10):

Overhead	7%
Profit	7%
AE fee	8.5%
General Conditions	5%
Contingency	2%

Completed Estimate Summary

Figure 16.11 is the completed estimate with mark-ups calculated. The cost represents the estimator's anticipation of the total project costs in a specific location at an assumed future date. Dividing this figure by the square foot area will produce a projected square foot cost:

$$\frac{\$3,490,096}{30,000 \text{ S.F.}} = \$116.34 \text{ per square foot}$$

Dividing by the building volume will produce a projected cubic foot cost:

$$\frac{\$3,490,096}{360,000 \text{ C.F.}} = \$9.69 \text{ per cubic foot}$$

ASSEMBLY NUMBER	DESCRIPTION	QTY	UNIT	TOTAL COST		COST PER S.F.
				UNIT	TOTAL	
DIVISION E	**EQUIPMENT & FURNISHINGS**					
E1010-110-0300	Office Safe, 1 hr Rating, 34" x 20" x 20"	1	EA	$ 1,575.00	$ 1,575	
	Subtotal, Division E, Equipment & Furnishings				$ 1,575	$ 0.05
DIVISION G	**SITEWORK**					
RG1010-011	Clear & Grub	1	ACRE	$ 7,100.00	$ 7,100	
G1030-805-1330	Utility trenching, backhoe, 0 to 1 slope, 4' deep	300	L.F.	$ 4.29	$ 1,287	
G1030-815-1460	Pipe bedding, 2' wide trench, pipe <8" dia.	300	L.F.	$ 1.57	$ 471	
G3020-110-4150	Drainage/sewage piping, conc., non-reinf., 6" dia.	100	L.F.	$ 11.38	$ 1,138	
G3020-110-4160	Drainage/sewage piping, conc., non-reinf., 8" dia.	160	L.F.	$ 13.16	$ 2,106	
G3060-110-3110	Gas service piping, steel, sched. 40, PE, 3" dia.	25	L.F.	$ 12.60	$ 315	
G3010-110-4110	Water distribution piping, copper, type K, 3" dia.	25	L.F.	$ 15.10	$ 378	
G3010-10-4130	Water distribution piping, copper, type K, 4" dia.	25	L.F.	$ 22.70	$ 568	
G3030-210-4620	Manhole/catchbasin, concrete, CIP, 4' ID riser, 4' deep	2	Ea.	$ 2,360.00	$ 4,720	
G3030-210-4640	Manhole/catchbasin, concrete, CIP, 4' ID riser, 6' deep	1	Ea.	$ 3,175.00	$ 3,175	
G2010-210-1800	Roadway, blt. Conc. Paving, 3" thick, 20' wide	260	L.F.	$ 78.00	$ 20,280	
G2020-210-1520	Parking Lot	60	CAR	$ 571.00	$ 34,260	
Allowance	Site Improvements (Grass, Trees, etc.)	6,800	S.F.	$ 2.00	$ 13,600	
	SUBTOTAL, DIVISION G, SITEWORK				$ 89,398	$ 2.98
DIVISION H	**GENERAL CONDITIONS**					
	Overhead @ 7%					
	Profit @ 7%					
	AE Fee @ 8.5%					
	General Conditions @ 5%					
	Contingency @ 2%					

Figure 16.10

Preliminary Estimate Cost Summary

PROJECT	TOTAL AREA	SHEET NO.
LOCATION	TOTAL VOLUME	ESTIMATE NO.
ARCHITECT	COST PER S.F.	DATE
OWNER	TOTAL VOLUME	NO. OF STORIES
QUANTITIES BY	EXTENSIONS BY	CHECKED BY

DIV	DESCRIPTION	SUBTOTAL COST	COST/S.F.	PERCENTAGE
A	SUBSTRUCTURE	$ 75,819	$ 2.53	3.3%
B10	SHELL: SUPERSTRUCTURE	$ 293,896	$ 9.80	12.9%
B20	SHELL: EXTERIOR CLOSURE	$ 363,262	$ 12.11	15.9%
B30	SHELL: ROOFING	$ 36,095	$ 1.20	1.6%
C	INTERIOR CONSTRUCTION	$ 444,011	$ 14.80	19.4%
D10	SERVICES: CONVEYING	$ 107,700	$ 3.59	4.7%
D20	SERVICES: PLUMBING	$ 91,271	$ 3.04	4.0%
D30	SERVICES: HVAC	$ 380,175	$ 12.67	16.6%
D40	SERVICES: FIRE PROTECTION	$ 59,380	$ 1.98	2.6%
D50	SERVICES: ELECTRICAL	$ 342,071	$ 11.40	15.0%
E	EQUIPMENT & FURNISHINGS	$ 1,575	$ 0.05	0.1%
F	SPECIAL CONSTRUCTION	$ -	$ -	0.0%
G	SITEWORK	$ 89,398	$ 2.98	3.9%

	BUILDING SUBTOTAL	$ 2,284,653

Sales Tax __ % x Subtotal /2 $ -

General Conditions _5_ % x Subtotal $ 114,233

Subtotal "A" $ 2,398,886

Overhead _7_ % x Subtotal "A" $ 167,922

Subtotal "B" $ 2,566,808

Profit _7_ % x Subtotal "B" $ 179,677

Subtotal "C" $ 2,746,485

Location Factor _115_ % x Subtotal "C"   Localized Cost $ 3,158,458
(Boston, MA)

Architects Fee _8.5%_ x Localized Cost = $ 268,469
Contingency _2_ % x Localized Cost = $ 63,169

Project Total Cost $ 3,490,096

Square Foot Cost $3,490,096/ 30,000 S.F. =   S.F. Cost $ 116.34
Cubic Foot Cost $3,490,096/ 360,000 C.F. =   C.F. Cost $ 9.69

Figure 16.11

Chapter 17

Overview/Estimate Analysis

When the estimate is complete, check the calculated square foot costs against known data for similar buildings. Consult contractor data, recent local projects, and published data, such as *Means Square Foot Costs*. This publication, updated annually, lists square foot costs for various building types. *Means Building Construction Cost Data* lists individual breakdown costs for foundations, substructures, superstructures, and so forth. Either type of cost data can be useful to check the accuracy of an assemblies square foot estimate.

Breakdown of Square Foot Costs

Means Square Foot Costs includes many building types and classifications. Figure 17.1 shows typical square foot costs for low-rise office buildings with a range of square foot areas and different framing and exterior wall systems. At the bottom of Figure 17.1, a section titled "Common Additives" lists various building elements that provide greater detail for tailoring the estimate. Figure 17.2 shows breakdown costs for a three-story, 20,000 S.F. office building with a 12' story height.

Square foot cost tables such as Figures 17.1 and 17.2 allow users to use certain categories or components from the listings, and substitute others. For instance, one may wish to customize the foundation, while using the costs from the tables for the remaining components. The foundation cost may have to be revised if piles or caissons are required because of poor soil conditions.

The models in *Means Square Foot Costs* do not show structural bay spacing or superimposed floor and roof loads. Loads in the models are commonly used in assemblies square foot estimates unless there is an unusual loading condition. The bay sizes provided are typical for each building type. Bay sizes can vary widely, however. Such changes could vary the superstructure costs by as much as 20%.

Cost Comparisons

Compare the assemblies square foot estimate costs developed in the preceding chapters for the example three-story office building to similar components in a three-story office building as analyzed in *Means Square Foot Costs* (Figure 17.1). The components in this example are factored for a 30,000 square foot building in lieu of the 20,000 square foot building shown in the model (Figure 17.2). The factor was developed as follows using Figure 17.1.

Costs per square foot of floor area

Exterior Wall	S.F. Area	5000	8000	12000	16000	20000	35000	50000	65000	80000
	L.F. Perimeter	220	260	310	330	360	440	490	548	580
Face Brick with Concrete Block Back-up	Wood Joists	153.75	131.20	118.30	109.45	104.80	95.65	91.20	88.95	87.20
	Steel Joists	154.00	131.40	118.55	109.70	105.00	95.85	91.40	89.15	87.40
Glass and Metal Curtain Wall	Steel Frame	148.80	128.35	116.70	108.80	104.65	96.45	92.55	90.55	89.00
	R/Conc. Frame	152.35	131.90	120.25	112.35	108.15	100.00	96.10	94.10	92.55
Wood Siding	Wood Frame	126.10	108.60	98.65	92.10	88.60	81.80	78.60	76.95	75.70
Brick Veneer	Wood Frame	137.95	117.35	105.60	97.65	93.40	85.15	81.20	79.20	77.65
Perimeter Adj., Add or Deduct	Per 100 L.F.	25.45	15.90	10.60	8.00	6.35	3.65	2.55	2.00	1.55
Story Hgt. Adj., Add or Deduct	Per 1 Ft.	4.20	3.05	2.45	1.95	1.70	1.20	.95	.80	.65
For Basement, add $21.35 per square foot of basement area										

The above costs were calculated using the basic specifications shown on the facing page. These costs should be adjusted where necessary for design alternatives and owner's requirements. Reported completed project costs, for this type of structure, range from $41.60 to $162.45 per S.F.

Common additives

Description	Unit	$ Cost	Description	Unit	$ Cost
Clock System			Smoke Detectors		
20 room	Each	12,700	Ceiling type	Each	149
50 room	Each	30,800	Duct type	Each	405
Closed Circuit Surveillance, One station			Sound System		
Camera and monitor	Each	1375	Amplifier, 250 watts	Each	1650
For additional camera stations, add	Each	745	Speaker, ceiling or wall	Each	145
Directory Boards, Plastic, glass covered			Trumpet	Each	271
30" x 20"	Each	565	TV Antenna, Master system, 12 outlet	Outlet	236
36" x 48"	Each	1025	30 outlet	Outlet	151
Aluminum, 24" x 18"	Each	425	100 outlet	Outlet	144
36" x 24"	Each	535			
48" x 32"	Each	745			
48" x 60"	Each	1600			
Elevators, Hydraulic passenger, 2 stops					
1500# capacity	Each	42,225			
2500# capacity	Each	43,425			
3500# capacity	Each	47,225			
Additional stop, add	Each	3650			
Emergency Lighting, 25 watt, battery operated					
Lead battery	Each	289			
Nickel cadmium	Each	655			

Note: *The estimate in this figure is organized using UniFormat rather than UNIFORMAT II.*

Figure 17.1

Model costs calculated for a 3 story building with 12' story height and 20,000 square feet of floor area

			Unit	Unit Cost		Cost Per S.F.		% of Sub-Total
A. Substructure								
1010	Footings and Foundation	Poured concrete; strip and spread footings and 4' foundation wall	S.F. Ground	$	8.19	$	2.73	
1030	Slab on Grade	4" reinforced concrete with vapor barrier and granular base	S.F. Slab	$	3.32	$	1.11	
2010	Excavation and Backfill	Site preparation for slab and trench for foundation wll and footing	S.F. Ground	$	1.08	$	0.36	5.3%
B. Shell								
	B10 Superstructure							
1010	Elevated Floor Construction	Open web steel joists, slab form, concrete, columns	S.F. Floor	$	31.41	$	7.44	
1020	Roof Construction	Metal deck, open web steel joists, columns	S.F. Roof	$	3.90	$	1.30	11.1%
	B20 Exterior Closure							
2010	Exterior Wall Construction	Face brick with concrete block backup 80% of wall	S.F. Wall	$	20.00	$	10.91	
2020	Exterior Windows	Aluminum outward projecting 20% of wall	Each	$	493.00	$	2.78	
2030	Exterior Doors	Aluminum and glass, hollow metal	Each	$	2,043.00	$	0.62	18.2%
	B30 Roofing							
3010	Roof Covering	Built-up tar and gravel with flashing; perlite w/EPS composite insulation	S.F. Roof	$	4.11	$	1.37	
3020	Roof Openings & Specialties	Gravel stop	S.F. Roof	$	6.92	$	0.12	1.9%
C. Interior Construction								
1010	Partitions	Gypsum board on metal studs, toilet partitions) S.F. of Floor/L.F. Partition	S.F. Partition	$	3.11	$	2.11	
1020	Interior Doors	Single leaf hollow metal 200 S.F. Floor/Door	Each	$	534.00	$	2.67	
1030	Interior Fittings & Accessories	N/A	-		-		-	
2010	Stairs	Concrete filled metal pan	Flight	$	5,025.00	$	1.76	
3010	Wall Finish	60% vinyl wall covering, 40% paint	S.F. Surface	$	4.10	$	2.43	
3020	Floor Finish	60% carpet, 30% vinyl composition tile, 10% ceramic tile	S.F. Floor	$	5.31	$	5.31	
3030	Ceiling Finish	Mineral fiber tile on concealed zee bars	S.F. Ceiling	$	3.63	$	3.63	22.8%
D. Building Services								
	D10 Conveying							
1010	Elevators & Lifts	Two hydraulic passenger elevators	Each	$	67,000.00	$	6.70	
1020	Escalators & Moving Walks	N/A	-		-		-	8.5%
	D20 Plumbing							
2010	Plumbing	Toilet and service fixtures, supply and drainage 1 Fixture/1320 S.F. Floor	Each	$	2,125.00	$	1.61	2.1%
	D30 HVAC							
3020	Heating Systems	Included in D3030	-		-		-	
3030	Cooling Systems	Multizone unit gas heating, electric cooling	S.F. Floor	$	11.75	$	11.75	15.0%
	D40 Fire Protection							
4010	Sprinklers	N/A	-		-		-	
4020	Stand Pipes	Standpipes and hose systems	S.F. Floor	$	0.60	$	0.60	0.8%
	D50 Electrical							
5010	Electric Service & Distribution	1000 ampere service, panel board and feeders	S.F. Floor	$	2.82	$	2.82	
5020	Lighting & Wiring	Fluorescent fixtures, receptacles, switches, A.C. and misc. power	S.F. Floor	$	7.78	$	7.78	
5030	Communications & Security	Alarm systems and emergency lighting	S.F. Floor	$	0.61	$	0.61	14.3%
E. Equipment & Furnishings								
1010	Commercial Equipment	N/A	-		-		-	
1020	Institutional Equipment	N/A	-		-		-	
1030	Vehicular Equipment	N/A	-		-		-	0.0%
F. Special Construction								
1020	Integrated Construction	N/A	-		-		-	
1040	Special Facilities	N/A	-		-		-	0.0%
G. Site Work								
1030	Site Earthwork	N/A	-		-		-	
2010	Roadways	N/A	-		-		-	
2020	Parking Lots	N/A	-		-		-	0.0%
				Sub-Total		$	78.52	100%
	CONTRACTOR FEES (General Requirements:10%)			25%		$	19.63	
	ARCHITECT FEES			7%		$	6.85	
			Total Building Cost			**$ 105.00**		

Figure 17.2

Model 3-Story Office with 12' story height:

20,000 S.F.	=	$105.00/S.F.
35,000 S.F.	=	$95.85/S.F.
1/3 ($105.00-$95.85) + $95.85	=	$98.90

Interpolating for a 30,000 square foot building:

30,000/S.F. @ $98.90/S.F. = .94 Cost Multiplier
20,000/S.F. @ $105.00/S.F.

The .94 multiplier, in all probability, would not be proportioned evenly over all components of the estimate. For purposes of comparison, mark up each component in Figure 17.2 by the size multiplier .94.

This is an example of how the numbers are derived:

Footings and Foundations: $2.73/S.F. × .94 = $2.57/S.F.

Component Comparison

To complete the analysis, check those components that represent a large proportion of the total project cost and those that can have a wide cost range. Checking the major cost factors in the project is an efficient way to ensure the accuracy of the estimate as a whole. Major variances from the model should be checked and revised if a suitable explanation is not found.

In the model building from *Means Square Foot Costs*, the share of each division toward the total building cost is the Contribution Percentage. The differential is the difference, in percent, between the model building and the three-story office building example in this book.

In addition to the cost comparison made in Figures 17.3 through 17.6 (between a model building from *Means Square Foot Costs* and the three-story office building example in this book), the following items should be considered when summarizing square foot costs:

System	Means S.F. Costs $/S.F.	% of S.T.	System S.F. Est. $/S.F.	% of S.T.
Substructure	$ 4.20	5.35	$ 2.53	3.46
Superstructure	8.74	11.13	9.80	13.39
Exterior Closure	14.31	18.22	12.11	16.55
Roofing	1.49	1.90	1.20	1.64
Interior Construction	17.91	22.81	14.80	20.23
Elevators	6.70	8.53	3.59	4.91
Mechanical	13.96	17.78	17.69	24.18
Electrical	11.21	14.28	11.40	15.58
Equipment & Furnishings	—	—	0.05	0.07
Subtotals	$78.52		$73.17	

Figure 17.3

Superstructure	Means S.F. Costs $/S.F.	Systems S.F. Est. $/S.F.	Remarks
Column Fireproofing	$1.83	$2.17	Larger Bay Size
Elevated Floors	5.61	6.68	Larger Bay Size
Roof	1.30	1.51	Larger Bay Size
	$8.74/S.F.	$8.66/S.F.	
Exterior Closure	**Means S.F. Costs $/S.F.**	**Systems S.F. Est. $/S.F.**	**Remarks**
Walls	$10.91	$4.66	Concrete Panel vs. Brick on Block
Doors	0.62	0.20	
Windows	2.78	6.79	Aluminum Window Wall in Lieu of Individual Steel Sash.
	$14.31/S.F.	$11.65/S.F.	

Owner or Architect's choice of fascia for aesthetics or code compliance.

Figure 17.4

Interior Construction	Means S.F. Costs $/S.F.	Systems S.F. Est. $/S.F.	Remarks
Partitions & Toilet Partitions	$2.11	$2.50	10' Floor to Floor vs. 12'
Interior Doors	2.67	1.03	200 S.F. vs. 500 S.F.
Wall Finishes	0.86	0.90	
Floor Finishes	5.31	5.72	
Ceiling Finish	3.63	2.69	Concealed Min. Fiber Tile (12" × 12") vs. Fibrous Glass (24" × 48")
Int. of Ext. Wall	1.57	0.64	
	$16.15/S.F.	$13.48/S.F.	

Adjust Doors:
$$\frac{20,000 \text{ S.F.}}{200 \text{ S.F./door}} = \frac{100 \text{ doors} \times \$534/\text{door}}{20,000 \text{ S.F.}} = \$2.67/\text{S.F.}$$

Adjust ceiling price by using lower price:

$$\text{Adjustment} \quad \frac{\$2.69/\text{S.F.}}{\$0.94/\text{S.F.}}$$

$16.15/S.F. − $0.94/S.F. = 15.21/S.F. versus $13.48/S.F.

Elevator	Means S.F. Costs $/S.F.	Systems S.F. Est. $/S.F.	Remarks
Elevator	$6.70	$3.59	100 FPM vs. 125 FPM

Figure 17.5

- **Foundation** square foot costs vary greatly, including soil capacities and the use of piles or caissons. Substructure depends on the selected floor system and the bay sizes.
- **Site preparation** is dependent on the existing contours, soil, rock, and vegetation conditions.
- **Superstructure** depends on the selected floor system and the bay sizes. In the comparison mentioned above, this part of the work contributes an average of 11% to the building cost and has a 2% differential.
- **Exterior closure** has the greatest influence on square foot costs. It is the most visible component with wide variations in material choice and cost. This portion of the estimate should be carefully evaluated. Exterior closure in the comparison mentioned above averages an 18% contribution to the total building cost, with a 2% differential.
- **Roofing** square foot costs are proportional to the number of stories in the building, with one-story buildings having the highest percent per square foot contribution.
- **Interior construction** also has a large percentage contribution to the total building cost. The variance of finishes in cost, appearance, and durability should be carefully evaluated. The individual items of interior construction do not represent a large percentage of the total building cost. Rather, it is the accumulation of many items that cause this division of the estimate to be so costly. Interior construction in the comparison mentioned above (Figure 17.3) has a 23% average contribution and a differential of 3%.
- **Mechanical** work represents a large contribution to the total building cost. There can be substantial variance with sophisticated heating, cooling, and control systems. The choice of plumbing fixtures will affect the overall cost, but not to the extent of the HVAC systems.
- **Electrical** work, like mechanical, is a major contributor to the overall square foot cost of the project. Service requirements and the choice of fixtures are variables in the estimate that must be resolved.

Mechanical	Means S.F. Costs $/S.F.	Systems S.F. Est. $/S.F.	Remarks
Plumbing	$1.61	$3.04	Additional fixtures for handicapped accessibility.
Fire Protection	0.60	1.98	Assemblies S.F. includes a wet pipe sprinkler.
HVAC	11.75	12.67	
	$13.96/S.F.	$17.69/S.F.	

Figure 17.6

Estimating Accuracy

Accuracy in an assemblies square foot estimate depends on several factors, including:

- Reliability and source of cost information.
- Specific design information available.
- Time available to create the estimate.

Sufficient time to accomplish an estimate is always a problem to be addressed. Whether it's a ballpark budget estimate or a detailed estimate for bidding purposes, there is rarely adequate time available to make sure that every reasonable factor has been considered and taken into account.

When preparing an estimate for bidding, think the project through from start to finish and use detailed unit prices. A preliminary budget, on the other hand, does not warrant the same amount of time or precision. Therefore, assemblies square foot estimating is a more logical approach. While it saves valuable time during the project's initial decision-making stages, square foot assemblies estimating still has potential pitfalls. Extra time is invested in the costlier portions of the estimate. A further analysis of the three-story office building helps to reinforce this point.

Estimate Summary

What follows is a summary of the assemblies square foot cost estimate before general conditions are added for the three-story office building.

Comparison

Figure 17.7 from *Means Assemblies Cost Data* identifies a percentage of total cost for each building system within the Means cost index model. The figures shown represent the "average" percentage contribution of each building system component to the total cost for the model commercial building costing over $1,000,000 and more. The weighted average index is calculated from over 66 materials and equipment types and 21 building trades. The table identifies the areas where the estimator should spend more time.

General: The following information on current city cost indexes is calculated for over 930 zip code locations in the United States and Canada. Index figures for both material and installation are based upon the 30 major city average of 100 and represent the cost relationship on July 1, 2000.

In addition to index adjustment, the user should consider:

1. productivity
2. management efficiency
3. competitive conditions
4. automation
5. restrictive union practices
6. unique local requirements
7. regional variations due to specific building codes

The weighted average index is calculated from about 72 materials and equipment types and 21 building trades.

If the systems component distribution of a building is unknown, the weighted average index can be used to adjust the cost for any city.

Labor, Material and Equipment Cost Distribution for Weighted Average Listed by System Division

Division No.	Building System	Percentage	Division No.	Building System	Percentage
A	Substructure	6.6 %	D10	Services: Conveying	3.7%
B10	Shell: Superstructure	18.4 %	D20-40	Mechanical	20.1
B20	Exterior Closure	12.2 %	D50	Electric	11.9
B30	Roofing	2.7	E	Equipment	2.0
C	Interiors	17.5 %	G	Sitework	4.9
				Total weighted index (Div. A-G)	100.0%

Figure 17.7

To compare the "average" index figures with those just calculated for the three-story office building, see Figure 17.8.

At first glance, the percentages appear to be in general agreement for the most significant sub-divisions: Superstructures, Exterior Closure, Interiors, and Services. Differences exist because the buildings being compared most likely do not correspond in some areas of design. If a difference of more than 5-10% occurs in any section, the estimator may wish to examine this part of the estimate more closely.

One of the fundamental "rules" of estimating is to spend the most time and pay closest attention to the parts of the project with the greatest potential cost. Unfortunately, the easiest part of the estimate, or that which relates to the estimator's specialty, tends to receive the most attention and time. Unless there are unusual circumstances involved in a project, the most likely order of cost significance is:

1. Mechanical
2. Superstructures
3. Exterior Closure
4. Interiors
5. Electrical
6. Substructures
7. Conveying
8. Sitework
9. Roofing
10. Special Construction

The above guidelines are only a general reference in determining priorities. There will always be exceptions.

Division Number	System	Estimate % of S.T.	Table % of S.T.
A	Substructure	3.3	6.6
B10	Superstructure	12.9	18.4
B20	Exterior Closure	15.9	12.2
B30	Roofing	1.6	2.7
C	Interiors	19.4	17.5
D10	Conveying	4.7	3.7
D20-40	Mechanical	23.2	20.1
D50	Electrical	15.0	11.9
E	Equipment & Furnishings	0.1	2.0
G	Building Sitework	3.9	4.9

Figure 17.8

A Final Word

Using published historical or model square foot costs for preliminary "guesstimates" is acceptable when the project is similar to the model. The accuracy of cost data is proportional to how well the project corresponds to the specifications of the cost model. If the published cost models do not correspond well, it is possible to produce realistic building costs, tailored to the particular project and its scope, in just a few hours using square foot assemblies estimating. This method is by far the most desirable for the preparation of preliminary budget estimates. Once you are familiar with it, this method can speed up the estimating process and produce a credible estimate of projected costs.

It is during the preliminary design stage that the single most significant event occurs: preparation of the initial budget. The cost figure chosen is critical, as it can determine whether the project is deemed feasible or not. If it is too high, the project will not move forward. If it is too low, problems will follow, and the project may stop during the design phase due to lack of funds. Unfortunately, the first cost figure written down or discussed is usually the one that someone (often the decision-maker) latches onto. No amount of discussion will cause those involved in the project to forget it, even when the scope or use shifts, or changes are made in design that significantly affect the final cost.

Don't be caught in the position of having to live with a construction budget based on a vague or "off-the-cuff" figure that is too low. Estimating building construction costs is an unusual business where the estimator is often cast in the role of either prophet or fool. Take control by making use of all the tools and advantages available. Square foot assemblies estimates look at a building's construction, element by element, in a logical and factually based progression. Costs are developed using a rational and verifiable thought process. If used wisely and with creativity, the process may provide advantages in terms of time and accuracy. Assemblies square foot estimating may be just the right edge needed in a fast-paced construction industry where accurate preliminary cost estimates have taken on a new significance.

Appendix

Using UNIFORMAT II for Facilities Planning & Design

*by Robert P. Charette, PE, CVS
and Anik Shooner, Architect*

Widely used in the construction industry, the UNIFORMAT classification was revised by the American Society of Testing Materials (ASTM) in 1993 and designated as UNIFORMAT II. It defines building elements as "major components common to most buildings that perform a given function, regardless of the design specification, construction method, or materials used." In practice, an element is any distinct part of a whole, such as the assemblies referred to throughout this book. In terms of project management, elements constitute the Work Breakdown Structure (W.B.S.) of a building.

The ASTM UNIFORMAT II classification system is particularly well suited to square foot estimating. The systems format is relatively easy to use since it reflects the instinctive way that designers and builders think about a building construction project. This approach makes it possible to define each building system, determine how each fits into the overall scheme, and establish realistic program budgets that can be monitored during design. This chapter provides some historical background information on the UNIFORMAT classification system, its evolution, and efficient use to increase design team performance.

The Importance of Standardization

Good coordination and communication among all team members is necessary to control the scope and costs of a building project. Team performance is greatly enhanced when the information exchanged is timely, reliable, and in a consistent format. This chapter addresses this objective as it relates to the early phases of project design.

During the program phase, the owner defines the project content, quality, cost, and schedule for the designers. These criteria will be used to measure the design team's performance. When action needs to be taken to keep the project on track, it must be timely to avoid any delays in the design schedule, and it must be kept within the allocated professional fees and the established budget.

The owner and design team leader must ensure that information is conveyed effectively to ensure clear communication among all participants. Potential problems or misunderstandings can be dealt with quickly in this environment. The design documents and estimates must be appropriate and consistent from project to project. Historically,

287

documentation presented to the owner is based on either the owner's own guidelines or on in-house procedures developed by consultants. In either case, it is important that the documentation not only be of a high quality, but also standardized to avoid costly inefficiencies. Lack of standardization not only reduces efficiency, but constrains the opportunities for value engineering because of a lack of timely information.

New Tools to Improve Communication, Coordination, and Productivity During Design

Two relatively recent publications by ASTM and CSI/CSC (Construction Specifications Institute/Construction Specifications Canada) contribute to improving communication, coordination, and productivity during design:

- ASTM Standard E1557-01 "Standard Classification for Building Elements and Related Sitework—UNIFORMAT II."
- CSI/CSC Practice FF/180 "Preliminary Project Descriptions and Outline Specifications."

The ASTM document (Figure A.1) provides a standard classification of building elements and related sitework that is universal and has many applications, including program and design estimates, specifications, building condition evaluation, and capital building budgeting.

The CSI/CSC document (Figure A.2) outlines the application of the classification of building elements for "concept/schematic phase specifications," referred to as "Preliminary Project Descriptions (PPD)"

 Designation: E 1557 – 01

Standard Classification for Building Elements and Related Sitework—UNIFORMAT II[1]

This standard is issued under the fixed designation E 1557; the number immediately following the designation indicates the year of original adoption or, in the case of revision, the year of last revision. A number in parentheses indicates the year of last reapproval. A superscript epsilon (ϵ) indicates an editorial change since the last revision or reapproval.

1. Scope

1.1 This standard establishes a classification of building elements and related sitework. Elements, as defined here, are major components common to most buildings. Elements usually perform a given function, regardless of the design specification, construction method, or materials used. The classification serves as a consistent reference for analysis, evaluation, and monitoring during the feasibility, planning, and design stages of buildings. Using UNIFORMAT II ensures consistency in the economic evaluation of buildings projects over time and from project to project. It also enhances reporting at all stages in construction—from feasibility and planning through the preparation of working documents, construction, maintenance, rehabilitation, and disposal.

1.2 This classification applies to buildings and related site work. It excludes specialized process equipment related to a building's functional use but does include furnishings and equipment.

1.3 The Classification incorporates three hierarchical levels described as Levels 1, 2, and 3. Appendix X1 presents a more detailed suggested Level 4 classification of sub-elements.

1.4 UNIFORMAT II is an elemental format similar to the original UNIFORMAT[2] elemental classification. UNIFORMAT II differs from the original UNIFORMAT, however, in that it takes into consideration a broader range of building types and has been updated to categorize building elements as they are in current building practice.

Copyright © ASTM, 100 Barr Harbor Drive, West Conshohocken, PA 19428-2959, United States

Extracted, with permission, from E 1557-97—Standard Classification for Building Elements and Related Sitework—UNIFORMAT II, copyright ASTM, 100 Barr Harbor Drive, West Conshohocken, PA 19428. A complete copy of the standard may be purchased from ASTM: phone: 610-832-9585, fax: 610-832-9555, e-mail: service@astm.org, Web site: www.astm.org.

Figure A.1

in CSI/CSC terminology. The latter expression is more correct, because at the concept phase "specifications" as such are not really produced.

Applying these two documents allows elemental design estimates and preliminary project descriptions to be linked in a common reference structure and provides a framework for more efficient, effective, and predictive planning and decision making. All parties can save time and money due to improved coordination, an increase in programming and design team performance, and a significant reduction in costly delays related to project scope "creep" and cost overruns.

Elemental Classifications

Background

The development of the first elemental classification is attributed to the British Ministry of Education immediately after World War II. At that time, quantity surveyors developed an "elemental classification" to control the cost of the accelerated post-war school expansion program. From the UK, the methodology was exported to other British Commonwealth countries such as Canada, South Africa, and Singapore, who adapted the classification to their needs. It was exported from Canada to the U.S. in the 1970s, and resulted in the adoption of the UNIFORMAT classification by the American Institute of Architects (AIA) and the U.S. General Services Administration (GSA). In 1993, the original UNIFORMAT classification was revised by ASTM and promulgated as Standard E1557–93–UNIFORMAT II, following four years of task group meetings with members from other organizations such as CSI/CSC, the American Association of Cost Engineers (AACE), the Tri Services Committee, and the GSA. At that time, CSI also recommended that Preliminary Project Descriptions at the schematic phase be structured according to a modified version of the UNIFORMAT elemental classification.

Manual of Practice

Construction Documents Fundamentals and Formats Module
Preliminary Project Descriptions and Outline Specifications

FF/180
Preliminary Project Descriptions and Outline Specifications

Well-prepared documents improve coordination, communication, and productivity during all phases of a project. Two written documents recommended for use during the initial design phases are "Preliminary Project Descriptions" and "Outline Specifications." This chapter describes these documents and how they apply to a conventional project delivered under a lump-sum, single prime contract. This chapter does not discuss multiple prime contracts, construction management, or design-build delivery methods.

Source: Reprinted from the Construction Specifications Institute (CSI) *Manual of Practice*, (FF Module, 1992 edition) with permission from CSI.

Figure A.2

Current Use

Today, elemental classifications are used throughout the world, primarily for design estimates. In Canada, all federal and provincial government departments call for mandatory elemental design estimates. In the U.S., the GSA and the Military Services have been the main proponents of UNIFORMAT, though a few state public works and education departments call for it as mandatory.

Additional applications for UNIFORMAT II are being developed. These include design-build contractual documents, layering of CAD drawings, and facility condition assessments. Since ASTM and CSI support the use of UNIFORMAT in North America, it is anticipated that the classification will gain widespread acceptance as a common standard, increasing efficiency and saving the construction industry significant sums of money.

For additional background information on elemental classifications, refer to the National Institute of Standards and Technology (NIST) Special Publication 6389: "UNIFORMAT II Elemental Classification for Building Specifications, Cost Estimating, and Cost Analysis." The 109-page report can be downloaded from the Internet Web site: *http://www.bfrl.nist.gov/oae/publications/nistirs/6389.html*

(Note: Excerpts from this document follow this chapter.)

Elemental Program and Design Estimates

To illustrate the concept of an elemental unit cost, consider the exterior wall of a building. In a conventional unit price estimate, each component would be quantified and priced without any relation to function (e.g., brick, vapor retarder, insulation, block back-up, and furring). The elemental approach uses a single composite rate per square foot. This rate includes all component costs, thus greatly simplifying the estimating task. Elemental cost data is available in *Means Assemblies Cost Data*, *Means Square Foot Costs*, and *Yardsticks for Costing*.

At the program phase, estimates or cost plans can be prepared using cost modeling techniques. Developing a model based on gross floor area and an appropriate massing configuration provides quantities for most building elements. This allows a realistic program estimate to be prepared, though elements are not defined other than by performance and levels of quality (e.g., the structure could be steel or concrete, and the exterior enclosure curtain wall or brick with block back-up).

With such a model, unit costs may be allocated to elements based on cost manuals, historical cost data from similar projects, and the level of quality called for in the facilities program or permissible within the owner's anticipated cost for the building. For example, even without a design, the exterior wall for an insurance company's head office could be set at $65.00/S.F., an anticipated high level of quality, versus $15.00/S.F., which would be appropriate for an industrial-type building.

Such realistic program estimates will permit a Design-to-Cost (DTC) approach for each discipline that can readily be monitored element by element. Designers must commit to the targeted cost, fully aware that cost issues are of prime importance to the team and the client. This approach will also facilitate the conducting of an effective Value Engineering (VE) workshop at this stage, which would be particularly cost effective if the budget exceeds the anticipated project cost. It is at the program phase that major budget problems should be resolved by ensuring that the program and budget are compatible, thus avoiding costly redesigns and delays after design has been initiated.

In a similar fashion, designers should, as a matter of routine, validate with cost modeling techniques any given budget to which they will be contractually committed. If the projected cost is higher than the budget, and the difference is validated, the program scope and quality levels can be reviewed to reach an agreement on the budget before time has been expended on design work—a saving for everyone that fosters a harmonious designer-client relationship and increased design efficiency.

At the schematic phase of a project, estimates must be prepared to monitor and control costs from the schematic phase through the completion of construction documents to ensure that the program budget is respected. This task is greatly facilitated using elemental estimate analytic parameters that allow effective cost analysis. In other words, all elements are expressed in quantity per unit GFA (gross floor area), cost per unit of measurement of the element, or a percentage of total cost. Such in-depth analysis would not be possible with costs based on trade or MasterFormat.

For Value Engineering workshops, elemental estimates:

- Serve as a checklist for all the elements of a project.
- Provide unit costs and quantities for all elements.
- Increase the team's comprehension of the project.
- Allow directed brainstorming of alternatives.
- Allow the rapid calculation of cost differentials for proposed alternatives, whatever their scope.

As a general rule, function-oriented design estimates permit effective cost analysis during the program and design phases of a project, and as a result, significantly increase the performance and return on investment for Value Engineering studies.

At the completion of the construction documents phase, design estimates may be converted to trade estimates or MasterFormat unit price estimates to accelerate bid analysis.

A typical UNIFORMAT II schematic design cost estimate summary is presented in Figure A.3.

Elemental Preliminary Project Descriptions (Schematic Design Phase)

The quality of project documentation at the schematic design and design development phases varies greatly. Few organizations have adopted standard guidelines that adequately define what should be presented. The absence of such documentation is often reflected in inaccurate design estimates that suffer from lack of information, misunderstandings between clients and designers on the nature of the project, delays in decision-making as to baseline systems selection, and difficulty in coordinating design team tasks. In most cases, designers know what they plan to incorporate in their design, but do not document this information for other team members because it is not mandatory. In general, no two consultants submit schematic or design development specifications in the same way, making it more difficult for the project manager and owner to analyze submittals as the design progresses.

UNIFORMAT II BUILDING ELEMENTAL COST SUMMARY

ELEMENTS Levels 2 & 3		RATIO QTY/GFA	ELEMENT COST			ELEMENT AMOUNT	COST per UNIT GFA	% Trade Cost
			UM	QUANTITY	UNIT PRICE			
A10	**FOUNDATIONS**					87 081	1,19	2,54%
A1010	Standard Foundations	0,25	FPA	18 200	1,93	35 211	0,48	
A1020	Special Foundations	0,00		0		0	0,00	
A1030	Slab on Grade	0,25	SF	18 200	2,85	51 870	0,71	
A20	**BASEMENT CONSTRUCTION**					65 813	0,90	1,92%
A2010	Basement Excavation	0,25	SF	18 200	1,43	26 026	0,36	
A2020	Basement Walls	0,04	SF	3 200	12,43	39 787	0,54	
B10	**SUPERSTRUCTURE**					669 695	9,15	19,57%
B1010	Floor Construction	0,75	SF	55 000	10,25	563 641	7,70	
B1020	Roof Construction	0,26	SF	18 900	5,61	106 054	1,45	
B20	**EXTERIOR CLOSURE**					435 496	5,95	12,73%
B2010	Exterior Walls	0,17	SF	12 100	7,47	90 421	1,24	
B2020	Exterior Windows	0,34	SF	25 060	13,50	338 310	4,62	
B2030	Exterior Doors	0,00	LVS	6	1127,50	6 765	0,09	
B30	**ROOFING**					48 558	0,66	1,42%
B3010	Roof Coverings	0,26	SF	18 900	2,47	46 752	0,64	
B3020	Roof Openings	0,00	EA	3	602,00	1 806	0,02	
C10	**INTERIOR CONSTRUCTION**					171 549	2,34	5,01%
C1010	Partitions	0,41	SF	29 940	4,23	126 654	1,73	
C1020	Interior Doors	0,00	EA	66	296,00	19 536	0,27	
C1030	Specialities	1,00	SF	73 200	0,35	25 359	0,35	

Figure A.3

292

CSI has recognized this situation, being fully aware that MasterFormat is not suitable for describing a project at such an early stage. This is because the final selection of products and materials has not been made, or is likely to change during design. As a result, CSI now recommends in their Practice FF/180 that Preliminary Project Descriptions (schematic phase specifications) be structured according to an elemental classification that will describe a project in terms of building systems or elements. The use of the Preliminary Project Description based on UNIFORMAT II links the specification and the estimate in a common reference structure and provides a framework for more efficient and productive planning and decision-making. Though baseline systems have been identified and their cost estimated at the schematic phase, changes are still possible within the budget allocated as the design progresses.

In Practice FF/180, CSI also recommends the use of a condensed MasterFormat specification referred to as "Outline Specifications" at the preliminary design phase. This approach is gaining acceptance in North America, where CSI, AIA, and HPT-Buildwrite offer such computerized specifications at a modest cost. The use of this type of specification at the design development phase pushes the decision-making up-front, reduces the risk of future misunderstandings, and accelerates the preparation of final working drawings and MasterFormat specifications. Though an Outline Specification is proposed at the design development phase in Practice FF/180, an elemental specification could also be considered to complement drawings that are less detailed than normal. HPT-Buildwrite also offers software for developing elemental specifications.

Designers may not be called on contractually to submit Preliminary Project Descriptions and Outline Specifications. However, the benefits are such that owners should consider, if necessary, offering a supplementary fee for this service.

Benefits of Elemental Estimate and Preliminary Project Descriptions

The benefits of elemental estimates and preliminary project descriptions include:

- At the schematic design phase, the project manager, owner, and end users are presented with a clear description of what is proposed for each element of the project, i.e., a *clear scope definition*. This allows them to make early comments on any changes that may be required and, in effect, commit to accepting the proposed design. This early review and approval process minimizes future changes that are usually at the expense of the consultants and possible delays.
- The architect is reassured that all engineering disciplines have taken the time to consider alternatives for all systems and have selected the most appropriate baseline systems at the schematic design phase. This pushes the decision-making up-front, and minimizes the possibility of major unexpected changes as the design proceeds. Similarly, the architect also must reflect on each architectural element and document his or her intentions. The classification, in essence, becomes a checklist for all building and sitework elements to monitor design and facilitate technical reviews.

- The cost manager, having been given descriptions of all building elements at the initial schematic design phase, is now in a position to provide a relatively accurate cost plan, element by element, that conforms exactly to the standard classification descriptions provided. The Preliminary Project Description saves the cost manager and design team members time as design data is collected in writing instead of in interviews with each consultant. As a result, the cost manager can now spend time more effectively, analyzing costs rather than estimating.
- The presentation of the project description and estimated cost in the same format facilitates and accelerates the design review and approval process. It also allows the taking of corrective actions related to scope or cost at the earliest possible time, without consuming a disproportionate amount of design fees, and without significant delays in the design schedule.
- Since the client and designers have a common understanding of the scope and cost of the project at the schematic design phase, the design development phase may be initiated with fewer uncertainties. As a result, the schedule may be accelerated. Redesigning should also be minimized, adding to the profitability for designers. This understanding among design team members and stakeholders should also create a positive environment that encourages teamwork.
- Communication, coordination, and productivity are improved significantly at all phases of a project. Many projects run into trouble because of costly redesigns resulting from scope "creep" and cost overruns. These problems are the result of a lack of effective communication among the client, project managers, and designers at the program and schematic design phases, when most problems should have been identified and addressed.

Other Applications of the UNIFORMAT II Classification

The UNIFORMAT II classification has other practical applications when standardized reporting is desirable during the design and operation life of the building. Some of these include:

- Performance specifications for design-build
- Preliminary construction schedules and cash flow projections
- Cost risk analysis (ASTM Standard E 1496)
- Life cycle costing analysis
- Value Engineering function/cost models
- Building condition evaluation
- Long-term capital replacement budgeting
- A checklist for brainstorming VE function/system alternatives
- Program technical requirements
- A checklist for technical reviews
- Layering of CAD drawings
- Reserve funds
- Filing of technical literature for assemblies
- Capital, maintenance, and repair cost databases
- Financial evaluation of buildings
- Defining the scope of contracts for each consultant
- Classifying construction details

Conclusion Improving communication and coordination among project team members, the client, and other stakeholders will lead to a significant increase in design team performance. This can be achieved in a consistent manner from project to project through the application of the ASTM UNIFORMAT II standard classification of building elements for estimates and simple elemental design specifications prepared as outlined in CSI Practice FF/180. These documents link the technical program, design estimates, and specifications in a common reference structure that provides a framework for more efficient, effective, and productive planning and decision making.